Lecture Notes in Computer Science 12840

More information about this subseries at http://www.springer.com/series/7409

Ken Barker · Kambiz Ghazinour (Eds.)

Data and Applications Security and Privacy XXXV

35th Annual IFIP WG 11.3 Conference, DBSec 2021
Calgary, Canada, July 19–20, 2021
Proceedings

 Springer

Editors
Ken Barker
University of Calgary
Calgary, AB, Canada

Kambiz Ghazinour ⓘ
State University of New York at Canton
Canton, NY, USA

ISSN 0302-9743   ISSN 1611-3349 (electronic)
Lecture Notes in Computer Science
ISBN 978-3-030-81241-6   ISBN 978-3-030-81242-3 (eBook)
https://doi.org/10.1007/978-3-030-81242-3

LNCS Sublibrary: SL3 – Information Systems and Applications, incl. Internet/Web, and HCI

This Springer imprint is published by the registered company Springer Nature Switzerland AG
The registered company address is: Gewerbestrasse 11, 6330 Cham, Switzerland

Preface

This volume contains the papers selected for presentation at the 35th Annual IFIP WG-11.3 Conference on Data and Applications Security and Privacy (DBSec 2021) that was supposed to take place during July 19–20, 2021, in Calgary, Canada. While the conference was held on the dates as scheduled, due to the COVID-19 situation it was held virtually, but we do look forward to gathering together in 2022 for the next DBSec!

In response to the call for papers for this edition, 45 submissions were received, and all submissions were evaluated on the basis of their significance, novelty, and technical quality. The Program Committee, comprising over 40 members, performed an excellent job, with the help of additional reviewers, of reviewing submissions through a careful anonymous process (three or more reviews per submission). The Program Committee's work was carried out electronically, yielding intensive discussions. Of the submitted papers, 15 full papers and 8 short papers were selected for presentation at the conference.

The success of DBSec 2021 depended on the voluntary effort of many individuals, and there is a long list of people who deserve special thanks. We would like to thank all the members of the Program Committee and all the external reviewers, for all their hard work in evaluating the papers and for their active participation in the discussion and selection process. We are very grateful to all the people who readily assisted and ensured a smooth organization process, in particular Sara Foresti (IFIP WG11.3 Chair) for her guidance and support, Khosro Salmani as Publicity Chair, and Leanne Wu for helping with other arrangements for the conference. EasyChair made the conference review and proceedings process run very smoothly.

Last but certainly not least, thanks to all the authors who submitted papers and all the conference attendees. We hope you find the proceedings of DBSec 2021 interesting, stimulating, and inspiring for your future research.

July 2021

Ken Barker
Kambiz Ghazinour

Organization

Program and General Chair

Ken Barker — University of Calgary, Canada

IFIP WG-11.3 Chair

Sara Foresti — Università degli Studi di Milano, Italy

Publicity Chair

Khosro Salmani — Mount Royal University, Canada

Publication Chair

Kambiz Ghazinour — State University of New York at Canton, USA

Local Arrangement Chair

Leanne Wu — University of Calgary, Canada

Program Committee

Adam J. Lee	University of Pittsburgh, USA
Andreas Schaad	WIBU-Systems, Germany
Anoop Singhal	NIST, USA
Ayesha Afzal	Air University, USA
Brad Malin	Vanderbilt University, USA
Catherine Meadows	NRL, USA
Charles Morisset	Newcastle University, UK
Costas Lambrinoudakis	University of Piraeus, Greece
Csilla Farkas	University of South Carolina, USA
Edgar Weippl	University of Vienna, Austria
Ehud Gudes	Ben-Gurion University, Israel
Fabio Martinelli	IIT-CNR, Italy
Frédéric Cuppens	Polytechnique Montréal, Canada
Giovanni Di Crescenzo	Perspecta Labs, USA
Giovanni Livraga	University of Milan, Italy

Günther Pernul	Universität Regensburg, Germany
Indrajit Ray	Colorado State University, USA
Indrakshi Ray	Colorado State University, USA
Jaideep Vaidya	Rutgers University, USA
Javier Lopez	University of Malaga, Spain
Kambiz Ghazinour	State University of New York at Canton, USA
Kui Ren	State University of New York at Buffalo, USA
Lingyu Wang	Concordia University, Canada
Maryam Majedi	University of Toronto, USA
Martin Olivier	University of Pretoria, South Africa
Nicola Zannone	Eindhoven University of Technology, the Netherlands
Nora Cuppens-Boulahia	Polytechnique Montréal, Canada
Pierangela Samarati	Università degli Studi di Milano, Italy
Sabrina De Capitani di Vimercati	Università degli Studi di Milano, Italy
Sara Foresti	Università degli Studi di Milano, Italy
Scott Stoller	Stony Brook University, USA
Shamik Sural	IIT, Kharagpur, India
Silvio Ranise	FBK-Irst and University of Trento, Italy
Sjouke Mauw	University of Luxembourg, Luxembourg
Sokratis Katsikas	Open University of Cyprus, Cyprus
Stefano Paraboschi	Universita di Bergamo, Italy
Steven Furnell	University of Nottingham, UK
Vijay Atluri	Rutgers University, USA
Vijay Varadharajan	The University of Newcastle, Australia
Wendy Hui Wang	Stevens Institute of Technology, USA
Yingjiu Li	University of Oregon, USA
Yuan Hong	Illinois Institute of Technology, USA

Additional Reviewers

Artsiom Yautsiukhin
Bingyu Liu
Giacomo Iadarola
Han Wang
Hossein Shirazi
Stefano Berlato
Tahir Ahmad
Xihui Chen
Yunior Ramírez-Cruz

Contents

Access Control

Differential Privacy

DPNeT: Differentially Private Network Traffic Synthesis with Generative Adversarial Networks

Liyue Fan[(⊠)] and Akarsh Pokkunuru

University of North Carolina at Charlotte, Charlotte, NC 28223, USA
{liyue.fan,apokkunu}@uncc.edu

Abstract. High quality network traffic data can be shared to enable knowledge discovery and advance cyber defense research. However, due to its sensitive nature, ensuring safe sharing of such data has always been a challenging problem. Current approaches for sharing networking data present several limitations to balance privacy (e.g., information leakage) and utility (e.g., availability and usefulness). To overcome those limitations, we develop DPNeT, a network traffic synthesis solution that generates high-quality network flows and satisfies (ϵ, δ)-differential privacy. We adopt generative adversarial networks (GANs) to capture the characteristics of real network flows and a similarity-preserving embedding model for mixed-type attributes. Furthermore, we propose new techniques to improve the outcome of differentially private learning and provide the privacy analysis of the overall solution. Through a comprehensive evaluation with large-scale network flow data, we demonstrate that our solution is capable of producing realistic network flows.

Keywords: Differential privacy · Generative adversarial networks · Network flow generation

1 Introduction

Sharing fine-grained network traffic data has enabled numerous research studies for knowledge discovery and cyber security applications, such as in anomaly detection [16] and cyber attack classification [4,29]. However, network traffic data is highly sensitive, e.g., with Internet protocol (IP) addresses and port numbers, etc., which may be used by adversaries to infer private information, e.g., a specific website visited by the user. In the worst case, home and commercial networks may be attacked [5,27]. Therefore, it is imperative to protect the privacy of individuals and organizations in the published network data. In order to enhance the privacy for sharing network traffic data, many anonymization techniques have been proposed, [3,20,30] to name a few. However, it has been shown that inference attacks may still be launched against anonymized network traces [6,13]. Moreover, it is challenging to quantify the quality of the anonymized data.

© IFIP International Federation for Information Processing 2021
Published by Springer Nature Switzerland AG 2021
K. Barker and K. Ghazinour (Eds.): DBSec 2021, LNCS 12840, pp. 3–21, 2021.
https://doi.org/10.1007/978-3-030-81242-3_1

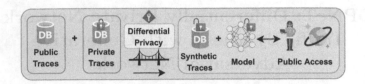

Fig. 1. DPNeT overview.

Recently, generative adversarial networks (GANs) [10] have been adopted to generate realistic network traffic data, e.g., for sequence of packet sizes [25], network flows [23], and traffic morphing [14]. However, it has been shown that deep learning models are subject to various attacks, e.g., inferring membership in the training set [26] and reconstructing training data [9]. Similarly, GAN models do not provide guarantees on what the generated data may reveal about real, sensitive training data. In fact, [12] successfully devised membership inference attacks against target GANs in both white-box and black-box access settings.

To provide rigorous privacy in network traffic synthesis, we propose to adopt differential privacy [7] when training GAN models with sensitive data. Our solution, dubbed DPNeT, builds on recent advances in training generative models (i.e., Wasserstein GAN [11]), and is able to produce *realistic* synthetic data that exhibits high *similarity* to the training data. As shown in Fig. 1, the trained models as well as the synthetic traces can be widely shared for research and educational purposes. The specific contributions of our work are as follows:

1. We are the first to develop a differentially private solution based on GANs for synthesizing flow-level network data. The privacy of the training examples is protected via differentially private optimization [1]. To ensure the quality of the synthetic data, we adopt the state-of-the-art methodology for training generative models, as well as an advanced embedding model to preserve similarities between mixed-type feature values.

2. To address the challenges in private learning, such as noisy or non-convergence, we propose two improvement techniques: decaying the clipping bound over epochs and privately selecting the best models across all training epochs. Our empirical analysis shows that decaying the clipping bound outperforms the standard option, i.e., no decay. Furthermore, we show that private model selection significantly improves the quality of the synthetic network flows, compared to the model obtained at the last epoch.

3. We provide privacy analysis results for both GAN training and model selection. In short, we implement Moments Accountant to account for the privacy loss during GAN model training; we further analyze the sensitivity and the privacy guarantees for selecting up to K models. Overall, we show that the DPNeT solution achieves (ϵ, δ)-differential privacy.

4. We conduct an extensive evaluation with large-scale network flows. We quantify the similarity and realism of the synthetic network flows using distributional measures and domain knowledge tests, respectively. Our results examine the impact of privacy and demonstrate the effectiveness of our proposed improvements.

The rest of the paper is organized as follows: we discuss related work in Sect. 2 and describe fundamental concepts such as differential privacy and GANs in Sect. 3. In Sect. 4, we provide a full description of DPNeT as well as privacy analysis results. In Sect. 5, we present and discuss empirical evaluation results. Finally, in Sect. 6 we conclude the paper with brief discussions on future work.

2   Related Work

Network Trace Anonymization. Anonymization techniques for network traffic data have been extensively studied in the last decades. For instance, IP addresses can be obfuscated with prefix-preserving pseudonyms [30] or bucketization [21]. Other features, such as timestamps and ports, can be shifted or suppressed [15,20]. However, it has been shown that classic anonymization methods are prone to inference attacks [3,6,13]. A recent study [19] proposed to create multiple views of the dataset based on the assumed adversarial knowledge, which is not suitable for our setting, i.e., sharing data widely. [17] proposed a differential privacy based solution for analyzing network traces; however, the sharing of private flows has not been discussed.

Network Traffic Synthesis with GANs. With rapid advancements in deep learning and generative models, a few research studies have proposed to incorporate generative adversarial networks (GANs) [10] for generating network traffic data. For instance, Shahid et al. [25] proposes to generate synthetic packet sizes for IoT applications. Unfortunately, the synthesis of other network traffic features was not addressed. Two recent works aim to synthesize network traffic flows for data sharing (Ring et al. [23]) and bypass internet censorship (Li et al. [14]). They are most relevant to this paper, especially Ring et al. [23]. However, GAN models do not provide privacy guarantees on what the generated data may reveal about real, sensitive training data. For instance, [12] successfully devised membership inference attacks against target GANs in both white-box and black-box access settings.

Differential Privacy and Machine Learning. The vulnerabilities of deep learning, e.g., membership inference [12,26], demonstrate great needs for strong privacy protection for the underlying training data. To this end, differential privacy has been applied to learning deep models [1] to combat such inference attacks. Recent studies adopt the framework in [1] to train GAN models [2,28] and we have conducted a survey study on those approach in [8]. Based on our survey results, most of the existing studies focus on generating image data, e.g., MNIST, where GAN models can learn input data distributions accurately; and very few studies attempt to publish mixed-type data, with features as challenging as IP addresses. Furthermore, all studies report difficulties and utility loss encountered with differentially private learning, e.g., the generator and discriminator may converge to a noisy equilibrium or do not converge. Our solution aims to address the shortcomings of existing approaches, and the proposed improvements can be applied to general differentially private learning tasks.

3   Preliminaries

3.1   Differential Privacy

The privacy model adopted in our work is differential privacy [7]. By definition, a randomized algorithm A is (ϵ, δ)-differentially private if for any two databases D, D' differ in at most a single record and any subset $S \in Range(A)$:

$$\Pr[A(D) \in S] \leq \exp(\epsilon) \Pr[A(D') \in S] + \delta \tag{1}$$

Here $\epsilon > 0$ is known as the privacy budget, which bounds the difference between output probabilities of two neighboring databases D, D'. In addition, $\delta \in [0, 1]$ accounts for the probability of *bad events* that might lead to a privacy breach. An advantage of DP is its resistance to post-processing [7], i.e., any computation performed on the output of a DP mechanism would not incur additional privacy cost. Other benefits of DP include the lightweight computation compared to crypto-based mechanisms and ease of control over the information leakage with the help of ϵ, δ parameters. Typically, smaller ϵ and δ values indicate stronger privacy protection, and vice versa. There exists a trade-off between preserving privacy and maintaining data utility.

In particular, we are interested in applying DP to deep learning, in order to protect the privacy of training examples. As shown in [1], it can be achieved by sanitizing the gradients during neural network optimization, which ultimately limits the overall influence of any training example on the model. A privacy accountant, i.e., Moments Accountant [1], has been proposed to account for differential privacy across training epochs, which provides stronger estimates of privacy loss compared to other composition theorems [7].

A key question to address in applying (ϵ, δ)-DP is the choice of privacy parameters. It is often seen that $\epsilon > 1$ in private deep learning, e.g., $\epsilon = 8$ as in [1] and up to $\epsilon = 96$ in some studies surveyed in [8]. In this study, we would like to provide privacy protection in deep learning applications without incurring significant utility loss. As a result, we consider $\epsilon \leq 20$. As for δ, to provide individual privacy in case of bad events, we set $\delta = \frac{1}{|D|}$ as recommended in [7].

3.2   Generative Adversarial Networks

Generative adversarial networks (GANs) [10] have become the state-of-the-art method to learn generative models, and has demonstrated superior performance in producing synthetic data that have similar characteristics as real data. A survey analysis for differentially private GANs can be found here [8]. GANs consist of two components, i.e., a generator G and a discriminator D. The problem is formulated as a minimax two-player game with the following objective [10]:

$$\min_{G} \max_{D} \mathbb{E}_{x \sim p_{data}}[\log D(x)] + \mathbb{E}_{z \sim p_z}[\log(1 - D(G(z)))]. \tag{2}$$

The generator G learns to capture the original data distribution p_{data} by mapping a latent distribution p_z. Specifically, G takes as input a random noise

z and generates synthetic data. On the other hand, the discriminator D learns to discriminate between samples drawn from p_{data}, i.e., x, and those generated by G, i.e., $G(z)$. D takes a sample as input and returns a score representing whether it is real or synthetic. By generating samples that appear to come from the original data distribution, the goal of the generator is to fool the discriminator. The generator and discriminator are trained simultaneously through an adversarial process: the more the generator improves the quality of synthetic data, the harder it is for the discriminator to distinguish between original and synthetic samples. Wasserstein GAN [11] minimizes the earth-mover distance (i.e., Wasserstein-1 distance) between p_z and p_{data}, and allows more stable and faster training by penalizing the norm of gradient of the discriminator. For those considerations, we will adopt Wasserstein GAN in DPNeT to generate synthetic network flows.

4   DPNeT Solution

Overview. DPNeT for network flow synthesis entails three main steps:

1 Embedding the input network flow records to numerical vectors;
2 Training GAN models using the input vectors and generating synthetic vectors;
3 Decoding the synthetic data to produce network flow records.

To generate private synthetic network flows, we utilize two separate deep learning architectures: an advanced *embedding* model, and the generator and discriminator of *GANs*. Step 1 and 3 rely on the embedding network which transforms network flow records to numerical vectors while preserving data semantics. Step 2 learns to mimic the input data distribution with GAN models, in a differentially private setting.

4.1   DPNeT: Embedding

Network flows are mixed-type data, consisting of both numerical features and categorical features (e.g., IP address, port number, packet size, and flag). To learn from the input data, categorical feature values are often converted into numerical values, e.g., via one-hot encoding. Ring et al. [22] have shown that the standard encoding methods do not capture the relationships between mixed-type features. They modified the Word2Vec model [18] to learn "word" embeddings for network flow features. We adopt a similar architecture for the embedding model in DPNeT.

Specifically, we consider 8 features of each flow, namely Source IP and Port, Destination IP and Port, Protocol, Packets, Bytes, and Duration. As a result, we are able to compute the similarities between Source IP and Source Port, or between Packets and Duration. For each flow, 13 bi-grams are constructed as the training data for the embedding model. Each bi-gram contains an input word, i.e., one of the 8 features, and a context word (expected output), i.e., among a set of features identified by domain experts for each input feature [23]. The embedding network is a simple neural network with a single hidden layer, which contains 20 neurons in our solution. After the embedding network is trained, it is utilized to produce numerical feature vectors for GANs training set (Step 1 in Overview), with the weights at the hidden layer as feature vectors of the words (i.e., network flow attributes). Given a synthetic feature vector generated by GANs, we retrieve the most similar word in the vocabulary based on cosine similarity in the embedding space (Step 3 in Overview), and output the synthetic network flow records.

In DPNeT, we train the embedding model using a *public* dataset that is disjoint from the sensitive training set for the GAN models. Our consideration is three-fold: (1) utilizing a public dataset enables accurate learning of feature similarities. For instance, we can fine-tune the parameters iteratively without incurring any privacy loss. (2) Thanks to the accuracy, the training set of the embedding model needs not be very large. Additionally, it is easier to sanitize the vocabulary on a smaller dataset, e.g., ensuring certain IP addresses or protocols are not present. (3) The embedding network should learn to encode general network flows independent of GAN training data, e.g., specific network traffic patterns that only occur in the sensitive training data. This also produces an embedding model that can be deployed for other applications.

By design, the vocabulary of the embedding model may not include all words in the GAN training set. We propose to impute the sensitive training data, i.e., replacing the feature values that do not appear in the embedding model's vocabulary with the mode of the feature. Our empirical evaluation confirms the feasibility of using disjoint training sets for the embedding model and GAN models.

4.2   DPNeT: GAN Training

The center of DPNeT is to learn GAN models with differential privacy. The embeddings obtained in Step 1 are concatenated with the remaining features in the network flows, e.g., class, and are provided to the GAN models as training data. To achieve differential privacy, we adopt the deep learning framework proposed in [1] and present the pseudocode in Algorithm 1.

We train our GAN models on N examples for E epochs, on a randomly selected mini-batch of size B with a sampling probability $q = B/N$. For each training example in a batch, the discriminator computes the gradients $g_j(x_k)$ w.r.t the model parameters and the DP mechanism sanitizes the gradients with clipping and perturbation. The clipping of gradient norm is upper bounded by a hyper-parameter γ. Additionally, the variance of the additive Gaussian noise is controlled through σ. The choice of σ and γ values is essential to achieve a balance between data quality and privacy protection. Finally, the discriminator parameters are updated after the completion of each batch. The generator learns to produce network flows through iterative updates using a traditional gradient based optimization approach. The discriminator is trained for E_D epochs per generator epoch, which helps alleviate mode collapse issues.

At the end of each epoch, we save the intermediate generator model G_i, which will be utilized for model selection purposes. Optionally, the clipping bound γ may decay according to a few proposed decay functions; we refer readers to Sect. 4.3. The privacy loss will be estimated by the Moments Accountant. After E epochs, the algorithm outputs the final generator model G and total privacy spent ϵ_1.

Algorithm 1. DPNeT Training Procedure

Input: training examples $x_1 \cdots x_N$, batch size B, noise σ, $\delta = \frac{1}{N}$, learning rate α, number of epochs E, number of discriminator epochs E_D, clipping bound γ

  for $i = 1 \cdots E$ **do**

    if $i > 1$ **then**

      For each generator parameter θ compute:

      $g_i \leftarrow Adam \left(\triangledown \frac{1}{\theta} \sum_{i=1}^{\theta} -D(G(Z_i)) \right)$

      Update generator: $\theta_{G_{(i+1)}} \leftarrow \theta_{G_{(i)}} + \alpha \; g_i$

    end if

    for $j = 1 \cdots E_D$ **do**

      Sample L with sampling probability $q = \frac{B}{N}$

      for each $x_k \in L$ **do**

        Compute gradients: $g_j(x_k) \leftarrow \triangledown_\theta W(\theta_j, x_k)$

        Clipping: $g_j(x_k) \leftarrow g_j(x_k)/max(1, \|g_j(x_k)\| /\gamma)$

      end for

      Perturbation: $g_j \leftarrow \frac{1}{|L|} \left(\sum_{k=1} g_j(x_k) + \mathcal{N}(0, (\sigma \, \gamma)^2 I) \right)$

      $\theta_{D_{(j+1)}} \leftarrow \theta_{D(j)} + \alpha \; g_j$

    end for

    Save the generator as G_i

    Optionally, decay the clipping bound γ (see Sec. 4.3)

    Estimate privacy ϵ_1 using Moments Accountant

  end for

Output: Generator G and total privacy spent ϵ_1

Implementation. In DPNeT, we adopt the MLP architecture for both the generator (5 hidden layers) and discriminator (4 hidden layers) as suggested in [23]. Every hidden layer has 1024 units. The choice of hidden layer activation is *ReLU* and *leaky ReLU* with a negative slope of 0.2 for the generator and discriminator, respectively. We use a linear activation for the output layers in both networks.

Privacy Analysis. As the generator update is solely based on the discriminator (i.e., post-processing of DP outputs), it is sufficient to apply clipping and perturbation to training the discriminator [8]. Using Moments Accountant [1], we can bound the moments of a mechanism's privacy loss and prove the (ϵ, δ)-differential privacy guarantee. In practice, we implement the Moments Accountant to estimate Algorithm 1's privacy loss ϵ_1 with the following: δ, the batch size, the total number of examples, the noise multiplier, and the number of training steps performed. Hence, *Algorithm* 1 *achieves (*ϵ_1*, *δ*)-differential privacy.* The intermediate models, i.e., G_i where $i < E$, incur less privacy loss due to small numbers of training steps required. We plot in Fig. 2a an empirical analysis of the privacy estimates over the training epochs.

4.3 Improvements in DPNeT

While Algorithm 1 incorporates the state-of-the-art techniques (i.e., GANs) for generating synthetic data, the application of differential privacy may introduce new challenges, due to clipping the gradient norm and adding perturbation noise. As a result, the quality of the generated data may be affected by privacy. Below, we describe the proposed improvements in the DPNeT solution, with the goal of overcoming the limitations of differentially private deep learning.

Clipping Bound Decay. The clipping bound γ is an important factor in Algorithm 1. Although not influential on the privacy accountant, the act of clipping changes the gradient estimation. Specifically, when γ is too small, the average gradient may point to a different direction than the true gradient; on the other hand, increasing γ would require higher noise to be added, as the noise distribution is based on $\sigma\gamma$. Our idea is to decay the value of γ over the course of training, such that the model is able to learn quickly initially (i.e., with higher γ values) and accurately in the final stages (i.e., with lower γ values).

To that end, we propose three types of decay functions to control the speed of decay and investigate the impact on private learning. Our approach is based on commonly used decay functions, namely linear, exponential, and logarithmic:

$$\text{Linear decay}: \ \gamma(i) = C - i \left(\frac{C - C'}{E - 1} \right) \tag{3}$$

$$\text{Exponential decay}: \ \gamma(i) = C \ d^{(i-1)} \tag{4}$$

$$\text{Logarithmic decay}: \ \gamma(i) = \frac{C}{1 + \log(i)} \tag{5}$$

$$\text{No decay}: \ \gamma(i) = C \tag{6}$$

In the above functions, i indicates the current epoch where $i \in [1, E]$; d is the exponential decay rate; C is the initial value for the clipping bound and C' is the final value. The design ensures that all decay functions start with $\gamma(1) = C$, including the option of no decay. We choose the other parameters such that at the final epoch, the three decay functions arrive at similar γ values, i.e., around C'. In our empirical evaluation, we show for each decay option how γ values change throughout the training process (see Fig. 4).

Private Model Selection. We also develop an effective strategy to privately select the best generative models among all models obtained throughout the training process. The rationale is that as DP introduces noise in training, the discriminator and the generator may converge toward a noisy equilibrium or do not converge. As a result, the generator obtained at the last epoch may not be optimal in quality. Therefore, in our approach, we consider models saved throughout the training epochs with the goal of choosing the best among them. Furthermore, we need to ensure differential privacy in the model selection process as the private training data is utilized to evaluate the goodness of each model. Our approach is inspired by [2], where a classification accuracy score was utilized to select the best model. However the goal of DPNeT is to produce network flows that can be broadly used, including but not limited to classification applications. Thus, a generic quality measure for the generated data would be much more beneficial. We achieve this with the L1-distance between histograms of real data and synthetic data. We choose histograms in our algorithm for the low sensitivity.

Algorithm 2. Private Model Selection

Input: ground truth flows X_{real}, synthetic flows X_{gen}, number of features n, number of epochs E, number of models to select K, privacy budget ϵ_m for each selection

$\ell_{dist} = \{\}$

for $i = 1 \cdots E$ **do**

    $X_{gen}^i \leftarrow N$ synthetic flows by model obtained at epoch i

    Compute the following score based on L1 distance between histograms:

    $\ell_i = \frac{1}{n} \sum_{j=1}^{n} ||\mathrm{HIST}(X_{real}[j]) - \mathrm{HIST}(X_{gen}^i[j])||_1$

    $\ell_{dist}.insert(\ell_i)$

end for

for $t = 1 \cdots K$ **do**

    Add noise every score: $\ell_{noisy} = \ell_{dist} + \mathrm{Laplace}(0, \frac{1}{\epsilon_m})$

    Sort ℓ_{noisy} and pick the epoch i_t with the smallest noisy score

    $\ell_{dist}.remove(\ell_{i_t})$

end for

Output: K epochs $\{i_t | t = 1 \cdots K\}$ with the best models

---

As shown in Algorithm 2, we aim to choose K best models among all models obtained over E epochs, where the trained generator model can be saved at the end of each epoch. For each epoch i, we compute L1-based score to measure the similarity between the training data distribution and the synthetic data distribution, averaged across all features. Note that HIST in Algorithm 2 generates histograms for a given dataset (e.g., X_{real} or X_{gen}^i) and a certain feature (i.e., indexed by j). All scores are perturbed with noise drawn from a Laplace distribution with 0 mean and $\frac{1}{\epsilon_m}$ scale. The model that corresponds to the lowest noisy score is picked in each iteration until all K models have been selected. The selected models (which may not include the final epoch model) will be used to generate synthetic data. We further propose a *mixture* strategy to combine synthetic flows from K models, which is demonstrated to be superior in our evaluation.

Privacy Analysis. Recall that intermediate generator models, i.e., G_i where $i \leq E$, are saved as part of Algorithm 1. Using G_i to generate synthetic data X_{gen}^i for Algorithm 2 does not incur additional privacy loss, thanks to the post-processing property of DP. The privacy result for Algorithm 2 is as follows:

1 the global sensitivity of ℓ_i is 1, for any i,
2 each iteration of model selection is ϵ_m-differentially private,
3 Algorithm 2 is $K\epsilon_m$-differentially private.

Suppose two neighboring datasets X_{real} and X'_{real}, where $X_{real} = X'_{real} \bigcup \{x^*\}$ and x^* is one flow record. The global sensitivity of ℓ_i can be analyzed with the Minkowski's inequality and the fact that by removing x^* the histograms of $X_{real}[j]$ for any attribute j can change by at most 1:

$$||\ell_i - \ell_i'||_1 \leq \frac{1}{n} \sum_{j=1}^{n} ||\text{HIST}(X_{real}[j]) - \text{HIST}(X'_{real}[j])||_1 \leq 1. \qquad (7)$$

Given $\Delta \ell_i = 1$ ($\forall i$), we can prove that each model selection with Laplace perturbation is ϵ_m-differentially private[1]. It follows that selecting K models with Algorithm 2 satisfies $K\epsilon_m$-differential privacy. Finally, we state the overall privacy guarantee of DPNeT.

Theorem 1. DPNeT satisfies (ϵ, δ)-differential privacy, where $\epsilon = \epsilon_1 + K\epsilon_m$ and $\delta = 1/N$.

Proof. Algorithm 1 satisfies (ϵ_1, δ)-DP where ϵ_1 can be estimated by Moments Accountant and $\delta = 1/N$. Algorithm 2 satisfies $(K\epsilon_m, 0)$-DP. By composition, the overall DPNeT solution satisfies $(\epsilon_1 + K\epsilon_m, \delta)$-DP.

5   Evaluation Results

In this section, we present our methodology for empirical evaluation and discuss results obtained.

Dataset. We utilize a large-scale network flow dataset CIDDS-001 [24] in our evaluation. CIDDS-001 was captured within an emulated small business environment, which contains four weeks of flow-based network traffic. This dataset is publicly available[2], with around 32 *million* labelled network flows consisting of both anomalous and normal behaviors. For each flow, we adopt 11 relevant features, namely: *Date_first_seen, Duration, Protocol, Source_IP, Source_Port, Destination_IP, Destination_Port, Packets, Bytes, Flags, Class,* and preprocess

---

[1] Proof omitted for brevity; it is similar to the proof of Report Noisy Max [7].
[2] http://www.hs-coburg.de/cidds.

them as suggested [23]. We then randomly subsample 2% of the entire dataset, which is then partitioned into two disjoint portions consisting of $N = 287435$ examples each. One partition is treated as *public* and utilized for training the embedding model, while the other partition remains *private* and used for training GAN models. Note that, since the two partitions are disjoint, some feature values in the GAN training data may not be present in the embedding model, e.g., unseen values. In those cases, we impute those feature values with the mode of their respective columns.

Metrics. For all our experiments, we utilize the following metrics to evaluate the quality of the synthetic data, in *similarity* and *realism*. To evaluate similarity, we measure the distance (e.g., Euclidean) between the probability distributions of the training data and the generated data for each attribute. To evaluate realism, we adopt 7 domain knowledge tests to check the consistency between attributes within each flow. For instance, one of them (Test 3) checks: "if the flow describes normal user behavior and the source port or destination port is 80 (HTTP) or 443 (HTTPS), the transport protocol must be TCP". The rationale as well as the detailed design of those domain knowledge tests can be found in [23]. For each test, we report the percentages of flows in the data that pass the test.

Hyper-parameters. The embedding model is trained for 500 epochs to accurately encode feature values. To achieve a trade-off between privacy and quality, the differentially private GAN models are trained for 50 epochs and we report the average results obtained in 3 separate runs. Other hyper-parameters for training the differentially private GAN models are: the training data size $N = 287435$, privacy parameter $\delta = \frac{1}{N}$, batch size $B = 2048$, initial clip norm $\gamma = 0.03$, and learning rate $\alpha = 0.0005$. For every generator iteration, discriminator is trained for $E_D = 5$ iterations. The privacy budget for training differentially private GANs is accounted by the Moment Accountant technique [1] as we vary the noise multiplier σ. For private model selection, we allocate $\epsilon_2 = 0.25$ for selecting top-5 models, i.e., spending 0.05 privacy budget for each round. We compare our results with the non-private baseline.

5.1   Impact of Privacy

We study the effect of privacy on the quality of generated data. As we vary the noise multiplier σ in Algorithm 1, we apply the Moment Accountant to track the privacy budget ϵ_1 spent over the training epochs. We report the intermediate results in Fig. 2a and ϵ_1 after 50 epochs in Table 1. Note that small ϵ values indicate stronger privacy protection. When setting $\sigma = 3$, we achieve $\epsilon_1 = 3.06$ for training the GAN model for 50 epochs; reducing the noise parameter degrades the privacy protection, e.g., $\epsilon_1 = 20.79$ when $\sigma = 0.8$. We also examine the impact of privacy on training quality. Figure 2b plots the loss in Wasserstein distance at the end of each epoch. Table 1 reports the final loss after 50 epochs, i.e., W_{dist}. We observe that in comparison to the non-private model, different privacy inflicts instability in the training process; stronger privacy leads to a higher level of instability, e.g., when $\sigma = 0.8$. Due to the added noise, a differentially private model may converge to a noisy equilibrium. As privacy is relaxed, the performance of differentially private models gradually approaches that of the non-private model.

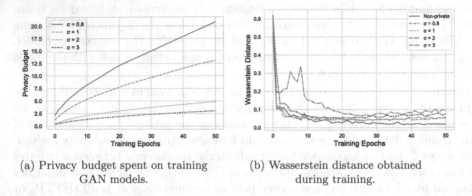

(a) Privacy budget spent on training GAN models.

(b) Wasserstein distance obtained during training.

Fig. 2. Impact of noise (σ) on GAN model training (best viewed in color). (Color figure online)

Table 1. Impact of privacy and Domain Knowledge Test Accuracy (in %) on real and synthetic data: ϵ_1 - privacy budget spent on training GAN models; ϵ - total privacy budget including training GAN models and model selection; W_{dist} - Wasserstein distance obtained at the last training epoch.

Privacy	ϵ_1/ϵ	W_{dist}	Domain Knowledge Test - Accuracy in %							
			Test 1	Test 2	Test 3	Test 4	Test 5	Test 6	Test 7	**Avg.**
Real data	∞/∞	0	100	100	100	100	100	100	100	**100**
Non-private	∞/∞	0.048	99.56	99.94	99.95	99.65	99.82	99.59	99.98	**99.64**
$\sigma = 0.8$	20.79/21.04	0.061	96.79	99.19	99.63	94.53	99.91	77.33	91.85	**94.18**
$\sigma = 1$	13.1/13.35	0.064	96.92	98.79	99.47	94.17	99.97	65.37	90.42	**92.16**
$\sigma = 2$	4.88/5.13	0.084	91.85	98.06	99.47	90.12	100.0	65.92	84.72	**90.02**
$\sigma = 3$	3.06/3.31	0.124	94.7	98.37	99.51	91.15	100.0	3.37	89.51	**82.37**

We conduct the domain knowledge tests on real/synthetic data and Table 1 reports the accuracy results, i.e., the percentage of flows that pass each test. We observe that both private and non-private GAN models are able to produce realistic network flows, e.g., highly accurate for Test 3 and Test 5. Test 6 appears to be the most challenging to capture by the private models. The content of Test 6 is: "if the flow represents a netbios message (destination port is 137 or 138), the source IP addresses must be internal (192.168.XXX.XXX) and the destination IP address must be an internal broadcast (192.168.XXX.255)." As it examines a specific type of flows and involves multiple features, even introducing a small amount of noise ($\sigma = 0.8$) inflicts a drop in performance. We also report the average accuracy among all 7 tests. It can be seen that the accuracy degrades gradually from real data, to synthetic data generated by the non-private model, to synthetic data generated by differentially private models.

Table 2. Impact of privacy and data quality: Euclidean distance between synthetic data distribution and training data distribution is reported.

Feature	Non-private	$\sigma = 0.8$	$\sigma = 1$	$\sigma = 2$	$\sigma = 3$
Day	0.009	0.032	0.043	0.047	0.082
Time	0.008	0.007	0.006	0.006	0.008
Duration	0.74	0.73	0.725	0.715	0.713
Protocol	0.004	0.003	0.002	0.007	0.012
Src IP	0.067	0.131	0.113	0.099	0.104
Src Pt	0.01	0.019	0.016	0.037	0.039
Dst IP	0.063	0.118	0.118	0.098	0.09
Dst Pt	0.01	0.017	0.016	0.024	0.036
Packets	0.016	0.044	0.055	0.068	0.074
Bytes	0.018	0.047	0.055	0.072	0.083
Flags	0.006	0.074	0.098	0.103	0.139
Class	0.003	0.004	0.006	0.005	0.011
Avg.	**0.079**	**0.103**	**0.104**	**0.107**	**0.116**

We report in Table 2 the similarity between real and synthetic data via the Euclidean distance computed for the probability distributions of each feature. The average similarity among all features is reported in the last row of the table. As can be seen, synthetic data generated by the non-private model is most similar to the real data; stronger privacy (i.e., higher σ values) would inflict a higher distance from the real distributions.

Qualitative Evaluation. To better understand the quality of the synthetic data, we plot the distributions of real and synthetic network flows in Fig. 3. Figure 3a shows the percentage of flows per hour in each dataset. We can observe that the non-private model can accurately capture variations in the real data distribution, and the private models are able to capture high-level trends while missing local variations. Similarly we observe in Fig. 3b that the distribution of Packet feature is well preserved by the non-private model. As the noise increases, higher distortions in the probability distribution can be observed.

5.2   Impact of Clipping Bound Decay

Here we study the effect of decaying the gradient norm bound on the quality of the generated data. We set $\sigma = 1$ in the following experiments and train GAN models with different decay functions proposed in Sect. 4.3. The resulting clipping bounds are depicted in Fig. 4. It can be seen that the clipping bounds reduce much more quickly when exponential decay and logarithmic decay are adopted, although all three decay functions reach similar values at the 50th epoch. In contrast, the no decay option adopts a constant clipping bound throughout all training epochs.

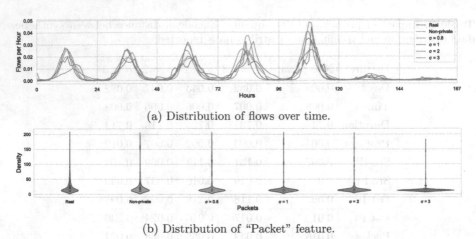

(a) Distribution of flows over time.

(b) Distribution of "Packet" feature.

Fig. 3. Distributions of real and synthetic flows (best viewed in color). (Color figure online)

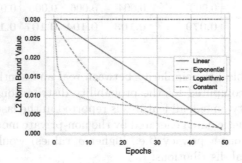

Fig. 4. Clipping bound decay: we set $C = 0.03$, $C' = 0.001$, and $d = 0.94$ such that all clipping bounds initialize at C and decay accordingly over 50 epochs, with the exception of Constant.

Table 3 and Table 4 report the distributional similarity and the accuracy of domain knowledge tests, respectively, using synthetically generated data. As expected, having a constant clipping bound (i.e., no decay), would force more noise to be added and thus reduce the quality of the learned model, i.e., higher Euclidean distance and lower domain knowledge test accuracy. It can be seen that linear decay is competitive in distributional similarity, and performs the best in domain knowledge tests. We believe that it is beneficial to gradually reduce the clipping bound over training epochs to achieve a trade-off between learning and privacy. Thus, we propose to adopt linear decay in the DPNeT solution.

Table 3. Impact of decay functions and data quality: Euclidean distance between synthetic data distribution and training data distribution is reported.

Feature	No decay	Exponential	Logarithmic	Linear
Duration	0.742	**0.717**	0.731	0.725
Protocol	0.003	0.004	0.01	**0.002**
Src IP	0.116	**0.103**	0.118	0.113
Src Pt	**0.016**	0.019	0.022	**0.016**
Dst IP	0.121	0.120	0.127	**0.118**
Dst Pt	0.02	**0.016**	**0.016**	**0.016**
Packets	0.056	**0.052**	0.056	0.055
Bytes	0.06	0.057	0.056	**0.055**
Flags	0.113	**0.095**	0.097	0.098
Class	0.007	0.007	**0.005**	0.006
Avg.	0.125	**0.119**	0.124	0.120

Table 4. Impact of decay functions and domain knowledge test accuracy (in %).

Domain test	No decay	Exponential	Logarithmic	Linear
Test 1	96.73	96.39	**97.09**	96.92
Test 2	98.72	**99.07**	98.97	98.79
Test 3	99.53	**99.59**	99.36	99.47
Test 4	92.55	93.01	**94.17**	**94.17**
Test 5	**99.98**	99.71	99.95	99.97
Test 6	45.05	50.03	53.56	**65.37**
Test 7	89.4	91.29	**91.36**	90.42
Avg.	88.85	89.87	90.64	**92.16**

5.3   Private Model Selection

Here we examine the effect of model selection using Algorithm 2. For each run of differentially private GAN training, we choose the best $K = 5$ epoch models and allocate $\epsilon_m = 0.05$ privacy budget for each selected model. We name the selected models Nosiy Best, Noisy 2nd, etc. $N = 287435$ synthetic flows are generated from each model; in addition, we create a *mixture* of size N by randomly subsampling from each model. The model obtained at the 50th epoch is also included as a baseline, since it may not be selected by the algorithm.

Table 5 and Table 6 report the distributional similarity and the accuracy of domain knowledge tests, respectively. Since the algorithm utilizes an L1-based score, we report the L1 distance between the probability distributions of real data and synthetic data in Table 5. We observe that our private model selection approach is highly beneficial: the average L1 distance monotonically increases from Noisy Best to Noisy 5th. The last epoch shows a much higher average L1 distance to the real data distributions, confirming that the final model may not be optimal. We also observe that the private model selection is affected by perturbation, i.e., Noisy 5th shows the best quality in Flags and Class. A very important observation is that the mixture strategy is superior to all other models, exhibiting the highest similarity to real data in many features.

Table 5. Impact of model selection on data quality: L1 distance between synthetic data distribution and training data distribution is reported.

Feature	Noisy best	Noisy 2nd	Noisy 3rd	Noisy 4th	Noisy 5th	Last epoch	Mixture
Duration	1.484	1.532	1.511	1.519	1.513	1.608	**1.408**
Protocol	0.008	0.008	0.016	0.012	0.016	0.015	**0.003**
Src IP	0.867	0.847	0.854	0.849	0.882	1.031	**0.708**
Src Pt	0.830	0.857	0.920	0.923	0.895	1.014	**0.723**
Dst IP	0.872	0.853	0.905	0.874	0.895	1.068	**0.708**
Dst Pt	0.849	0.862	0.912	0.880	0.924	1.008	**0.717**
Packets	0.212	0.276	0.251	0.257	0.232	0.277	**0.202**
Bytes	0.713	0.759	0.731	0.768	0.786	0.922	**0.636**
Flags	0.267	0.251	0.219	0.225	**0.215**	0.219	0.219
Class	0.025	0.011	0.009	0.025	**0.007**	0.024	0.009
Avg.	0.613	0.626	0.633	0.633	0.637	0.718	**0.533**

Table 6. Domain knowledge test accuracy (in %) for models selected with Algorithm 2.

Domain test	Noisy best	Noisy 2nd	Noisy 3rd	Noisy 4th	Noisy 5th	Last epoch	Mixture
Test 1	97.68	94.20	**98.34**	96.57	97.46	97.46	96.92
Test 2	98.86	98.87	**99.23**	98.25	98.76	98.28	98.79
Test 3	99.37	99.41	**99.77**	99.28	99.52	99.39	99.47
Test 4	95.08	94.10	93.31	91.86	**96.06**	94.02	94.17
Test 5	**100.00**	100.00	100.00	100.00	99.55	100.00	99.97
Test 6	45.26	**66.79**	61.78	60.52	18.32	35.03	65.37
Test 7	90.02	90.56	**90.67**	90.66	90.18	90.55	90.42
Avg.	89.47	91.99	91.87	91.02	85.69	87.82	**92.16**

Similarly in Table 6, Noisy 5th and Last Epoch show lower average test accuracy, compared to other models. The mixture shows the highest average test accuracy among all models. Both tables demonstrate the advantage of our model selection approach and creating a diverse set of synthetic data using *mixture*.

6   Conclusion and Future Work

In this paper, we have described DPNeT, a differentially private solution for generating high quality synthetic network flow data. Privacy of the sensitive training data is protected by training GAN models with differential privacy. We have also proposed novel approaches for clipping bound decay and private model selection. We have demonstrated their effectiveness in improving the quality of synthetic data through comprehensive empirical evaluations. Our approaches may be applied to other differentially private deep learning tasks, e.g. classification.

We identify several directions for future work. Firstly, we observe that fine tuning of hyper parameters, such as learning rate and number of discriminator epochs, is essential to circumvent issues such as mode collapse and non-convergence. In the future, effective parameter tuning methods can be explored for differentially private solutions. Secondly, it is desirable to investigate the usefulness of the synthetic network flow data in domain specific applications, e.g., anomaly detection. Future work can study the performance of anomaly detection models trained with synthetic data. Thirdly, as our solution builds on an embedding model trained with public data, it is important to study the efficacy of the solution when public data may come from a different distribution.

Acknowledgements. The authors would like to thank the anonymous reviewers for their suggestions and comments. This work has been supported in part by NSF CNS-1949217, NSF CNS-1951430, and UNC Charlotte. The opinions, findings, and conclusions or recommendations expressed in this material are those of the authors and do not necessarily reflect the views of the sponsors.

References

1. Abadi, M., et al.: Deep learning with differential privacy. In: Proceedings of the 2016 ACM SIGSAC Conference on Computer and Communications Security, pp. 308–318 (2016)
2. Beaulieu-Jones, B.K., et al. : Privacy-preserving generative deep neural networks support clinical data sharing. Circul. Cardiovasc. Qual. Outcomes **12**(7), e005122 (2019)
3. Brekne, T., Årnes, A., Øslebø, A.: Anonymization of IP traffic monitoring data: attacks on two prefix-preserving anonymization schemes and some proposed remedies. In: Danezis, G., Martin, D. (eds.) PET 2005. LNCS, vol. 3856, pp. 179–196. Springer, Heidelberg (2006). https://doi.org/10.1007/11767831_12
4. Chawla, N.V., Lazarevic, A., Hall, L.O., Bowyer, K.W.: SMOTEBoost: improving prediction of the minority class in boosting. In: Lavrač, N., Gamberger, D., Todorovski, L., Blockeel, H. (eds.) PKDD 2003. LNCS (LNAI), vol. 2838, pp. 107–119. Springer, Heidelberg (2003). https://doi.org/10.1007/978-3-540-39804-2_12
5. Chen, Y., Trappe, W., Martin, R.P.: Detecting and localizing wireless spoofing attacks. In: 2007 4th Annual IEEE Communications Society Conference on Sensor, Mesh and Ad Hoc Communications and Networks, pp. 193–202. IEEE (2007)
6. Coull, S.E., et al.: Playing devil's advocate: inferring sensitive information from anonymized network traces. Ndss **7**, 35–47 (2007)

7. Dwork, C., Roth, A., et al.: The algorithmic foundations of differential privacy. Found. Trends Theor. Comput. Sci. **9**(3–4), 211–407 (2014)
8. Fan, L.: A survey of differentially private generative adversarial networks. In: The AAAI Workshop on Privacy-Preserving Artificial Intelligence (2020)
9. Fredrikson, M., Jha, S., Ristenpart, T.: Model inversion attacks that exploit confidence information and basic countermeasures. In: Proceedings of the 22nd ACM SIGSAC Conference on Computer and Communications Security, pp. 1322–1333. ACM (2015)
10. Goodfellow, I., et al.: Generative adversarial nets. In: Advances in Neural Information Processing Systems, pp. 2672–2680 (2014)
11. Gulrajani, I., Ahmed, F., Arjovsky, M., Dumoulin, V., Courville, A.C.: Improved training of Wasserstein gans. In: Advances in Neural Information Processing Systems, pp. 5767–5777 (2017)
12. Hayes, J., Melis, L., Danezis, G., De Cristofaro, E.: Logan: membership inference attacks against generative models. Proc. Privacy Enhan. Technol. **2019**(1), 133–152 (2019)
13. King, J., Lakkaraju, K., Slagell, A.: A taxonomy and adversarial model for attacks against network log anonymization. In: Proceedings of the 2009 ACM Symposium on Applied Computing, pp. 1286–1293 (2009)
14. Li, J., Zhou, L., Li, H., Yan, L., Zhu, H.: Dynamic traffic feature camouflaging via generative adversarial networks. In: 2019 IEEE Conference on Communications and Network Security (CNS), pp. 268–276. IEEE (2019)
15. Li, Y., Slagell, A., Luo, K., Yurcik, W.: Canine: a combined conversion and anonymization tool for processing netflows for security. In: International Conference on Telecommunication Systems Modeling and Analysis. vol. 21 (2005)
16. Lippmann, R.P., et al.: Evaluating intrusion detection systems: the 1998 Darpa off-line intrusion detection evaluation. In: Proceedings DARPA Information Survivability Conference and Exposition (DISCEX 2000). vol. 2, pp. 12–26. IEEE (2000)
17. McSherry, F., Mahajan, R.: Differentially-private network trace analysis. ACM SIGCOMM Comput. Commun. Rev. **40**(4), 123–134 (2010)
18. Mikolov, T., Sutskever, I., Chen, K., Corrado, G.S., Dean, J.: Distributed representations of words and phrases and their compositionality. In: Advances in Neural Information Processing Systems, pp. 3111–3119 (2013)
19. Mohammady, M., Wang, L., Hong, Y., Louafi, H., Pourzandi, M., Debbabi, M.: Preserving both privacy and utility in network trace anonymization. In: Proceedings of the 2018 ACM SIGSAC Conference on Computer and Communications Security, pp. 459–474 (2018)
20. Pang, R., Allman, M., Paxson, V., Lee, J.: The devil and packet trace anonymization. ACM SIGCOMM Comput. Commun. Rev. **36**(1), 29–38 (2006)
21. Riboni, D., Villani, A., Vitali, D., Bettini, C., Mancini, L.V.: Obfuscation of sensitive data in network flows. In: 2012 Proceedings IEEE INFOCOM, pp. 2372–2380. IEEE (2012)
22. Ring, M., Dallmann, A., Landes, D., Hotho, A.: Ip2vec: learning similarities between IP addresses. In: 2017 IEEE International Conference on Data Mining Workshops (ICDMW), pp. 657–666. IEEE (2017)
23. Ring, M., Schlör, D., Landes, D., Hotho, A.: Flow-based network traffic generation using generative adversarial networks. Comput. Secur. **82**, 156–172 (2019)
24. Ring, M., Wunderlich, S., Grüdl, D., Landes, D., Hotho, A.: Flow-based benchmark data sets for intrusion detection. In: Proceedings of the 16th European Cnference on Cyber Warfare and Security, pp. 361–369 (2017)

25. Shahid, M.R., Blanc, G., Jmila, H., Zhang, Z., Debar, H.: Generative deep learning for internet of things network traffic generation. In: 2020 IEEE 25th Pacific Rim International Symposium on Dependable Computing (PRDC), pp. 70–79 (2020). https://doi.org/10.1109/PRDC50213.2020.00018
26. Shokri, R., Stronati, M., Song, C., Shmatikov, V.: Membership inference attacks against machine learning models. In: 2017 IEEE Symposium on Security and Privacy (SP), pp. 3–18. IEEE (2017)
27. Son, S., Shmatikov, V.: The hitchhiker's guide to DNS cache poisoning. In: International Conference on Security and Privacy in Communication Systems, pp. 466–483. Springer (2010)
28. Torkzadehmahani, R., Kairouz, P., Paten, B.: DP-CGAN: differentially private synthetic data and label generation. In: The IEEE Conference on Computer Vision and Pattern Recognition (CVPR) Workshops (2019)
29. Wright, C., Monrose, F., Masson, G.M.: Hmm profiles for network traffic classification. In: Proceedings of the 2004 ACM Workshop on Visualization and Data Mining for Computer Security, pp. 9–15 (2004)
30. Xu, J., Fan, J., Ammar, M., Moon, S.B.: On the design and performance of prefix-preserving ip traffic trace anonymization. In: Proceedings of the 1st ACM SIGCOMM Workshop on Internet Measurement, pp. 263–266 (2001)

Comparing Local and Central Differential Privacy Using Membership Inference Attacks

Daniel Bernau[1(✉)], Jonas Robl[1], Philip W. Grassal[2], Steffen Schneider[3], and Florian Kerschbaum[4]

[1] SAP, Karlsruhe, Germany
{daniel.bernau,jonas.robl}@sap.com
[2] Heidelberg University, Heidelberg, Germany
philip-william.grassal@iwr.uni-heidelberg.de
[3] Procure.AI, London, UK
steffen.schneider@procure.ai
[4] University of Waterloo, Waterloo, Canada
florian.kerschbaum@uwaterloo.ca

Abstract. Attacks that aim to identify the training data of neural networks represent a severe threat to the privacy of individuals in the training dataset. A possible protection is offered by anonymization of the training data or training function with differential privacy. However, data scientists can choose between local and central differential privacy, and need to select meaningful privacy parameters ϵ. A comparison of local and central differential privacy based on the privacy parameters furthermore potentially leads data scientists to incorrect conclusions, since the privacy parameters are reflecting different types of mechanisms.

Instead, we empirically compare the relative privacy-accuracy trade-off of one central and two local differential privacy mechanisms under a white-box membership inference attack. While membership inference only reflects a lower bound on inference risk and differential privacy formulates an upper bound, our experiments with several datasets show that the privacy-accuracy trade-off is similar for both types of mechanisms despite the large difference in their upper bound. This suggests that the upper bound is far from the practical susceptibility to membership inference. Thus, small ϵ in central differential privacy and large ϵ in local differential privacy result in similar membership inference risks, and local differential privacy can be a meaningful alternative to central differential privacy for differentially private deep learning besides the comparatively higher privacy parameters.

Keywords: Anonymization · Membership inference · Neural networks

J. Robl and P. W. Grassal—Contributed equally to this research.
P. W. Grassal and S. Schneider—This work was done during an internship at SAP.

K. Barker and K. Ghazinour (Eds.): DBSec 2021, LNCS 12840, pp. 22–42, 2021.
https://doi.org/10.1007/978-3-030-81242-3_2

1   Introduction

Neural networks have successfully been applied to a wide range of learning tasks, each requiring its own specific set of training data, architecture and hyperparameters to achieve meaningful classification accuracy and foster generalization. In some learning tasks data scientists have to deal with personally identifiable or sensitive information, which results in two challenges. First, legal restrictions might not permit collecting, processing or publishing certain data, such as National Health Service data [5]. Second, membership inference (MI) [20,31,38] and model inversion attacks [15,16] are capable of identifying and reconstructing training data based on information leakage from a trained, published neural network model. A mitigation to both challenges is offered by anonymized deep neural network training with differential privacy (DP). However, a data scientist can choose between two categories of DP mechanisms: local DP (LDP) [40] and central DP (CDP) [9]. LDP perturbs the training data before any processing takes place, whereas CDP perturbs the gradient update steps during training. The degree of perturbation, which affects the accuracy of the trained neural network on test data, is calibrated for both DP categories by adjusting their respective privacy parameter ϵ. Choosing ϵ too large will unlikely mitigate privacy attacks such as MI, and setting ϵ too small will significantly reduce model accuracy. Balancing the privacy-accuracy trade-off is a challenging problem especially for data scientists who are not experts in DP. Furthermore, data scientists might rule out LDP when designing differentially private neural networks due to concerns raised by the comparatively higher privacy parameter ϵ in LDP. In this work, we compare the empirical privacy protection under the white-box MI attack of Nasr et al. [31] for LDP and CDP mechanisms for learning problems from diverse domains: consumer preferences, face recognition and health data. The MI attack indicates a lower bound on the inference risk whereas DP formulates an upper bound [24,43,44], but in practice even high privacy parameters such as experienced in LDP may already offer protection. This work makes the following contributions:

- Comparing LDP and CDP by the average precision of their MI precision-recall curve as privacy measure, and showing that under this measure LDP and CDP have similar privacy-accuracy trade-offs despite vastly different ϵ.
- Showing that CDP mechanisms are not achieving a consistently better privacy-accuracy trade-off on various datasets and reference models. The trade-off rather depends on the specific dataset.
- Analyzing the relative privacy-accuracy trade-off and showing that it is not constant over ϵ, but that for each data set there are ranges where the relative trade-off is greater for protection against MI than accuracy.

Section 2 revisits differential privacy and Sect. 3 formulates our approach for comparing LDP and CDP under membership inference. We describe evaluation datasets in Sect. 4. Findings are presented in Sect. 5 and discussed in Sect. 6. Related work and conclusions are provided in Sect. 7 and Sect. 8.

2   Differential Privacy

DP [8] anonymizes a dataset $\mathcal{D} = \{d_1, \ldots, d_n\}$ by perturbation and can be either enforced centrally to a function $f(\mathcal{D})$, or locally to each entry $d \in \mathcal{D}$.

2.1   Central DP

In central DP the aggregation function $f(\cdot)$ is evaluated and perturbed by a trusted server. Due to perturbation, it is no longer possible for an adversary to confidently determine whether $f(\cdot)$ was evaluated on $\mathcal{D}$, or some neighboring dataset $\mathcal{D}'$ differing in one element. Assuming that every participant is represented by one element, privacy is provided to participants in $\mathcal{D}$ as their impact on $f(\cdot)$ is limited. *Mechanisms* $\mathcal{M}$ fulfilling Definition 1 are used for perturbation of $f(\cdot)$ [9]. We refer to the application of a mechanism $\mathcal{M}$ to a function $f(\cdot)$ as *central differential privacy*. CDP holds for all possible differences $\|f(\mathcal{D}) - f(\mathcal{D}')\|_2$ by adapting to the global sensitivity of $f(\cdot)$ per Definition 2.

Definition 1 ((ϵ, δ)-central differential privacy). *A mechanism* $\mathcal{M}$ *gives* (ϵ, δ)-*central differential privacy if* $\mathcal{D}, \mathcal{D}' \subseteq \mathcal{DOM}$ *differing in at most one element, and all outputs* $\mathcal{S} \subseteq \mathcal{R}$

$$\Pr[\mathcal{M}(\mathcal{D}) \in \mathcal{S}] \leq e^\epsilon \cdot \Pr[\mathcal{M}(\mathcal{D}') \in \mathcal{S}] + \delta$$

Definition 2 (Global ℓ_2 Sensitivity). *Let* $\mathcal{D}$ *and* $\mathcal{D}'$ *be neighboring. The global* ℓ_2 *sensitivity of a function* f, *denoted by* Δf, *is defined as*

$$\Delta f = max_{\mathcal{D}, \mathcal{D}'} \|f(\mathcal{D}) - f(\mathcal{D}')\|_2.$$

Definition 3 (Gaussian Mechanism [10]). *Let* $\epsilon \in (0, 1)$ *be arbitrary. For* $c^2 > 2ln(\frac{1.25}{\delta})$, *the Gaussian mechanism with parameter* $\sigma \geq c\frac{\Delta f}{\epsilon}$ *gives* (ϵ, δ)-*CDP, adding noise scaled to* $\mathcal{N}(0, \sigma^2)$.

For CDP in deep learning we use differentially private versions[1] of two standard gradient optimizers: SGD and Adam [27]. We refer to these CDP optimizers as DP-SGD and DP-Adam. A CDP optimizer represents a differentially private training mechanism $\mathcal{M}_{nn}$ that updates the weight coefficients θ_t of a neural network per training step $t \in T$ with $\theta_t \leftarrow \theta_{t-1} - \alpha(\tilde{g})$, where $\tilde{g} = \mathcal{M}_{nn}(\partial loss/\partial \theta_{t-1})$ denotes a Gaussian perturbed gradient and α is some scaling function on $\tilde{g}$ to compute an update, i.e., learning rate or running moment estimations. Differentially private noise is added by the Gaussian mechanism of Definition 3 [1]. After T update steps, $\mathcal{M}_{nn}$ outputs a differentially private weight matrix θ which is used by the prediction function $h(\cdot)$ of a neural network.

---

[1] We used Tensorflow Privacy: https://github.com/tensorflow/privacy.

A CDP gradient optimizer bounds the sensitivity of the computed gradients by clipping norm $\mathcal{C}$ based on which the gradients get clipped before perturbation. Since weight updates are performed iteratively during training a composition of $\mathcal{M}_{nn}$ is required until the training step T is reached and the final private weights θ are obtained. For CDP we measure privacy decay under composition by tracking σ of the Gaussian mechanism. After training we transform and compose σ under Renyi DP [29], and transform the aggregate again to CDP. We choose this accumulation method over other composition schemes [1, 25] due to the tighter bound for heterogeneous mechanism invocations.

2.2   Local DP

We refer to the perturbation of entries $d \in \mathcal{D}$ as local differential privacy [40]. LDP is the standard choice when the server which evaluates a function $f(\mathcal{D})$ is untrusted. We adapt the definitions of Kasiviswanathan et al. [26] to achieve LDP by using local randomizers $\mathcal{LR}$. In the experiments within this work we use a local randomizer to perturb each record $d \in \mathcal{D}$ independently. Since a record may contain multiple correlated features (e.g., items in a preference vector) a local randomizer must be applied sequentially which results in a linearly increasing privacy loss. A series of local randomizer executions per record composes a local algorithm according to Definition 5. ϵ-local algorithms are ϵ-local differentially private [26], where ϵ is a summation of all composed local randomizer guarantees. We perturb low domain data with randomized response [41], a (composed) local randomizer. By Eq. (1) randomized response yields $\epsilon = \ln(3)$ LDP for a one-time collection of values from binary domains (e.g., $\{\text{yes}, \text{no}\}$) with two fair coins [12]. That is, retention of the original value with probability $\rho = 0.5$ and uniform sampling with probability $(1 - \rho) \cdot 0.5$.

$$\epsilon = \ln \left(\frac{\rho + (1 - \rho) \cdot 0.5}{(1 - \rho) \cdot 0.5} \right) = \ln \left(\frac{\Pr[\text{yes}|\text{yes}]}{\Pr[\text{yes}|\text{no}]} \right). \tag{1}$$

Definition 4 (Local differential privacy). *A local randomizer (mechanism) $\mathcal{LR} : \mathcal{DOM} \rightarrow \mathcal{S}$ is ϵ-local differentially private, if $\epsilon \geq 0$ and for all possible inputs $v, v' \in \mathcal{DOM}$ and all possible outcomes $s \in \mathcal{S}$ of $\mathcal{LR}$*

$$\Pr[\mathcal{LR}(v) = s] \leq e^\epsilon \cdot \Pr[\mathcal{LR}(v') = s]$$

Definition 5 (Local Algorithm). *An algorithm is ϵ-local if it accesses the database $\mathcal{D}$ via $\mathcal{LR}$ with the following restriction: for all $i \in \{1, \dots, |\mathcal{D}|\}$, if $\mathcal{LR}_1(i), \dots, \mathcal{LR}_k(i)$ are the algorithms invocations of $\mathcal{LR}$ on index i, where each $\mathcal{LR}_j$ is an ϵ_j-local randomizer, then $\epsilon_1 + \dots + \epsilon_k \leq \epsilon$.*

Definition 6 (Laplace Mechanism [10]). *Given a numerical query function $f : DOM \rightarrow \mathbb{R}^k$, the Laplace mechanism with $\lambda = \frac{\Delta_f}{\epsilon}$ is an ϵ-differentially private mechanism, adding noise scaled to $Lap(\lambda, \mu = 0)$.*

In our evaluation we also look at image data for which we rely on the local randomizer by Fan [14] for LDP image pixelization. The randomizer applies the Laplace mechanism of Definition 6 with scale $\lambda = \frac{255 \cdot m}{b^2 \cdot \epsilon}$ to each pixel, thus fulfilling Definition 4. Parameter m represents the neighborhood in which LDP is provided. Full neighborhood for an image dataset would require that any picture can become any other picture. In general, providing DP or LDP within a large neighborhood will require high ϵ values to retain meaningful image structure. High privacy will result in random black and white images. Within this work we consider the use of LDP and CDP for deep learning along a generic data science process (e.g., CRISP-DM [42]). In such a processes the dataset $\mathcal{D}$ of a data owner $\mathcal{DO}$ is (i) transformed, and (ii) used to learn a model function $h(\cdot)$ (e.g., classification), which (iii) afterwards is deployed for evaluation by third parties. In the following $h(\cdot)$ will represent a neural network. DP is applicable at every stage in the data science process. In the form of LDP by perturbing each record $d \in \mathcal{D}$, while learning $h(\cdot)$ centrally with a CDP gradient optimizer, or to the evaluation of $h(\cdot)$ by federated learning with voting. We leave learning with more than two parties, such as used in PATE [33] with CDP or amplification by shuffling for LDP [11] as future work. However, independent of the stage of application, the privacy-accuracy trade-off is of particular interest. We follow the evaluation of regularization techniques that apply noise to the training data to foster generalization [17,18,28] and measure utility by the test accuracy of $h(\cdot)$.

3   White-Box MI Attack

Membership inference (MI) attacks aim at identifying the presence or absence of individual records in the training data of data owner $\mathcal{DO}$. MI attacks are of particular importance for members of the training dataset when the nature of the training dataset is revealing sensitive information. For example, a medical training dataset containing patients with different types of cancer, or a training dataset that is used to predict the week of pregnancy based on the shopping cart [21]. A related attack building upon MI is attribute inference [44] where individual records are partially known and specific attribute values shall be inferred. In this work we solely consider MI since protection against MI offers protection against attribute inference. In specific, we consider white-box MI by Nasr et al. [31] which is stronger than previously suggested black-box MI attacks (e.g., Shokri et al. [38]). The MI attack assumes an honest-but-curious adversary $\mathcal{A}$ with access to a trained prediction function $h(\cdot)$, knowledge about the hyperparameters and DP mechanisms that were used for training. We refer to the trained prediction function as *target model* and the training data of $\mathcal{DO}$ as $\mathcal{D}_{\text{target}}^{\text{train}}$. Given this accessible information $\mathcal{A}$ wants to learn a binary classifier, the *attack model*, that allows to classify data into members and non-members w.r.t. the target model training dataset with high accuracy. The accuracy of an MI attack model is evaluated on a balanced dataset including all members (target model training data) and an equal number of non-members (target model test data), which simulates the worst case where $\mathcal{A}$ tests membership for all training records. White-box MI exploits that an ML classifier such as a neural network (NN) tends to classify a record $d = (x, y)$ from its training dataset $\mathcal{D}_{target}^{train}$ with different confidence $p(\mathbf{x})$ given $h(\mathbf{x})$ for features $\mathbf{x}$ and true label y than a record $d \notin \mathcal{D}_{target}^{train}$. White-box MI makes two assumptions about $\mathcal{A}$. First, $\mathcal{A}$ is able to observe internal features of the ML model in addition to external features (i.e., model outputs). The internal features comprise observed losses $L(h(x; W))$, gradients $\frac{\delta L}{\delta W}$ and the learned weights W of $h(\cdot)$. Second, $\mathcal{A}$ is aware of a portion of $\mathcal{D}_{target}^{train}$ and $\mathcal{D}_{target}^{test}$. These portions were set to 50% by Nasr et al. [31] and will be the same within this work to allow comparison. Second, $\mathcal{A}$ extracts internal and external features of a balanced set of confirmed members and non-members. An illustration of the white-box MI attack is given in Fig. 1. Again, $\mathcal{A}$ is assumed to know a portion of $\mathcal{D}_{target}^{train}$ and $\mathcal{D}_{target}^{test}$ and generates attack features by passing these records through the trained target model. $\mathcal{A}$ trains a binary classification attack model per target variable $y \in Y$ to map $p(\mathbf{x})$ to the indicator "in" or "out". The set $(L(h(x; W)), \frac{\delta L}{\delta W}, p(\mathbf{x}), y, \mathtt{in/out})$ serves as attack model training data, i.e., $\mathcal{D}_{attack}^{train}$. Thus, the MI attack model exploits the imbalance between predictions on $d \in \mathcal{D}_{target}^{train}$ and $d \notin \mathcal{D}_{target}^{train}$. Attack model accuracy is computed on features extracted from the target model likewise.

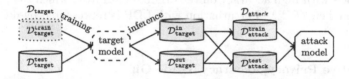

Fig. 1. White-box MI with attack features $(y^*, p(\mathbf{x}), L(h(x; W), y), \frac{\delta L}{\delta W})$. LDP perturbation on $\mathcal{D}_{target}^{train}$ (dotted) and CDP on target model training (dashed). Target model training is colored: training (violet) and validation (green). (Color figure online)

3.1   Evaluating CDP and LDP Under MI

DP has been shown to formulate a theoretical upper bound on the accuracy of MI adversaries [44], and thus the use of DP should impact the classification accuracy of $\mathcal{A}$. To illustrate the effect of the privacy parameter ϵ on the MI attack we focus on two questions related to the identifiability of training data within this work: "How many records predicted as **in** are truly contained in the training dataset?" (precision), and "How many truly contained records are predicted as **in**?" (recall). For analysis we use precision-recall curves which depict the precision and recall for various classification thresholds, and thus reflect the possible MI attack accuracies of $\mathcal{A}$. We compare the precision-recall curves by their average precision (AP) to assess the overall effect of DP on MI. The AP approximates the integral under the precision-recall curve as a a weighted mean of the precision P per threshold t and the increase in recall R from the previous threshold, i.e.: $AP = \sum_t (R_t - R_{t-1}) \cdot P_t$. We prefer this non-interpolated technique over interpolated calculations of the area under curve, since the precision-recall curve is not guaranteed to decline monotonically and thus the linear trapezoidal interpolation might yield an overoptimistic representation [7,13]. Good MI attack models will realize an AP of close to 1 while poor MI attack models will be close to the baseline of uniform random guessing, hence $AP = 0.5$. The data owner $\mathcal{DO}$ has two options to apply DP against MI within the data science process introduced in Sect. 2. Either in the form of LDP by applying a local randomizer to the training data and using the resulting $\mathcal{LR}(\mathcal{D}_{target}^{train})$ for training, or CDP with a differentially private optimizer on $\mathcal{D}_{target}^{train}$. A discussion and comparison of LDP and CDP purely based on the privacy parameter ϵ likely falls short and potentially leads data scientists to incorrect conclusions, since the privacy parameters are reflecting different types of mechanisms. Furthermore, data scientists give up flexibility w.r.t. applicable learning algorithms, if ruling out the use of LDP due to comparatively greater ϵ and instead solely investigating CDP (e.g., DP-SGD). We suggest to compare LDP and CDP by their concrete effect on the AP and the resulting privacy-accuracy trade-off. While we consider a specific MI attack our methodology is applicable to other MI attacks as well. Models that use CDP are represented by dashed lines in Fig. 1. In the LDP setup, the target model is trained with perturbed records from a local randomizer, i.e., $\mathcal{LR}(\mathcal{D}_{target}^{train})$. However, in order to increase his attack accuracy $\mathcal{A}$ needs to learn attack models with high accuracy on the original data from which the perturbed records stem, i.e., $\mathcal{D}_{target}^{train}$. Perturbation with LDP is represented by dotted lines in Fig. 1.

3.2 Relative Privacy-Accuracy Trade-Off

We calculate the relative privacy-accuracy trade-off for LDP and CDP as the relative difference between $\mathcal{A}$'s change in AP to $\mathcal{DO}$'s change in test accuracy. Let AP_{orig}, AP_{ϵ} be the MI APs and ACC_{orig}, ACC_{ϵ} be the test accuracies for the original and DP target model. Furthermore, let ACC_{base} be the baseline test accuracy of uniform random guessing $1/\mathbb{C}$, where $\mathbb{C}$ denotes the number of classes in the dataset, and AP_{base} be the baseline AP at 0.5. We fix ACC_{base}, AP_{base}, since $\mathcal{A}$ or $\mathcal{DO}$ would perform worse than uniform random guessing at lower values. Rearranging and bounding the cases where AP and ACC increases over ϵ yields:

$$\varphi = \frac{(AP_{orig} - AP_{\epsilon})/(AP_{orig} - AP_{base})}{(ACC_{orig} - ACC_{\epsilon})/(ACC_{orig} - ACC_{base})}$$

$$\varphi = \frac{\max(0, AP_{orig} - AP_{\epsilon}) \cdot (ACC_{orig} - ACC_{base})}{\max(0, ACC_{orig} - ACC_{\epsilon}) \cdot (AP_{orig} - AP_{base})}$$

$$\varphi = \min\left(2, \frac{\max(0, (AP_{orig} - AP_{\epsilon}) \cdot (ACC_{orig} - ACC_{base}))}{\max(0, (ACC_{orig} - ACC_{\epsilon}) \cdot (AP_{orig} - AP_{base}))}\right)$$

To avoid φ from approaching infinitely large values when the accuracy remains stable while AP decreases significantly, and the undefined case of $ACC_{orig} \leq ACC_{\epsilon}$, we bound φ at 2. In consequence, when the relative gain in privacy (lower AP) exceeds the relative loss in accuracy, it applies that $1 < \varphi \leq 2$, and $0 \leq \varphi < 1$ when the loss in test accuracy exceeds the gain in privacy. Hence, φ quantifies the relative loss in accuracy and the relative gains in privacy for a given privacy parameter ϵ and captures the relative privacy-accuracy trade-off as a ratio which we seek to maximize.

4   Datasets and Learning Tasks

We consider four datasets for experiments. The datasets have been used in related work on MI and face recognition. The reference datasets are mostly unbalanced w.r.t. the amount of training data per training label, a characteristic that we found to benefit MI attacks. Each dataset is also summarized in Table 1 and the distributions for the two unbalanced datasets Texas Hospital Stays and Purchases Shopping Carts are provided in the appendix.

Texas Hospital Stays. The Texas Hospital Stays dataset [38] is an unbalanced dataset and consists of high dimensional binary vectors representing patient health features. Each record within the dataset is labeled with a procedure. The learning task is to train a fully connected neural network for classification of patient features to a procedure and we do not try to re-identify a known individual, and fully comply with the data use agreement for the original public use data file. We train and evaluate models for a set of most common procedures $\mathbb{C} \in \{100, 150, 200, 300\}$. Depending on the number of procedures the dataset comprises $67,330$–$89,815$ records and $6,170$–$6,382$ features. To allow comparison with related work [31,38], we train and test the target model on $n = 10,000$ records respectively.

Purchases Shopping Carts. This dataset is also unbalanced and consists of binary vectors with 600 features that represent customer shopping carts [38]. However, a significant difference to the Texas Hospital Stays dataset is that the number of features is almost 90% lower. Each vector is labeled with a customer group. The learning task is to classify shopping carts to customer groups by using a fully connected neural network. The dataset is provided in four variations with varying numbers of labels $\mathbb{C} \in \{10, 20, 50, 100\}$ and comprises $38,551$–$197,324$ records. We sample $n = 8,000$ records each for training and testing the target model. Again, this methodology ensures comparability with related work [31,38].

Labeled Faces in the Wild. The Labeled Faces in the Wild (LFW) dataset contains labeled images each depicting a specific person with a resolution of 250×250 pixels (i.e., features) [22]. The dataset has a long distribution tail w.r.t. to the number of images per label. We thus focus on learning the topmost classes $\mathbb{C} \in \{20, 50, 100\}$ with 1906, 2773 and 3651 overall records respectively. We start our comparison of LDP and CDP from a pre-trained VGG-Very-Deep-16 CNN faces model [34] by keeping the convolutional core, exchanging the dense layer at the end of the model and training for LFW grayscale faces.

For LDP, we apply differentially private image pixelization within neighborhood $m = \sqrt{250 \times 250}$ and avoid coarsening by setting $b = 1$. We transform all images to grayscale before LDP and CDP training.

Skewed Purchases. We specifically crafted this balanced dataset[2] to mimic a transfer learning task, i.e., the application of a trained model to novel data which is similar to the training data w.r.t. format but following a different distribution. This situation arises for Purchases Shopping Carts, if for example not enough high-quality shopping cart data for a specific retailer are available yet. Thus, only few high-quality data (e.g., manually crafted examples) can be used for testing and large amounts of low quality data from potentially different distributions for training (e.g., from other retailers). In effect the distribution between train and test data varies for this dataset. Similar to Purchases Shopping Carts the dataset consists of $200,000$ records with 600 features and is analyzed for $\mathbb{C} \in \{10, 20, 50, 100\}$ labels. However, each vector x in the training dataset X is generated by using two independent random coins to sample a value from $\{0, 1\}$ per position $i = 1, \ldots, 600$. The first coin steers the probability $\Pr[x_i = 1]$ for a fraction of 600 positions per record x. We refer to these positions as indicator bits (*ind*) which indicate products frequently purchased together. The second coin steers the probability $\Pr[x_i = 1]$ for a fraction of $600 - (\frac{600}{|\mathbb{C}|})$ positions per record. We refer to these positions as noise bits (*noise*) that introduce scatter in addition to *ind*. We let $\Pr_{ind}[x_i = 1] = 0.8 \wedge \Pr_{noise}[x_i = 1] = 0.2, \forall x \in X_{train}$ and $\Pr_{ind}[x_i = 1] = 0.8 \wedge Pr_{noise}[x_i = 1] = 0.5, \wedge x \in X_{test}, 1 \leq i \leq 600$. The difference in information entropy between test and train data is ≈ 0.3.

5   Experiments

We perform an experiment which compares the privacy-accuracy trade-off for LDP and CDP by MI AP instead of privacy parameter ϵ per dataset. The results of each experiment are visualized by three sets of figures. First, we compare the relative privacy-accuracy trade-off φ resulting from test accuracy and MI AP over ϵ. We present this information for CDP per dataset in Figs. 2, 3, 4 and 5a, b, c and for LDP in Figs. 2, 3, 4 and 5d, e, f. The obtained information serves as basis to identify privacy parameters at which the MI AP is converging towards the baseline. Second, we state the precision-recall curves from which MI AP was calculated to illustrate the slope with which precision and recall are diverging from the baseline for LDP and CDP in Figs. 2, 3, 4 and 5g,h. Third, we compare the absolute privacy-accuracy trade-offs per dataset for both LDP and CDP in a scatterplot. We present this information in Figs. 2, 3, 4 and 5i. For each dataset the model training stops once the test data loss is stagnating (i.e., early stopping) or a maximum number of epochs is reached. This design avoids excessive overfitting and increases real-world relevance. For all executions of the experiment

---

[2] We provide this dataset along with all evaluation code on GitHub: https://github.com/SAP-samples/security-research-membership-inference-and-differential-privacy.

CDP noise is sampled from a Gaussian distribution (cf. Definition 3) with scale $\sigma = noise\ multiplier\ z \times clipping\ norm\ \mathcal{C}$. We evaluate increasing noise regimes per dataset by evaluating noise multipliers $z \in \{0.5, 2, 4, 6, 16\}$ and calculate the resulting ϵ at a fixed $\delta = \frac{1}{n}$. However, since batch size, dataset size and number of epochs are also influencing the Renyi differential privacy accounting a fixed z will inevitably result in different composed ϵ for different datasets. For LDP we use the same hyperparameters as in the original training and evaluate two local randomizers, namely randomized response and LDP image pixelization with the Laplace mechanism. For each randomizer we state the individual ϵ_i per invocation (i.e., per anonymized value). We apply randomized response to all datasets except LFW with a range of privacy parameter values $\epsilon_i \in \{0.1, 0.5, 1, 2, 3\}$ that reflect retention probabilities ρ from 5%–90% (cf. Eq. (1)). For LFW each pixel is perturbed with Laplace noise, and also investigate a wide range of resulting noise regimes by varying ϵ_i.

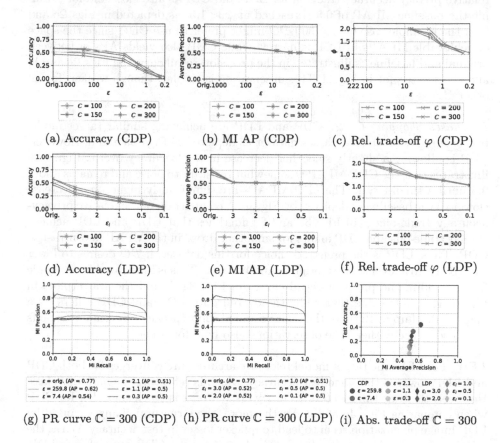

(a) Accuracy (CDP)   (b) MI AP (CDP)   (c) Rel. trade-off φ (CDP)

(d) Accuracy (LDP)   (e) MI AP (LDP)   (f) Rel. trade-off φ (LDP)

(g) PR curve $\mathbb{C} = 300$ (CDP) (h) PR curve $\mathbb{C} = 300$ (LDP) (i) Abs. trade-off $\mathbb{C} = 300$

Fig. 2. Texas Hospital Stays accuracy and privacy (error bars lie within points)

32   D. Bernau et al.

For sake of completeness we provide the resulting overall privacy parameters ϵ, z, hyperparameters and train accuracies for all datasets for LDP and CDP in Table 1 and 2 in the appendix. The experiment is repeated five times per dataset to stabilize measurements and we report mean values with error bars unless otherwise stated. Precision-recall curves depict all experiment data.

Texas Hospital Stays. For Texas Hospital Stays we observe that LDP and CDP are achieving very similar privacy-accuracy trade-offs under MI. The main difference in LDP and CDP is observable in a smoother decrease of target model accuracy for CDP in contrast to LDP, which are depicted in Figs. 2a and d. The smoother decay also manifests in a slower drop of MI AP for CDP in comparison to LDP as stated in Figs. 2b and e. Texas Hospital Stays represents an unbalanced high dimensional dataset and both factors foster MI. However, the increase in dataset imbalance by increasing $\mathbb{C}$ is negligible w.r.t. MI AP. The relative privacy-accuracy trade-off for LDP and CDP is also close and for example the baseline MI AP of 0.5 is reached at $\varphi \approx 1.5$, as depicted in Figs. 2c and f. In the example case of $\mathbb{C} = 300$ $\mathcal{DO}$ might prefer to use CDP, since the space of achievable MI APs in LDP is narrow while CDP also yields APs in between original and baseline as illustrated in the precision-recall curves in Figs. 2g and h, and the scatterplot in Fig. 2i. This observation is similar, though weaker, for all other $\mathbb{C}$.

Purchases Shopping Carts. CDP and LDP are achieving similar target model test accuracies on the Purchases dataset as depicted in Figs. 3a and d. However, LDP is allowing a slightly smoother decrease in test accuracy over ϵ. Figure 3b illustrates that the CDP MI AP is somewhat resistant to noise and remains above 0.5 until a small $\epsilon \approx 1$. The LDP MI APs are significantly higher and decrease slower to the baseline as depicted by Fig. 3e. A comparison of the relative privacy-accuracy trade-offs φ in Figs. 3c and f underlines that CDP and LDP achieve similar trade-offs and LDP allows for smoother drops in the MI AP in contrast to CDP. Thus, LDP is the preferred choice for this dataset, if $\mathcal{DO}$ desires to lower the MI AP to a level *between* original and baseline. This is illustrated for example for $\mathbb{C} = 50$ in the precision-recall curves in Figs. 3g, h and the scatterplots in Fig. 3i. It is noticeable that while the overall ϵ for LDP and CDP differs by a magnitude of up to 10 times the relative and absolute privacy-accuracy trade-offs are close to each other. The observations also hold for other $\mathbb{C}$.

LFW. For LFW the target model reference architecture converges for both CDP and LDP towards the same test accuracy, which is reflecting the majority class. However, the target model test accuracy decay over ϵ is much smoother for CDP when comparing Figs. 4a and d. Furthermore, the structural changes caused by LDP image pixelization seem to lead to quicker losses in test accuracy. W.r.t. the relative privacy-accuracy trade-off φ in Figs. 4c and f CDP outperforms LDP. At MI AP $= 0.5$ CDP achieves $\varphi \approx 1.5$ for all $\mathbb{C}$ while LDP yields $\varphi \approx 1.1$ for all $\mathbb{C}$. The $\varphi = 0$ observed at $\epsilon_i = 10000$ for $\mathbb{C} = 100$ is due to an actual increase in AP that is comparatively larger than the decrease in test accuracy.

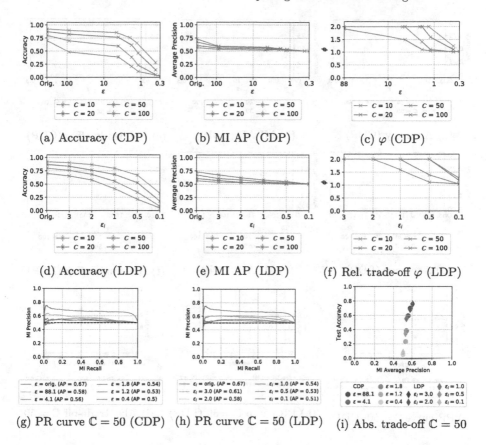

Fig. 3. Purchases accuracy and privacy (error bars lie within points)

The exemplary precision-recall curves for $\mathbb{C} = 50$ in Figs. 4g and h furthermore illustrate that CDP can already have a large effect on MI AP at high ϵ. In addition, we observe from Fig. 4i that CDP realizes a strictly better absolute privacy-accuracy trade-off under MI for $\mathbb{C} = 50$.

Skewed Purchases. The effects of dimensionality and imbalance of a dataset on MI have been addressed by related work [31,38]. However, the effect of a domain gap between training and test data which is found in transfer learning when insufficient high-quality data for training is initially available and reference data that potentially follows a different distribution has not been addressed. For this task we consider the Skewed Purchases dataset. Figures 5a and d show that the LDP test accuracy is in fact only decreasing at very small ϵ_i whereas CDP again gradually decreases over ϵ. This leads to a consistently higher test accuracy in comparison to CDP. W.r.t. the relative privacy-accuracy trade-off LDP outperforms CDP as depicted by φ in Figs. 5c and f. However, we observe several outliers. Most notably for CDP, the MI AP decreases for $\mathbb{C} = 100$ and

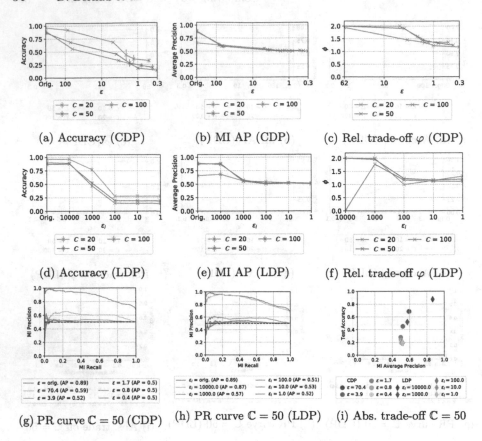

Fig. 4. LFW accuracy and privacy (error bars lie within points)

large ϵ values, but increases for small ϵ as shown in Fig. 5b. This is a consequence of the target model resorting to random guessing for test records. Similarly, for LDP the MI AP for $\mathbb{C} \in \{10, 100\}$ first decreases before recovering again as depicted in Fig. 5e. We reason about the cause of these outliers by analyzing the target model decisive confidence values. LDP generalizes the training data towards the test data, however, at $\epsilon_i = 1.0$ LDP leads to nearly indistinguishable test and train distributions. Thus, the decisive softmax confidence of the target model increases in comparison to smaller and larger ϵ_i. For $\mathbb{C} = 10$ the absolute privacy-accuracy trade-off is also favorable for LDP as depicted in Fig. 5i.

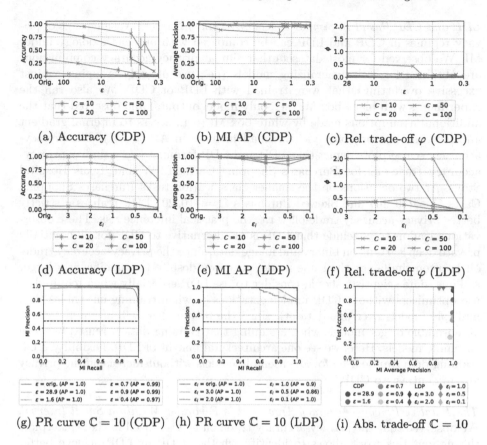

(a) Accuracy (CDP)   (b) MI AP (CDP)   (c) Rel. trade-off φ (CDP)

(d) Accuracy (LDP)   (e) MI AP (LDP)   (f) Rel. trade-off φ (LDP)

(g) PR curve $\mathbb{C} = 10$ (CDP)   (h) PR curve $\mathbb{C} = 10$ (LDP)   (i) Abs. trade-off $\mathbb{C} = 10$

Fig. 5. Skewed Purchases accuracy and privacy (error bars lie within points)

6   Discussion

Privacy Parameter ϵ Alone is Unsuited to Compare and Select and Compare DP Mechanisms. We consistently observed that while the theoretic upper bound on inference risk reflected by ϵ in LDP is higher by a factor of hundreds or even thousands in comparison to CDP, the practical protection against a white-box MI attack is actually not considerably weaker at similar model accuracy. For Texas Hospital Stays LDP mitigates white-box MI at an overall $\epsilon = 6382$ whereas CDP lies between $\epsilon = 0.9$ for $\mathbb{C} = 100$ and $\epsilon = 0.3$ for $\mathbb{C} = 300$. This observation at the baseline AP also holds for Purchases Shopping Carts where LDP $\epsilon = 60$ and CDP is between $\epsilon = 0.4$ for $\mathbb{C} = 10$ and $\epsilon = 0.3$ for $\mathbb{C} = 100$, and LFW (LDP $\epsilon = 62.5 \times 10^2$, CDP $\epsilon = 2.1$ to $\epsilon = 1.5$). Thus, we note that assessing privacy solely based on ϵ falls short. Given the results of the previous sections we rather encourage data scientists to also quantify privacy under an empirical attack such as white-box MI in addition to ϵ.

LDP and CDP Result in Similar Privacy-Accuracy Curves. A wide range of privacy regimes in CDP and LDP can be compared with our methodology under MI. We observed for most datasets that similar privacy-accuracy combinations are obtained for well generalizing models (i.e., use of early stopping against excessive overfitting) that were trained with LDP or CDP. We also ran the experiments with black-box MI (i.e., only model outputs) and observed that the additional assumptions made by white-box MI (e.g., access to internal gradient and loss information) only yield a small increase in AP (3–5%). The privacy-accuracy scatterplots depict that LDP and CDP formulate very similar privacy-accuracy trade-offs for Purchases Shopping Carts, LFW and Texas Hospital Stays. At two occasions on the smaller classification tasks Purchases Shopping Carts $\mathbb{C} = \{10, 20\}$ and Skewed Purchases $\mathbb{C} = \{10, 20\}$ LDP realizes a strictly better privacy-accuracy trade-off w.r.t. the practical inference risk. These observations lead us to conclude that LDP is an alternative to CDP for differentially private deep learning on binary and image data, since the privacy-accuracy trade-off is often similar at the same model accuracy despite the significantly larger ϵ. Thus, data scientists should consider to use LDP especially when required to use optimizers without CDP implementations or when training ensembles (i.e., multiple models over one dataset), since the privacy loss will accumulate over all ensemble target models when assuming that training data is reused between ensemble models. Here, we see one architectural benefit of LDP: flexibility. LDP training data can be used for all ensemble models without increasing the privacy loss in contrast to CDP.

The Relative Privacy-Accuracy Trade-Off is Favorable Within a Small Interval. We observed that the privacy-accuracy trade-off as visualized in the scatterplots throughout this work allows to identify whether CDP or LDP achieve better test accuracy at similar APs. However, the scatterplots do not reflect whether target model test accuracy is decreasing slower, similar or stronger than MI AP decreases over the privacy parameter ϵ. For this purpose we introduced φ. We found that φ allows to identify ϵ intervals in which the AP loss is stronger than the test accuracy loss for all datasets. On the high dimensional datasets Texas Hospital Stays and LFW CDP consistently achieves higher φ than LDP. In contrast, φ values are similar for LDP and CDP on Purchases, and superior for LDP on Skewed Purchases.

7   Related Work

Our work is related to DP in neural networks, attacks against the confidentiality of training data and performance benchmarking of neural networks.

CDP is a common approach to realize differentially private neural networks by adding noise to the gradients during model training. Fundamental approaches for perturbation with the differentially private gradient descent (DP-SGD) during model training were provided by Song et al. [39], Bassily et al. [4] and Shokri et al. [37]. Abadi et al. [1] formulated the DP-SGD that was used in this work.

Mironov [29] introduces Renyi DP for measuring the DP-SGD privacy loss over composition. Iyengar et al. [23] suggest a hyperparameter free algorithm for differentially private convex optimization for standard optimizers.

Fredrikson et al. [15,16] formulate model inversion attacks that use target model softmax confidence values to reconstruct training data per class. In contrast, MI attacks address the threat of identifying individual records in a dataset [3,36]. Yeom et al. [44] have demonstrated that the upper bound on MI risk for CDP can be converted into an expected bound for MI Advantage. We state MI precision and recall, arguing that in is the sensitive information. Jaymaran and Evans [24] showed that the theoretic MI upper bound and the achievable MI lower bound are far apart in CDP. We observe, that LDP can be an alternative to CDP as the upper and lower bounds are even farther apart from each other. Shokri et al. [32] formulate an optimal mitigation against their MI attack [38] by using adversarial regularization. By applying the MI attack gain as a regularization term to the objective function of the target model, a non-leaking behavior is enforced w.r.t. MI. While their approach protects against their MI adversary, DP mitigates any adversary with arbitrary background information. Carlini et al. [6] suggest *exposure* as a metric to measure the extent to which neural networks memorize sensitive information. Similar to our work, they apply DP for mitigation. We focus on attacks against machine learning models targeting identification of members of the training dataset. Abowd and Schmutte [2] describe an economic social choice framework to choose privacy parameter ϵ. We compare LDP and CDP mechanisms aside from ϵ. Rahman et al. [35] applied a black-box MI attack against DP-SGD models on CIFAR-10 and MNIST. They evaluate the severity of MI attack by the F1-score which results in numerically higher scores, but assumes out labels to be sensitive.

MLPERF [30] and DPBench [19] are frameworks for machine learning performance measurements and evaluation of DP. We focus on comparing the privacy-utility trade-off and apply the core principles of both benchmarks.

8 Conclusion

We compared LDP and CDP mechanisms for differentially private deep learning under a white-box MI attack. The comparison comprises the average precision of the MI precision-recall curve and the target model test accuracy to support data scientists in choosing among available DP mechanisms and selecting privacy parameter ϵ. Our experiments on diverse learning tasks show that neither LDP nor CDP yields a consistently better privacy-accuracy trade-off. While MI only yields a lower bound on MI whereas ϵ in DP yields an upper bound, we observed that the lower bounds for LDP and CDP are close at similar model accuracy despite large difference in their upper bound. This suggests that the upper bound is far from the practical susceptibility to MI attacks and that data scientists should also consider to apply LDP despite the large privacy parameter values. Especially, since LDP does not require privacy accounting when training multiple models and offers flexibility w.r.t. optimizers. We consider the relative privacy-accuracy trade-off for LDP and CDP as the ratio of losses in accuracy and privacy over ϵ, and show that it is only favorable within a small interval.

Acknowledgements. This work has received funding from the European Union's Horizon 2020 research and innovation program under grant agreement No. 825333 (MOSAICROWN).

Appendix

Neural network models and composed ϵ for LDP are provided in Table 1. We state hyperparameters, composed ϵ for CDP, and training accuracies in Table 2. Texas Hospitals Stays and Purchases Shopping Carts provided by Shokri et al. are unbalanced in terms of records per class, as shown in Figs. 6 and 7.

Table 1. Overview of datasets considered in evaluation.

Dataset	Model	LDP
Texas Hospital Stays [38]	Fully connected NN with three layers ($512 \times 128 \times \mathbb{C}$) [38]	$19,125 - 638$ ($6382 \times \epsilon_i$)
Purchases Shopping Carts [38]	Fully connected NN with two layers ($128 \times \mathbb{C}$) [38] (i.e., logistic regression)	$1800 - 60$ ($600 \times \epsilon_i$)
Labeled Faces in the Wild [22]	VGG-Very-Deep-16 CNN [34]	$62.5 \times 10^6 -$ $6,250$ ($250 \times 250 \times \epsilon_i$)
Skewed Purchases	Fully connected NN with two layers ($128 \times \mathbb{C}$) [38] (i.e., logistic regression)	$1,800 - 60$ ($600 \times eps_i$)

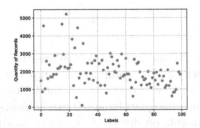

Fig. 6. Quantity of records per label for Purchases Shopping Carts

Table 2. Target Model training accuracy (from orig. to smallest ϵ), CDP ϵ values (from $z = 0.5$ to $z = 16$) and hyperparameters

C		Texas Hospital Stays				Purchases Shopping Carts				LFW			Skewed Purchases			
		100	150	200	300	10	20	50	100	20	50	100	10	20	50	100
LDP		0.86	0.92	0.83	0.81	0.99	1.0	1.0	0.99	1.0	1.0	1.0	1.0	1.0	1.0	1.0
		1.0	1.0	1.0	1.0	0.97	0.97	0.95	0.94	1.0	1.0	1.0	1.0	1.0	1.0	0.99
		1.0	1.0	1.0	1.0	0.88	0.85	0.86	0.90	1.0	0.96	1.0	1.0	1.0	1.0	0.97
		1.0	1.0	0.98	0.92	0.64	0.58	0.69	0.79	0.22	0.18	0.13	1.0	0.99	0.97	0.89
		0.99	0.95	0.86	0.72	0.58	0.47	0.62	0.75	0.24	0.17	0.13	0.93	0.98	0.9	0.80
		0.82	0.71	0.59	0.53	0.44	0.38	0.49	0.51	0.25	0.17	0.13	0.52	0.55	0.71	0.45
CDP		0.86	0.92	0.83	0.81	1.0	1.0	1.0	0.99	1.0	1.0	1.0	1.0	1.0	1.0	1.0
		0.74	0.75	0.69	0.62	0.95	0.91	0.82	0.63	0.99	0.87	0.79	1.0	1.0	0.97	0.58
		0.57	0.54	0.48	0.42	0.91	0.84	0.71	0.51	0.76	0.5	0.35	1.0	0.96	0.6	0.1
		0.35	0.31	0.26	0.22	0.80	0.69	0.46	0.27	0.44	0.28	0.25	0.92	0.8	0.25	0.02
		0.22	0.19	0.16	0.13	0.69	0.51	0.28	0.14	0.36	0.23	0.18	0.89	0.64	0.12	0.02
		0.05	0.04	0.03	0.02	0.28	0.14	0.05	0.02	0.32	0.19	0.13	0.66	0.24	0.03	0.01
ϵ		222.6	259.8	251.5	259.8	88.1	88.1	88.1	88.1	84.3	70.4	62.4	28.9	29.8	42.2	73.5
		6.3	6.6	7.3	7.4	4.6	4.1	4.1	4.1	4.8	3.9	3.4	1.6	1.7	3.5	2.1
		2.3	2.0	2.2	2.1	2.0	1.8	1.8	1.8	2.1	1.7	1.5	0.7	1.6	1.3	1.3
		0.9	1.1	1.0	1.1	1.3	1.2	1.2	1.2	1.3	0.8	1.0	0.9	0.9	0.7	0.6
		0.3	0.2	0.3	0.3	0.4	0.4	0.4	0.3	0.5	0.4	0.3	0.4	0.4	0.3	0.3
Learning rate	Orig.	0.01	0.01	0.01	0.01	0.001	0.001	0.001	0.001	0.001	0.001	0.001	0.001	0.001	0.001	0.001
	CDP	0.01	0.01	0.01	0.01	0.01	0.01	0.01	0.01	0.001	0.0008	0.0008	0.001	0.001	0.001	0.001
	LDP	0.01	0.01	0.01	0.01	0.001	0.001	0.001	0.001	0.001	0.001	0.001	0.001	0.001	0.001	0.001
Batch size	Orig.	128	128	128	128	128	128	128	128	32	32	32	100	100	100	100
	CDP	128	128	128	128	128	128	128	128	16	16	16	100	100	100	100
	LDP	128	128	128	128	128	128	128	128	32	32	32	100	100	100	100
Epochs	Orig.	200	200	200	200	200	200	200	200	30	30	30	200	200	200	200
	CDP	1000	1000	1000	1000	200	200	200	200	110	110	110	200	200	200	200
	LDP	200	200	200	200	200	200	200	200	30	30	30	200	200	200	200
Clipping Norm	CDP	4	4	4	4	4	4	4	4	3	3	3	4	4	4	4

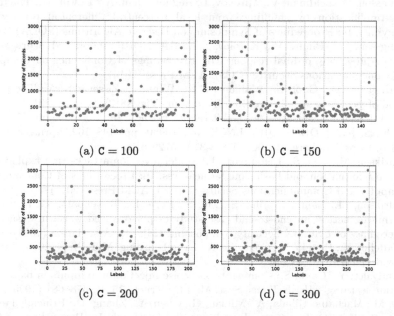

(a) C = 100   (b) C = 150

(c) C = 200   (d) C = 300

Fig. 7. The Quantity of records per Label for the Texas Hospital Stays Dataset

References

1. Abadi, M., et al.: Deep learning with differential privacy. In: Proceedings of Conference on Computer and Communications Security (CCS). ACM Press (2016)
2. Abowd, J.M., Schmutte, I.M.: An economic analysis of privacy protection and statistical accuracy as social choices. Am. Econ. Rev. **109**(1), 171–202 (2019)
3. Backes, M., Berrang, P., Humbert, M., Manoharan, P.: Membership privacy in microRNA-based studies. In: Proceedings of Conference on Computer and Communications Security (CCS). ACM Press (2016)
4. Bassily, R., Smith, A., Thakurta, A.: Private empirical risk minimization. In: Proceedings of Symposium on Foundations of Computer Science (FOCS). IEEE Computer Society (2014)
5. BBC News: Google DeepMind NHS app test broke UK privacy law (2017). https://www.bbc.com/news/technology-40483202
6. Carlini, N., Liu, C., Kos, J., Erlingsson, Ú., Song, D.: The secret sharer: measuring unintended neural network memorization and extracting secrets (2018)
7. Davis, J., Goadrich, M.: The relationship between precision-recall and ROC curves. In: Proceedings of Conference on Machine Learning (ICML). Omnipress (2006)
8. Dwork, C.: Differential privacy. In: Bugliesi, M., Preneel, B., Sassone, V., Wegener, I. (eds.) ICALP 2006. LNCS, vol. 4052, pp. 1–12. Springer, Heidelberg (2006). https://doi.org/10.1007/11787006_1
9. Dwork, C., Kenthapadi, K., McSherry, F., Mironov, I., Naor, M.: Our data, ourselves: privacy via distributed noise generation. In: Vaudenay, S. (ed.) EUROCRYPT 2006. LNCS, vol. 4004, pp. 486–503. Springer, Heidelberg (2006). https://doi.org/10.1007/11761679_29
10. Dwork, C., Roth, A.: The algorithmic foundations of differential privacy. Found. Trends Theoret. Comput. Sci. **9**(3–4), 211–407 (2014)
11. Erlingsson, U., Feldman, V., Mironov, I., Raghunathan, A., Talwar, K., Thakurta, A.: Amplification by shuffling: from local to central differential privacy via anonymity. In: Proceedings of Symposium on Discrete Algorithms (SODA) (2019)
12. Erlingsson, U., Pihur, V., Korolova, A.: RAPPOR: randomized aggregatable privacy-preserving ordinal response. In: Proceedings of Conference on Computer and Communications Security (CCS). ACM Press (2014)
13. Everingham, M., Gool, L., Williams, C.K., Winn, J., Zisserman, A.: The pascal visual object classes challenge. Int. J. Comput. Vis. **88**(2), 98–136 (2010)
14. Fan, L.: Image pixelization with differential privacy. In: Kerschbaum, F., Paraboschi, S. (eds.) DBSec 2018. LNCS, vol. 10980, pp. 148–162. Springer, Cham (2018). https://doi.org/10.1007/978-3-319-95729-6_10
15. Fredrikson, M., Jha, S., Ristenpart, T.: Model inversion attacks that exploit confidence information and basic countermeasures. In: Proceedings of Conference on Computer and Communications Security (CCS). ACM Press (2015)
16. Fredrikson, M., Lantz, E., Jha, S., Lin, S., Page, D., Ristenpart, T.: Privacy in pharmacogenetics: an end-to-end case study of personalized warfarin dosing. In: Proceedings of USENIX Security Symposium. USENIX Association (2014)
17. Goodfellow, I., Bengio, Y., Courville, A.: Deep Learning. MIT Press (2016). http://www.deeplearningbook.org
18. Grandvalet, Y., Canu, S.: Comments on "noise injection into inputs in back propagation learning". IEEE Trans. Syst. Man Cybernet. **25**(4), 678–681 (1995)
19. Hay, M., Machanavajjhala, A., Miklau, G., Chen, Y., Zhang, D.: Principled evaluation of differentially private algorithms using DPBench. In: Proceedings of Conference on Management of Data (SIGMOD). ACM Press (2016)

20. Hayes, J., Melis, L., Danezis, G., De Cristofaro, E.: LOGAN: membership inference attacks against generative models. Proc. Priv. Enhanc. Technol. (PoPETs) **2019**(1), 133–152 (2019)
21. Kashmir Hill: How Target Figured Out A Teen Girl Was Pregnant Before Her Father Did (2012). https://www.forbes.com/sites/kashmirhill/2012/02/16/how-target-figured-out-a-teen-girl-was-pregnant-before-her-father-did/
22. Huang, G.B., Ramesh, M., Berg, T., Learned-Miller, E.: Labeled faces in the wild: a database for studying face recognition in unconstrained environments. University of Massachusetts, Technical report (2007)
23. Iyengar, R., Near, J.P., Song, D., Thakkar, O.D., Thakurta, A., Wang, L.: Towards practical differentially private convex optimization. In: Proceedings of Symposium on Security and Privacy (S&P). IEEE Computer Society (2019)
24. Jayaraman, B., Evans, D.: Evaluating differentially private machine learning in practice. In: Proceedings of the USENIX Security Symposium. USENIX Association (2019)
25. Kairouz, P., Oh, S., Viswanath, P.: The composition theorem for differential privacy. IEEE Trans. Inf. Theory **63**(6), 4037–4049 (2017)
26. Kasiviswanathan, S.P., Lee, H.K., Nissim, K., Raskhodnikova, S., Smith, A.: What can we learn privately? SIAM J. Comput. **40**, 793–826 (2008)
27. Kingma, D.P., Ba, J.: Adam: a method for stochastic optimization. In: 3rd International Conference on Learning Representations, ICLR 2015, San Diego, CA, USA, 7–9 May 2015, Conference Track Proceedings. ICLR (2015)
28. Matsuoka, K.: Noise injection into inputs in back-propagation learning. IEEE Trans. Syst. Man Cybernet. **22**(3), 36–440 (1992)
29. Mironov, I.: Rényi differential privacy. In: Proceedings of Computer Security Foundations Symposium (CSF). IEEE Computer Society (2017)
30. MLPerf Website: MLPerf - Fair and useful benchmarks for measuring training and inference performance of ML hardware, software, and services (2018). https://mlperf.org/
31. Nasr, M., Shokri, R., Houmansadr, A.: Comprehensive Privacy Analysis of Deep Learning: Stand-alone and Federated Learning under Passive and Active White-box Inference Attacks (2018)
32. Nasr, M., Shokri, R., Houmansadr, A.: Machine learning with membership privacy using adversarial regularization. In: Proceedings of Conference on Computer and Communications Security (CCS). ACM Press (2018)
33. Papernot, N., Song, S., Mironov, I., Raghunathan, A., Talwar, K., Erlingsson, Ú.: Scalable private learning with pate (2018)
34. Parkhi, O.M., Vedaldi, A., Zisserman, A.: Deep face recognition. In: British Machine Vision Conference. BMVA Press (2015)
35. Rahman, M.A., Rahman, T., Laganière, R., Mohammed, N.: Membership inference attack against differentially private deep learning model. Trans. Data Priv. **11**, 61–79 (2018)
36. Sankararaman, S., Obozinski, G., Jordan, M.I., Halperin, E.: Genomic privacy and limits of individual detection in a pool. Nature Genetics **41**, 965–967 (2009)
37. Shokri, R., Shmatikov, V.: Privacy-preserving deep learning. In: Proceedings of Conference on Computer and Communications Security (CCS). ACM Press (2015)
38. Shokri, R., Stronati, M., Song, C., Shmatikov, V.: Membership inference attacks against ML models. In: Proceedings of Symposium on Security and Privacy (S&P). IEEE Computer Society (2017)

39. Song, S., Chaudhuri, K., Sarwate, A.D.: Stochastic gradient descent with differentially private updates. In: Proceedings of Conference on Signal and Information Processing. IEEE Computer Society (2013)
40. Wang, T., Blocki, J., Li, N., Jha, S.: Locally differentially private protocols for frequency estimation. In: Proceedings of USENIX Security Symposium. USENIX Association (2017)
41. Warner, S.L.: Randomized response: a survey technique for eliminating evasive answer bias. J. Am. Stat. Assoc. **60**(309), 63–69 (1965)
42. Wirth, R., Hipp, J.: Crisp-DM: towards a standard process model for data mining. In: Proceedings of Conference on Practical Applications of Knowledge Discovery and Data Mining. Practical Application Company (2000)
43. Yeom, S., Fredrikson, M., Jha, S.: The unintended consequences of overfitting: training data inference attacks (2017)
44. Yeom, S., Giacomelli, I., Fredrikson, M., Jha, S.: Privacy Risk in Machine Learning: Analyzing the Connection to Overfitting (2018)

Preventing Manipulation Attack in Local Differential Privacy Using Verifiable Randomization Mechanism

Fumiyuki Kato$^{(\boxtimes)}$, Yang Cao$^{(\boxtimes)}$, and Masatoshi Yoshikawa$^{(\boxtimes)}$

Department of Social Informatics, Graduate School of Informatics,
Kyoto University, Kyoto, Japan
fumiyuki@db.soc.i.kyoto-u.ac.jp, yang@i.kyoto-u.ac.jp,
yoshikawa.masatoshi.5w@kyoto-u.ac.jp

Abstract. Local differential privacy (LDP) has been received increasing attention as a formal privacy definition without a trusted server. In a typical LDP protocol, the clients perturb their data locally with a randomized mechanism before sending it to the server for analysis. Many studies in the literature of LDP implicitly assume that the clients honestly follow the protocol; however, two recent studies show that LDP is generally vulnerable under malicious clients. Cao et al. (USENIX Security '21) and Cheu et al. (IEEE S&P '21) demonstrated that the malicious clients could effectively skew the analysis (such as frequency estimation) by sending fake data to the server, which is called data poisoning attack or manipulation attack against LDP. In this paper, we propose secure and efficient verifiable LDP protocols to prevent manipulation attacks. Specifically, we leverage Cryptographic Randomized Response Technique (CRRT) as a building block to convert existing LDP mechanisms into a verifiable version. In this way, the server can verify the completeness of executing an agreed randomization mechanism on the client side without sacrificing local privacy. Our proposed method can completely protect the LDP protocol from output manipulation attacks, and significantly mitigates unexpected damage from malicious clients with acceptable computational overhead.

Keywords: Local differential privacy · Manipulation attack · Data poisoning · Verifiable computation · Oblivious transfer

1 Introduction

Today's data science has been very successful in collecting and utilizing large amounts of data. Useful data often includes personal information, and there are serious privacy concerns. In particular, recent data breaches [14,15] and strict rules [12,13] by the government significantly promote the concerns. Local differential privacy (LDP) [5,6] is a promising privacy-enhanced technique for collecting sensitive information. Each client perturbs sensitive data locally by a

K. Barker and K. Ghazinour (Eds.): DBSec 2021, LNCS 12840, pp. 43–60, 2021.
https://doi.org/10.1007/978-3-030-81242-3_3

randomized mechanism satisfying differential privacy. A server can run analysis such as frequency estimation based on the perturbed data without accessing the raw data. We can see the effectiveness and feasibility of LDP in recent production releases of the platformers such as Google [7], Apple [8], and Microsoft [9], which all utilize LDP for privacy-preserved data curation.

While many studies have been focusing on improving LDP's utility [10,11,16–19] in the literature, recent studies [1,2] report a vulnerability of LDP protocol and alert the lack of security. Specifically, [1,2] show that the malicious clients can manipulate the analysis, such as frequency estimation, by sending false data to the server. Malicious clients can skew the estimations effectively by considering that estimations are calculated by normalizing with randomization probability defined in the LDP protocol and can even control the estimations. Their studies significantly highlight the necessity of a secure LDP protocol to defend against malicious clients. The problematic point of protecting such an attack is that, in a general LDP protocol, others cannot verify the integrity of results without the original data. The randomization would provide data providers plausible deniability for their outputs.

To the best of our knowledge, no effective way in the literature can *completely* prevent manipulation attacks. Although Cao et al. [1] showed some of the countermeasures against malicious clients, their empirical results showed that preventing against *output manipulation attack* is still an open problem. Among their proposed methods, the one normalizing the estimated probability distribution was shown to be to some extent effective for input-manipulation (i.e., the attackers can falsely manipulate the raw input data but honestly execute the local randomized mechanism). However, the proposed countermeasures in [1] are not very effective for output-manipulation attack (i.e., the attackers can arbitrarily change the output of the local randomized mechanism). In addition, their detection-based countermeasures are based on the assumption of specific attack methods and may not be effective against arbitrary output-manipulation attacks. The authors concluded the need for more robust defenses against these attacks. Concurrently, Cheu et al. [2] also emphasize the same conclusion for manipulation attacks they call. There is another promising direction against an attacker who exploits the random mechanism of Differential privacy. Narayan et al. [3] propose an interesting scheme to prove integrity for executing correct randomization mechanisms for Differential privacy. However, their setting is different from ours since they focus on central DP with the data curator, who has the sensitive data, and the analyst, who creates the proof (in our setting, the client needs to prove their local execution).

To solve these problems, we design a novel verifiable LDP protocol based on Multi-Party Computation (MPC) techniques in this work. Our contributions are summarized below. First, we categorize the attacks of malicious clients into two classes, *output-manipulation* and *input-manipulation* (formally defined in Sect. 3). For input-manipulation attacks, efficient countermeasures have been provided in [1], but existing studies cannot prevent output-manipulation wholly and effectively. We analyze the effectiveness of output-manipulation compared to input-manipulation, highlight the importance of output-manipulation protection, and formalize the definition of *output-manipulation secure* LDP protocol. Second, we propose secure and efficient verifiable LDP protocols to prevent manipulation attack. Our protocols enable the server to verify the completeness of executing an agreed randomization mechanism on the client side without sacrificing local privacy. Specifically, we leverage Cryptographic Randomized Response Technique (CRRT) [4] as a building block to convert existing state-of-the-art LDP mechanisms including kRR [10], OUE [11], and OLH into output-manipulation secure LDP protocols with negligible utility loss. Our proposed secure protocols do not assume any specific attack, and work effectively against general output-manipulation, and thus is more potent than previously proposed countermeasures. Third, we conduct intensive experiments to test the performance of the proposed protocols. We demonstrate that the proposed methods can completely protect the LDP protocol from output manipulation attacks with acceptable computational overhead.

2   Background: Attacks on LDP Protocols

2.1   Local Differential Privacy

Differential privacy (DP) [6] is a rigorous mathematical privacy definition, which quantitatively evaluates the degree of privacy protection when we publish outputs about sensitive data in a database. DP is a central model where a trusted server collects sensitive data and releases differentially private statistical information to an untrusted third party. On the other hand, Local DP (LDP) is a local model, considering an untrusted server that collects clients' sensitive data. Clients perturb their data on their local environment and send only randomized data to the server to protect privacy.

In this work, we suppose server S collects data and aggregates them, and N clients c_i ($0 \leq i \leq N - 1$) send their sensitive data in a local differentially private manner. Each client has an item v which is categorical data, and the items have d domains and $v \in [0, d - 1](:= [d])$. Additionally, v_i denotes c_i's item. The clients randomize their data by randomization mechanism $\mathcal{A}$, and c_i send $\mathcal{A}(v_i) = y_i(\in D)$ to the server, where D is the output space of $\mathcal{A}$. The server estimates some statistics by $\mathcal{F}(y_0, ..., y_{N-1})$. In particular of this work, $\mathcal{F}_k$ corresponds to *frequency estimation* for item k (i.e., how many clients have chosen item k). The formal LDP definition is as follows:

Definition 1 (ϵ-local differential privacy (ϵ-LDP)). *A randomization mechanism $\mathcal{A}$ satisfies ϵ-LDP, if and only if for any pair of input values $v, v' \in [d]$ and for all randomized output $y \in D$, it holds that*

$$\Pr[\mathcal{A}(v) \in y] \leq e^{\epsilon} \Pr[\mathcal{A}(v') \in y].$$

Under a specific randomized algorithm $\mathcal{A}$, we want to estimate the frequency of any items. Wang et al. [11] introduce "pure" LDP protocols with nice symmetric property and a generic aggregation procedure to calculate the unbiased frequency estimations from given randomization probabilities. Let Support be a function that maps each possible output y to a set of input that y supports. Support is defined for each LDP protocol, and it specifies how the estimation can be computed under the LDP protocol. A formal definition of pure LDP is as follows:

Definition 2 (Pure LDP [11]). *A protocol given by $\mathcal{A}$ and Support is pure if and only if there exist two probability values $p > q$ such that for all v_1,*

$$\Pr[\mathcal{A}(v_1) \in \{o|v_1 \in \mathsf{Support}(o)\}] = p,$$
$$\forall_{v_2 \neq v_1} \Pr[\mathcal{A}(v_2) \in \{o|v_1 \in \mathsf{Support}(o)\}] = q. \tag{1}$$

where p, q are probabilities, and q must be the same for all pairs of v_1 and v_2.

While maximizing p and minimizing q make the LDP protocol more accurate, under ϵ-LDP it must be $\frac{p}{q} \leq e^\epsilon$. The important thing is that, in pure LDP protocol, we can simply estimate the frequency of item k as follows:

$$\mathcal{F}_k = \frac{\sum_i \mathbb{1}_{\mathsf{Support}(y^i)}(k) - Nq}{p - q} \tag{2}$$

We can interpret that this formula normalizes observed frequencies using probabilities p and q to adjust for randomization.

For frequency estimation under LDP, we introduce three state-of-the-art randomization mechanisms, kRR [10], OUE [11] and OLH [11]. These mechanisms includes three steps: (1) Encode is encoding function: $\mathcal{E} : v(\in [d]) \to v'(\in [g])$, (2) Perturbation is randomized function: $\mathcal{A} : v'(\in [g]) \to y(\in D)$, (3) Aggregation calculates estimations from all collected values: $\mathcal{F} : (y_0, ..., y_{N-1}) \to \mathbb{R}$. Formal proofs that each protocol satisfies ϵ-LDP can be found in [11].

k-ary Randomized Response (kRR) is an extension of Randomized Response [32] to meet ϵ-LDP. In particular, kRR provides accurate results in small item domains. This mechanism does not require any special encoding, and provides a identity mapping $\mathcal{E}(v) = v$ ($[g] = [d]$). Perturbation is as follows;

$$Pr[\mathcal{A}(v) = y] = \begin{cases} p = \dfrac{e^\epsilon}{e^\epsilon + d - 1}, & \text{if } y = v \\ q = \dfrac{1-p}{d-1} = \dfrac{1}{e^\epsilon + d - 1}, & \text{if } y \neq v \end{cases} \tag{3}$$

For aggregation, we can consider Support function as $\mathsf{Support}(v) = (v)$ and make this follow pure LDP protocol (Definition 2). Therefore, aggregation follows Eq. (2).

Optimized Unary Encoding (OUE) encodes item v into d-length bit vector and encode function is defined as $\mathcal{E}(v) = [0, ..., 0, 1, 0, ..., 0]$ where only single bit corresponding to v-th position is 1. Final output space is also d dimensional bit vector $\{0, 1\}^d$ (e.g. $\mathbf{y} = [1, 0, 1, 1, 0]$). Let i-th bit of output vector as y_i, perturbation is as follows;

$$Pr[y_i = 1] = \begin{cases} p = \dfrac{1}{2}, & \text{if } i = v \\ q = \dfrac{1}{e^\epsilon + 1}, & \text{if } i \neq v \end{cases} \tag{4}$$

These p and q minimize the variance of the estimated frequency in similar bit vector encoding (e.g. RAPPOR [7]). In aggregation step, we consider Support function as $Support(\mathbf{y}) = \{v | y_v = 1\}$, and also calculate using Eq. (2).

Optimized Local Hashing (OLH) employs hash function for dimensional reduction to reduce communication costs. It picks up H from a universal hash function family $\mathbb{H}$, and H maps $v \in [d]$ to $v' \in [g]$ where $2 \leq g < d$. Therefore, encode function is $\mathcal{E}(v) = H(v)$. Perturbation is the same as kRR, except that the input/output space is $[g]$. Then, p and q is defined as follows;

$$Pr[\mathcal{A}(x) = y] = \begin{cases} p = \dfrac{e^\epsilon}{e^\epsilon + g - 1}, & \text{if } y = H(v) \\ q = \dfrac{1}{g} \cdot p + \left(1 - \dfrac{1}{g}\right) \cdot \dfrac{1}{e^\epsilon + g - 1} = \dfrac{1}{g}, & \text{if } y \neq H(v) \end{cases} \tag{5}$$

In aggregation step, we consider Support function as $Support(\mathbf{y}) = \{v | v \in [d]$ and $y = H(v)\}$ and follow Eq. (2) using p and q.

2.2 Attacks on LDP Protocols

In this subsection, we introduce two important studies suggesting caution to the necessity of secure LDP protocols.

Targeted Attack. Cao et al. [1] focus on *targeted* attacks to LDP protocols, where the attacker tries to promote the estimated frequencies of a specific item set. Considering the attacker against the LDP protocols, M malicious clients, who can arbitrarily control local environments and send crafted data to the server, are injected by the attacker. (They call *data poisoning attacks*.) Attacker wants to promote r target items $T = \{t_1, ..., t_r\}$ in the frequency estimation. Cao et al. propose three attacks: Random perturbed-value attack (RPA), Random item attack (RIA), Maximal gain attack (MGA). The first two attacks are as baselines and MGA is an optimized attack. In RIA, malicious clients perform uniform random samplings of a value from the target item set. And then, following the LDP protocol, encoding and perturbation are performed and sent to the server. MGA is more complicated than others. It aims to maximize the attacker's overall gain G: sum of the expected frequency gains for the target items, $G = \sum_{t \in T} \mathbb{E}[\Delta f_t]$ where Δf_t represents the increase of estimated frequency of item t ($\forall t \in T$) from without attack to with attack. In MGA, the output item selection is performed according to the optimal solution maximizing the attacker's gain and sent to the server without perturbation.

Table 1. MGA can achieve the highest **gains** against all three protocols. $\beta = \frac{M}{N+M}$ and $f_T = \sum_{t \in T} f_t$ in the table. (The summary results of [11].)

	kRR	OUE	OLH
RPA (output-manipulation)	$\beta(\frac{r}{d} - f_T)$	$\beta(r - f_T)$	$-\beta f_T$
RIA (input-manipulation)	$\beta(1 - f_T)$	$\beta(1 - f_T)$	$\beta(1 - f_T)$
MGA (output-manipulation)	$\beta(1 - f_T) + \frac{\beta(d-r)}{e^\epsilon - 1}$	$\beta(2r - f_T) + \frac{2\beta r}{e^\epsilon - 1}$	$\beta(2r - f_T) + \frac{2\beta r}{e^\epsilon - 1}$

Table 2. Overall, output-manipulations are much more vulnerable than input-manipulation. The differences of both manipulations gain are calculated by output-manipulation gain − input-manipulation gain (resp. output-manipulation gain/input-manipulation gain) in Targeted (resp. Untargeted) Attack.

Targeted Attack [1]	kRR	OUE	OLH
	$+\left(\dfrac{\beta(d - r)}{e^\epsilon - 1} \right)$	$+\left(\beta(2r - 1) + \dfrac{2\beta r}{e^\epsilon - 1} \right)$	$+\left(\beta(2r - 1) + \dfrac{2\beta r}{e^\epsilon - 1} \right)$
Untargeted Attack [2]	$\times \Omega\left(\dfrac{\sqrt{d}}{\epsilon} \right)$		

Cao et al. describe the details of these three attacks against kRR, OUE, OLH in the frequency estimation and give theoretical analysis. The summary of the results is shown in Table 1. The table shows the overall gains of the three attacks against kRR, OUE, and OLH. MGA can achieve the highest gains for all protocols, clearly because MGA maximizes the gains. A notable point is a difference, summarized in Table 2, showing the difference in gains between MGA and RIA They respectively correspond to *output-manipulation* and *input-manipulation* (described later) in our work. Note that the difference is remarkable, especially under the higher privacy budget.

Untargeted Attack. Albert et al. [2] analyze manipulation attacks in LDP. Compared to Cao et al.'s work, their study mainly focuses on *untargeted* attacks. The attackers aim to skew the original distribution and degrade the overall estimation accuracy of the server.

They suggest for the LDP protocols that the architecture is inherently vulnerable to malicious clients' manipulations. They suppose a general manipulation attack: the attacker injects M users in N clients in the LDP protocol. These injected users can send arbitrary data sampled from carefully skewed distributions to the server without supposed perturbation. We consider this attacker model corresponds to MGA in [1] and *output-manipulation* (described later) in this paper. We should focus on one of their contributions: they show the general manipulation attack can skew the estimated distribution by $\Omega(\frac{M\sqrt{d}}{\epsilon N})$ in the frequency estimation, which causes more significant error than input-manipulation by about a $\frac{\sqrt{d}}{\epsilon}$ factor (Table 2). The difference is, for example, defined as l_1-norm of the original and skewed distribution.

Summary. We summarize these notable results in Table 2, showing how effective output-manipulation can attack compared to input-manipulation. The above two

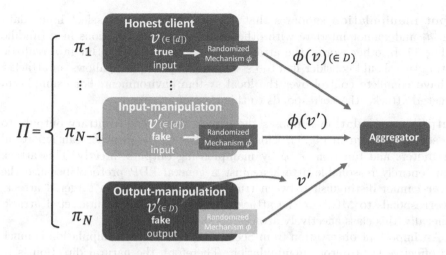

Fig. 1. From top to bottom, normal protocol, input-manipulation attack and output-manipulation attack against an LDP protocol.

previous studies' common conclusion is highlighting the great necessity of enforcing the correctness of users' randomization to defend the output-manipulation attacks.

3   Problem Statements

Firstly, we give some notations to LDP protocols, partially following the abovementioned in Sect. 2. We denote a single LDP protocol as π_i, where a client c_i sends sensitive data v to server S in ϵ-LDP manner. Encode and perturbation are denoted together as ϕ. ϕ is a probabilistic function (i.e., randomization mechanism) that takes $v \in [d]$ as input and output $y \in D$, such that output space $D = [d]$ if kRR, $D = \{0,1\}^d$ if OUE, $D = [g]$ if OLH. And we denote overall protocol including all clients as $\Pi = \{\pi_i | i \in [N]\}$.

3.1   Overview of Our Goal

An attacker against Π injects compromised users into the protocol to send many fake data to a central server. Note that such an attack results in manipulation against a single protocol π by each compromised user. Therefore, we consider security for π, and by protecting security for π, we can naturally protect security for Π. As for the attacker's capability, the attacker can access the implementation of ϕ because this is executed on clients' local, and he knows all parameters and functions including ϕ, ϵ, d, D and $Support(y)$, and employs this information to craft effective malicious outputs. However, in fact, there is little variation in the attacker's behavior because the server can easily deny the protocol if output $y \notin D$. Under such conditions, as shown in Fig. 1, we can observe that an attacker can carry out the following two classes of attacks:

Input-manipulation supposes that the attacker can only select input data $v \in [d]$ and cannot interfere with other parameters and functions in π (middle in Fig. 1). In other words, the attacker must send $y = \phi(v)$. But in a realistic setting, we should consider it a too strong assumption, as it allows an attacker to have complete control over the local system environment. For example, in targeted attack, RIA corresponds to this class of attacks.

Output-manipulation supposes the attacker can send arbitrary outputs to the server (bottom in Fig. 1). This corresponds to the attacker can ignore all parameters and functions ϵ, ϕ by manipulating outputs directly. This attack is an entirely reasonable attack against a general LDP protocol because the server cannot distinguish between true data or fake data. In targeted attack, it corresponds to MGA or the attacks proposed in [2] for untargeted attack. Generally, this class effectively attacks, as shown in Table 2.

An important observation from Sect. 2.2 is that input-manipulation is much less effective than output-manipulation. Therefore, the natural direction is to defend against output-manipulation and limit the attack to the range of input-manipulation to achieve secure LDP protocols. On the other hand, it is hard to prevent input-manipulation completely. These have been studied in the fields of game theory [26,27] or truth discovery [28], and we leave such a solution as future work.

Overall, our goal is to mitigate attacks against the LDP protocols by completely defending output-manipulation and limiting to input-manipulation. For this purpose, we consider enforcing the correct mechanism ϕ for protocol π. The key idea is to make the protocol verifiable against malicious clients from a server. In the rest of the paper, we refer to this property as *output-manipulation secure*. (It is also expressed simply as *secure*, and we call it as secure LDP protocol.)

Definition 3 (output-manipulation secure). *An LDP protocol π is output-manipulation secure if any malicious client cannot perform output-manipulation and can only perform input-manipulation against π.*

3.2 Security Definitions

In this subsection, we clarify what we should achieve for a secure LDP protocol. Similar to [4], security definitions of secure LDP protocol are consistent with a traditional secure two-party computation (2PC) protocol described in [25]. It considers an ideal world where we can employ Trusted Third Party (TTP) to execute arbitrary confidential computations indeed. And we aim to replace the TTP with a real-world implementation of cryptographic protocol $\pi = (c, S)$ between client c and server S. The protocol's flow when using a TTP is very simple. The client c sends input v to the TTP, and the TTP provides $y = \phi(v)$ to S. After all, c and S never receive any other information; S does not know v, and c does not know y. (S can estimate v from y and ϕ, but c's privacy should be guaranteed by LDP.)

While it is seemingly apparent that this ideal world's protocols will satisfy our requirements, let us review possible attacks closely. Goldreich [25] summarizes that there are just three types of attacks in malicious model 2PC against ideal world protocols; (1) denial of participation in the protocol; (2) fake input, not the true one; (3) aborting the protocol prematurely. We cannot hope to avoid these, but (1) and (3) cannot influence the estimation of original data distributions in LDP protocols. (2) is exactly input-manipulation described in the previous subsection. Thus, it is sufficient that the ideal world in 2PC is *output-manipulation secure* (see Definition 3) in the LDP protocols.

Considering the substituted cryptographic protocol $\pi = (c, S)$, let client c as prover $\mathcal{P}$ and server S as verifier $\mathcal{V}$ and $\pi = (\mathcal{P}, \mathcal{V})$. More specifically, we should guarantee secure LDP under the worst case that both $\mathcal{P}$ and $\mathcal{V}$ behave maliciously. The case where $\mathcal{P}$ is malicious is obvious, considering output-manipulation. Still, in the $\mathcal{V}$'s case, it is because, given the original scenario of LDP, we need to guarantee the privacy of $\mathcal{P}$. And, we assume $\mathcal{P}$ is a polynomial computational adversary and $\mathcal{V}$ is unbounded.

Following [4], the ideal world protocol can be substituted with protocol π if (a) for any prover algorithm $\mathcal{P}^*$, $\mathcal{V}$ who receives $\phi(v) = y$ accepts only when $\mathcal{P}^*$'s secret input is surely v, or otherwise halts with negligible error; (b) for any prover algorithm $\mathcal{P}^*$, y is indistinguishable from other categories; (c) for any verifier algorithm $\mathcal{V}^*$, v is indistinguishable from other categories. Additionally, we need to verify that the randomization function ϕ used in the protocol does indeed satisfy ϵ-LDP.

Let $\mathrm{view}_{\mathcal{P}}$ (resp. $\mathrm{view}_{\mathcal{V}}$) as the set of messages generated by the protocol that $\mathcal{P}$ (resp. $\mathcal{V}$) can observe And let k as a security parameter that increases logarithmically with cryptographic strength. Then, our security definitions are reduced as following three properties:

- **Verifiability**: This property corresponds to the above-mentioned (a). We consider the protocol is verifiable if it satisfies as follows;

$$Pr[\mathcal{V} \text{ does not halts } |y \leftarrow \phi(*)] = 1 \text{ and,} \qquad (6)$$

$$1 - Pr[\mathcal{V} \text{ halts } |y \leftarrow \mathcal{P}^*] < negl(k) \qquad (7)$$

where $negl(k)$ is negligible function in k, $y \leftarrow \phi(*)$ means y is obtained by correct execution of ϕ and $y \leftarrow \mathcal{P}^*$ means y is obtained by $\mathcal{P}^*$ other than correct $\phi(v)$.

- **Indistinguishability**: This property corresponds to (b) and (c). (b) satisfies if $\mathrm{view}_{\mathcal{P}^*}$ has indistinguishable distributions for any input category $v \in [d]$. Formally, we define this property as follows; for any adversary $\mathcal{P}^*$,

$$|Pr[\mathcal{P}^*(\mathrm{view}_{\mathcal{P}^*}, v) = y] - Pr[\mathcal{P}^*(v) = y]| < negl(k) \qquad (8)$$

where $negl(k)$ is negligible function in k. This means that a malicious client can use any information obtained from the protocol but only get negligible information about the final output of the server side. Similarly, (c) satisfies if, for any unbounded adversary $\mathcal{V}^*$,

$$|Pr[\mathcal{V}^*(\mathrm{view}_{\mathcal{V}^*}, y) = v] - Pr[v|y]| < negl(k) \qquad (9)$$

- **Local Differential Privacy**: The randomization mechanism ϕ in the given protocol must satisfy ϵ-LDP as shown in Definition 1. The verification of the correct execution is performed in Eq. (6).

4   Proposed Method

We design secure LDP protocols for kRR, OUE, and OLH, respectively, so that we defend output-manipulations completely. In our method, a major building block is Cryptographic Randomized Response Technique (CRRT) [4] which employs Pedersen's commitment scheme [21] for secure verifiability using additive homomorphic property, and Naor-Pinkas 1-out-of-n Oblivious Transfer (OT) technique [22] for tricks for a verifiable randomization mechanism. Overall, the proof of validity is based on disjunctive proof [23]. It is a lightweight interactive proof protocol based on a secret sharing scheme and, can perform witness-indistinguishable [31] proofs of knowledge (similar to zero-knowledge proofs). Combined with the security of the encryption scheme proposed in [4], it is possible to securely prove that the output value y is obtained by sampling from a probability distribution that satisfies the ϵ-LDP i.e. $y = \phi(v)$. For simplicity, we explain several phases separately in the following protocol description (Protocol 1, 2), but they can be done simultaneously in the actual implementation.

Before explanation of the protocols in detail, we introduce the following cryptographic setting. Assume that p and q are sufficiently large primes such that q divides $p - 1$, $\mathcal{Z}_p$ has a unique subgroup G of order q. q is the shared security parameter between $\mathcal{P}$ and $\mathcal{V}$. Security parameter k is $k = \log_2 q_{max}$ such that q_{max} is the maximum value of possible q. We select g and h as a public key. They are two generators of G and, their mutual logarithms $\log_g h$ and $log_h g$ are hard to compute. We use this public key in the following protocols.

4.1   Secure kRR

Protocol 1 shows the details of the secure version of kRR, an extension of CRRT [4] to satisfy LDP for multidimensional data. As a whole, in the setup phase, both $\mathcal{P}$ and $\mathcal{V}$ prepare the same parameters l, n, z from accuracy parameter $width$ and privacy budget ϵ by Algorithm 1. l, n, z identify an categorical probability distributions that satisfies LDP and we use it in 1-out-of-n OT for verifiable random sampling. In mechanism phase, $\mathcal{P}$ creates a vector $\boldsymbol{\mu}$ representing the categorical distribution containing n data where each data μ_i corresponds to one of the categories $[d]$. $width$ (i.e., n) is the size of the vector and decides a trade-off between accuracy to approximate LDP and overheads caused by the protocol. For proof P2, we use z^{μ_i} instead of μ_i. All z^{μ_i} is encrypted to y_i by an encryption scheme that combines Pedersen's commitment and OT. Only the μ_σ, where σ is pre-chosen by $\mathcal{V}$, can be decrypted correctly. Such a trick allows us to surely perform random sampling from vector $\boldsymbol{\mu}$ representing the categorical distribution. In the proof phase, two proofs are verified in the protocol. The first one is a disjunctive proof for each encrypted data y_i belonging to one of the categories $[d]$ (**P1**) . The second one also uses a disjunctive proof that the summation of the vector used as categorical distribution in the OT belongs to one of the possible values (**P2**). There are just d possible values for the summation of $\boldsymbol{\mu}$ (4.(a)).

Algorithm 1. DECIDESHAREDPARAMETERS

Input: ϵ, *width*
1: $i \leftarrow \lfloor \frac{e^\epsilon}{(d-1)+e^\epsilon} \rfloor$ // as an integer
2: **while** $i > 0$ **do**
3:    **if** $(width - i)$ divides $(d-1)$ **then**
4:      $g \leftarrow gcd(i, width, \frac{width-i}{d-1})$
5:      $l, n \leftarrow \frac{i}{g}, \frac{width}{g}$
6:      **break**
7:    **end if**
8:    $i \leftarrow i - 1$
9: **end while**
10: $z \leftarrow \max([l, \frac{n-l}{d-1}]) + 1$
Output: l, n, z

Here, we confirm Protocol 1 is secure. From the protocol, prover and verifier get $\mathrm{view}_{\mathcal{P}} = \{g^a, g^b, g^{ab-\sigma-1}, x_i, x\}$ and $\mathrm{view}_{\mathcal{V}} = \{w_i, y_i, com_i^{(j)}, c_i^{(j)}, h_i^{(j)},$ $com_j, c_i, h_i\}$ for all $i \in [n], j \in [d]$ respectively.

Firstly, we consider indistinguishability. Our encryption scheme (e.g., μ_i is encrypted to y_i) is the same as the one presented in [4], which has been shown that a protocol using the scheme is sufficiently indistinguishable for $\mathcal{P}^*$ and $\mathcal{V}^*$. That is, it is as hard for $\mathcal{P}^*$ to know about the σ, and also hard for $\mathcal{V}^*$ to guess the distribution of μ and input v. Considering the attacker views, for $\mathcal{P}^*$, calculating σ from $\mathrm{view}_{\mathcal{P}}$ is as hard as the Decisional Diffie Hellman (DDH) problem. And x and x_i are completely random integers. For $\mathcal{V}^*$, (w_i, y_i) of $\mathrm{view}_{\mathcal{V}}$ is indistinguishable by the security of the cryptographic scheme, and $(com_i^{(j)}, c_i^{(j)})$ is also indistinguishable because of the secret sharing scheme [23]. Verifiability is satisfied by proofs, P1 and P2. If both P1 and P2 are verified, $\mathcal{V}$ itself selects one value from the verified vector by OT. Then, for any operation by $\mathcal{P}^*$, $\mathcal{V}$ can confirm the correctness of the protocol. Hence, verifiability entirely depends on the protocol that proves the P1 and P2. We use disjunctive proofs and Eq. (6) and Eq. (7) are respectively satisfied by the completeness and soundness of the disjunctive proofs shown in [23]. Lastly, Algorithm 1 definitely generates l, n such that $\frac{l}{n} \leq \frac{e^\epsilon}{(d-1)+e^\epsilon}$ and $\frac{n-l}{d-1} \geq \frac{1}{(d-1)+e^\epsilon}$. Hence, because random sampling from μ is equivalent to kRR with $p = \frac{l}{n}, q = \frac{n-l}{d-1}$, at least ϵ-LDP is satisfied.

4.2   Secure OUE

We show the secure version of OUE protocol in Protocol 2. Unlike kRR, OUE sends a d-length bit vector where each i-th bit corresponds to that client likely has the item $i \in [d]$. In OUE, mechanism ϕ performs random bit flips with given constant probability independently for each bit. The Bernoulli distributions, which determine the probabilities of each flip, are approximated by a distribution of n-length bit vectors. As in the case of kRR, verifiable random sampling is achieved by a trick using Pedersen's commitment and OT. However, there are d distribution vectors since it needs for each category.

Protocol 1. secure kRR

Client c as prover $\mathcal{P}$ who holds an secret input $v \in [d]$ and server S as verifier $\mathcal{V}$. ϵ is privacy budget and $width$ is a parameter representing the degree of approximation.

1. **Setup phase.**
   (a) $\mathcal{P}$ and $\mathcal{V}$ run DecideSharedParameters($\epsilon, width$) and prepare l, n, z as shown in Algorithm 1. This is an algorithm for approximating integers l, n, z for given ϵ with as little degradation in accuracy as possible while still satisfying privacy protection.
   (b) $\mathcal{V}$ selects $\sigma \in [n]$. And $\mathcal{P}$ prepares a n-length random number vector $\boldsymbol{\mu} = (\mu_1, ..., \mu_n)$ where for all $1 \leq i \leq n$, $\mu_i \in [d]$, the vector satisfies $\#\{\mu_i | \mu_i \in \boldsymbol{\mu} \text{ and } \mu_i = v\} = l$ and for all $\{v' | v' \in [d] \setminus \{v\}\}$, $\#\{\mu_i | \mu_i = v'\} = \frac{n-l}{d-1}$ where $\#\{\cdot\}$ returns count on a set.

2. **Mechanism phase.**
   (a) $\mathcal{V}$ picks random $a, b \leftarrow \mathbb{Z}_q$ and sends g^a, g^b and $g^{ab-\sigma+1}$ to $\mathcal{P}$.
   (b) For all $i \in \{1, ..., n\}$, $\mathcal{P}$ performs the following subroutine; (1) Generate (r_i, s_i) at random; (2) Compute $w_i \leftarrow g^{r_i}(g^a)^{s_i} = g^{r_i + as_i}$ and $h_i \leftarrow (g^b)^{r_i}(g^{ab-\sigma+1}g^{i-1})^{s_i} = g^{(r_i+as_i)b+(i-\sigma)s_i}$; (3) Encrypt μ_i to y_i as $y_i \leftarrow g^{z\mu_i}h_i$. Then, send (w_i, y_i) to $\mathcal{V}$.
   (c) $\mathcal{V}$ computes w_σ^b where σ is what $\mathcal{V}$ choose at setup phase, and computes $g^{\mu_\sigma} \leftarrow \frac{y_\sigma}{h^{w_\sigma^b}}$. And then, find μ_σ from the result and g. Thus, $\mathcal{V}$ receives μ_σ as a randomized output from $\mathcal{P}$.

3. **Proof phase for P1.**
   (a) For all $j \in [d] \setminus \{\mu_i\}$, for all $i \in \{1, ..., n\}$, $\mathcal{P}$ generates challenge $c_i^{(j)}$ and response $s_i^{(j)}$ from $\mathbb{Z}_q$ and prepares commitments $com_i^{(j)} \leftarrow h^{s_i^{(j)}}/(y_i/g^{z^j})^{c_i^{(j)}}$. For $\{\mu_i\}$ and for all $i \in \{1, ..., n\}$, $\mathcal{P}$ generates $w_i \leftarrow \mathbb{Z}_q$ and let $com_i^{(\mu_i)} = h^{w_i}$. Then, send $com_i^{(j)}$ to $\mathcal{V}$, for all i, j.
   (b) $\mathcal{V}$ picks $x_i \leftarrow \mathbb{Z}_q$ for all $i \in \{1, ..., n\}$ and sends it to $\mathcal{P}$.
   (c) For all $i \in \{1, ..., n\}$, $\mathcal{P}$ computes $c_i^{(\mu_i)} = x_i - \sum_{j \in [d] \setminus \mu_i} c_i^{(j)}$ and $s_i^{(\mu_i)} = v_i c_i^{(\mu_i)} + w_i$. Then, send $c_i^{(j)}$ and $s_i^{(j)}$ for all i, j to $\mathcal{V}$.
   (d) Finally, $\mathcal{V}$ checks if $h^{s_i^{(j)}} = b(y_i/g^{z^j})^{c_i^{(j)}}$ for all $j \in [d]$ and $x_i = \sum_{j \in [d]} c_i^{(j)}$, for all $i \in 1, ..., n$. Otherwise halts.

4. **Proof phase for P2.**
   (a) For all $j \in [d] \setminus \{v\}$, $\mathcal{P}$ generates challenge c_j and response s_j from $\mathbb{Z}_q$ and prepares commitments $com_j \leftarrow h^{s_j}/(\prod_{i \in \{1,..,n\}} y_i/g^{Z_j})^{c_j}$ where $Z_j = \frac{n-l}{d-1}\left(\sum_{k \in [d] \setminus \{j\}} z^k\right) + lz^j$. And $\mathcal{P}$ generates $w \leftarrow \mathbb{Z}_q$ and let $com_v = h^w$. Then, send com_j to $\mathcal{V}$, for all $j \in [d]$.
   (b) $\mathcal{V}$ picks $x \leftarrow \mathbb{Z}_q$ and sends it to $\mathcal{P}$.
   (c) $\mathcal{P}$ computes $c_v = x - \sum_{j \in [d] \setminus \{v\}} c_j$ and $s_v = \left(\sum_{i \in 1, ..., n} v_i\right) c_v + w$. Then, send c_j and s_j for all j to $\mathcal{V}$.
   (d) Finally, $\mathcal{V}$ checks if $h^{s_j} = b(\prod_{i \in \{1,..,n\}} y_i/g^{Z_j})^{c_j}$ for all $j \in [d]$ and $x = \sum_{j \in [d]} c_j$. Otherwise halts.

Protocol 2. secure OUE

$\mathcal{P}, v \in [d], \mathcal{V}, width, \epsilon$ as with Protocol 1.

1. **Setup phase.**
   (a) $\mathcal{P}$ and $\mathcal{V}$ set l, n as $\lceil \frac{1}{1+e^{\epsilon}} \cdot width \rceil$ and $width$ itself respectively.
   (b) $\mathcal{V}$ selects d random numbers $\boldsymbol{\sigma} = \{\sigma_1, ..., \sigma_d\}$ where $1 \leq \sigma_j \leq n$. $\mathcal{P}$ prepares d n-length random bit vectors $\boldsymbol{\mu} = (\boldsymbol{\mu_1}, ..., \boldsymbol{\mu_n})$ such that $\boldsymbol{\mu_j} = (\mu_1^{(j)}, ..., \mu_n^{(j)})$ where all $\mu_i^{(j)} \in \{0, 1\}$, and the vector satisfies $\sum_i \mu_i^{(j)} = n - l$ if $j = v$ and $\sum_i \mu_i^{(j)} = l$ if $j \neq v$.
2. **Mechanism phase.**
   (a) $\mathcal{V}$ picks random $a_j, b_j \leftarrow \mathbb{Z}_q$ and sends g^{a_j}, g^{b_j} and $g^{a_j b_j - \sigma_j + 1}$ to $\mathcal{P}$ for all $j \in [d]$.
   (b) For all $j \in [d]$ and $i \in \{1, ..., n\}$, $\mathcal{P}$ performs the following subroutine;
      (1) Generate $(r_i^{(j)}, s_i^{(j)})$ at random; (2) Compute $w_i^{(j)} \leftarrow g^{r_i^{(j)}}(g^{a_j})^{s_i^{(j)}} = g^{r_i^{(j)} + a s_i^{(j)}}$ and $h_i^{(j)} \leftarrow (g^{b_j})^{r_i^{(j)}}(g^{a_j b_j - \sigma_j + 1}g^{i-1})^{s_i^{(j)}} = g^{(r_i^{(j)} + a s_i^{(j)})b_j + (i - \sigma_j)s_i^{(j)}}$;
      (3) Encrypt $\mu_i^{(j)}$ to $y_i^{(j)}$ as $y_i^{(j)} \leftarrow g^{\mu_i^{(j)}} h^{h_i^{(j)}}$. Then, $\mathcal{P}$ sends all $(w_i^{(j)}, y_i^{(j)})$ to $\mathcal{V}$.
   (c) For all $j \in [d]$, $\mathcal{V}$ computes $g^{\mu_{\sigma_j}^{(j)}} \leftarrow y_{\sigma_j}^{(j)}/h^{(w_{\sigma_j}^{(j)})^{b_j}}$. And then, find μ_{σ_j}. Thus, $\mathcal{V}$ receives $[\mu_{\sigma_1}, ..., \mu_{\sigma_d}]$ as a randomized output from $\mathcal{P}$.
3. **Proof phase for P1.**
   (a) For all $j \in [d]$, for all $i \in \{1, ..., n\}$, $\mathcal{P}$ generates challenge $c_{1-\mu_i^{(j)}, i}^{(j)}$ and response $s_{1-\mu_i^{(j)}, i}^{(j)}$ from $\mathbb{Z}_q$ and prepares commitments $com_{1-\mu_i^{(j)}, i}^{(j)} \leftarrow h^{s_{1-\mu_i^{(j)}, i}^{(j)}}/(y_i^{(j)}/g^{\mu_i^{(j)}})^{c_i^{(j)}}$. Generate $w_i^{(j)} \leftarrow \mathbb{Z}_q$ and compute $com_{(\mu_i^{(j)}), i}^{(j)} \leftarrow h^{w_i^{(j)}}$. Then, send $com_{\{0,1\}, i}^{(j)}$ to $\mathcal{V}$, for all i, j.
   (b) $\mathcal{V}$ picks $x_i^{(j)} \leftarrow \mathbb{Z}_q$ for all $j \in [d]$ and $i \in 1, ..., n$ and sends it to $\mathcal{P}$.
   (c) For all $j \in [d]$ and $i \in \{1, ..., n\}$, $\mathcal{P}$ computes $c_{\mu_i^{(j)}, i}^{(j)} = x_i^{(j)} - c_{1-\mu_i^{(j)}, i}^{(j)}$ and $s_{\mu_i^{(j)}, i}^{(j)} = v_i^{(j)} c_{\mu_i^{(j)}, i}^{(j)} + w_i^{(j)}$. Then, send $c_{\{0,1\}, i}^{(j)}$ and $s_{\{0,1\}, i}^{(j)}$ for all i, j to $\mathcal{V}$.
   (d) Finally, $\mathcal{V}$ checks if $h^{s_{\{0,1\}, i}^{(j)}} = b(y_i^{(j)}/g^{\{0,1\}})^{c_{\{0,1\}, i}^{(j)}}$ and $x_i^{(j)} = c_{0,i}^{(j)} + c_{1,i}^{(j)}$, for all $i \in \{1, ..., n\}$ and for all $j \in [d]$. Otherwise halts.
4. **Proof phase for P2. (Simplified because it is similar to P1.)**
   (a) $\mathcal{P}$ generates and sends all $com_{\{p,q\}}^{(j)}$ to $\mathcal{V}$.
   (b) $\mathcal{V}$ picks $x_j \leftarrow \mathbb{Z}_q$ for all $j \in [d]$ and sends it to $\mathcal{P}$.
   (c) $\mathcal{P}$ sends $c_{\{p,q\}}^{(j)}$ and $s_{\{p,q\}}^{(j)}$ for all j to $\mathcal{V}$
   (d) $\mathcal{V}$ checks if $h^{s_p^{(j)}} = b(\prod_{i \in 1, .., n} y_i^{(j)}/g^{n/2})^{c_p^{(j)}}$ and $h^{s_q^{(j)}} = b(\prod_{i \in 1, .., n} y_i^{(j)}/g^l)^{c_q^{(j)}}$ and $x_j = c_p^{(j)} + c_q^{(j)}$ for all $j \in [d]$. Otherwise halts.
5. **Proof phase for P3.**
   (a) $\mathcal{P}$ computes $h_{sum} \leftarrow \sum_{i,j} h_i^{(j)}$ and sends h_{sum} to $\mathcal{V}$.
   (b) $\mathcal{V}$ checks if $h^{h_{sum}} g^{n/2 + l(d-1)} = \prod_{i,j} y_i^{(j)}$. Otherwise halts.

In addition, each vector's distribution is one of two types: j-th vector such that secret input $v = j$ or otherwise (i.e., p or q in Eq. (4)). Thus, we perform independent OT and decide 0 or 1 for d categories and finally, get randomized output $[\mu_{\sigma_1}, ..., \mu_{\sigma_d}]$.

Then, similar to secure kRR, we must show that all Bernoulli distributions represented by d vectors are correct. Specifically, the proofs are that all elements of bit vectors μ are surely a bit (0 or 1) (**P1**) and distribution of the vectors are surely equivalent to either of p or q of Eq. (4) (**P2**). The number of p and q are 1 and $d - 1$ respectively (**P3**). If all these three proofs are verified, we can confirm the OUE protocol is simulated correctly. Like kRR's proofs, **P1** and **P2** are proved by d disjunctive proofs. **P3** is based on hardness of discrete logarithm problem. $\mathcal{P}$ cannot find h_{sum} in polynomial time without all correct $h_i^{(j)}$ that is used when encrypting $y_i^{(j)}$. While $\mathcal{P}$ has to release h_{sum}, this is information theoretically indistinguishable from $\mathcal{V}$ for each $h_i^{(j)}$ unless $n = d = 1$. Security statements for the secure OUE protocol are similar to secure kRR. For LDP, as we can see 1.(a) in Protocol 2, we set $q = l/n$ such that $\frac{l}{n} \geq \frac{1}{1+e^\epsilon}$.

4.3   Secure OLH

To make OLH output-manipulation secure, basically, we can use Protocol 1 except that it requires sharing of a hash function and using reduced output category space. As a first step, $\mathcal{V}$ generates and sends a seed s to $\mathcal{P}$ to initialize hash function $H_s : v \to v'$ where $v \in [d]$ and $v' \in [g]$. $\mathcal{V}$ and $\mathcal{P}$ use the same H_s as a hash function. We can apply Protocol 1 to achieve secure OLH by using category set $[g]$ instead of $[d]$ and sensitive input value v is handled as $v' = H_s(v)$. The rest of the steps are almost the same as kRR.

Even if $\mathcal{P}^*$, who does not use the hash function correctly, participates the protocol, $\mathcal{V}$ can easily detect it if it sends the output of a different output space, i.e. $y \notin [g]$. If attacker does not use a different output space, the attack can only be equivalent to input-manipulation because $\mathcal{V}$ verifies the correctness of the categorical distribution used in random sampling after applying hash function.

5   Evaluation

In this section, we evaluate and analyze the performance of our proposed protocols. The code in Python is available on github[1].

Experimental Setup. We use an HP Z2 SFF G4 Workstation, with 4-core 3.80 GHz Intel Xeon E-2174G CPU (8 threads, with 8 MB cache), 64 GB RAM. The host OS is Ubuntu 18.04 LTS. The client and server exchange byte data serialized by pickle (from Python standard library) over TCP. We use $\epsilon = 1.0$ and in OLH, set $g = d/2$ as the hashed space instead of $g = \lfloor e^\epsilon + 1 \rfloor$ for demonstration.

Parameter Generator. First, we analyze the approximated probability distribution generated by our proposed method. In secure kRR protocol, we

---

[1] https://github.com/FumiyukiKato/verifiable-ldp.

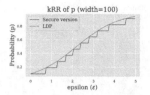 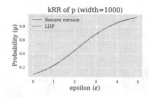 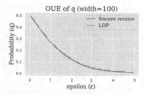

Fig. 2. In secure kRR, with a sufficiently large *width*, categorical distribution by Algorithm 1 can accurately approximate the LDP distributions (left and middle). In secure OUE, it is almost exact discrete approximation with relatively small *width* (right).

approximate the probability distribution where we generate data to satisfy LDP by Algorithm 1. Figure 2 shows how accurate the algorithm generates discrete distribution for $\epsilon = (0, 5]$ and for *width* $= \{100, 1000\}$. The red curve represents probability p for the normal mechanism, and the blue one represents the approximated one. When the *width* is small, there is a noticeable loss of accuracy due to approximation. However, with a sufficiently large *width*, the approximated p has a sufficiently small loss. As the *width* increases, the performance degrades, indicating that there is a trade-off between the accuracy of the probability approximation and the performance. This is true not only for kRR but also for OUE and OLH. For secure OUE, in the right-side of Fig. 2, we compare probability q because p is constant in OUE. It is almost exact discrete approximation with small *width*. This is due to the difference in the structure of the vectors that form the probability distribution, with OUE having a simpler structure.

Performance. We evaluate performances of our proposed method. Figure 3 shows total bandwidths, caused in communications of the total protocol, of each three methods for different category sizes. Generally, when increasing category size, total bandwidth also increases. While it increases linearly in OUE, there are fluctuations in kRR and OLH. This is because the probability value that Algorithm 1 approximates may have a smaller denominator (i.e., n) by reduction, which can make the distribution vector smaller. Overall, larger *width* generates almost linear increases in bandwidth. And for the same *width*, secure OUE causes larger communication overhead than others. However, as mentioned in the previous paragraph, secure OUE can approximate the probability distribution with high accuracy using smaller *width*. Hence, in particular, when the number of categories is large, secure OUE is considered to be more efficient by using smaller *width*. Figure 3 shows that, comparing kRR with *width* $= 1000$ and OUE with *width* $= 100$, many categories require several times more bandwidth. On the other hand, when the discretized probability distribution can be approximated with a small denominator by reduction, kRR and OLH show a very small bandwidth. When comparing kRR and OLH, OLH is smaller overall. This is due to the fact that the output space is reduced by hashing.

Figure 4 shows total execution time from the time the client sends the first request until the entire protocol is completed. Most of the characteristics are

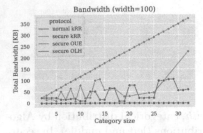

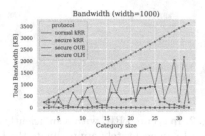

Fig. 3. With the same *width*, the communication costs of kRR and OLH are small. However, OUE can approximate LDP accurately with small widths (Fig. 2).

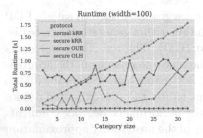

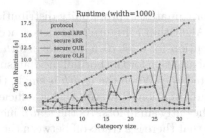

Fig. 4. The characteristics of runtime is similar to bandwidth. OLH takes a little longer because of the hashing.

similar to those of bandwidth. As the size of the proofs that need to be computed increases, the execution time is also expected to increase. The only difference is OLH, which takes extra time to execute the hash function. However, as the number of categories becomes larger, the influence becomes smaller.

Therefore, the overhead can be minimized by providing a privacy budget for optimal efficiency for kRR and OLH, and by using different methods for different *width*. The overhead is expected to increase as the number of categories increases, but since the limit on the number of categories is determined to some extent by the use of LDP, we do not think this is a major problem.

At the end, impressively, our method is algorithm-only, making it more feasible than alternatives that assume secure hardware [30] or TEE [29]. Nevertheless, overall, we believe the overhead is acceptable. We believe this is due to the fact that we use relatively lightweight OT techniques as a building block.

6   Conclusion

In this paper, we showed how we prevent malicious clients from attacking to LDP protocol. An important observation was the effectiveness of output-manipulation and the importance of protection against it. Our approach was verifiable randomization mechanism satisfying LDP. Data collector can verify the completeness of executing an agreed randomization mechanism for every possibly malicious data provider. Our proposed method was based on only lightweight cryptography Hence, we believe it has high feasibility and can be implemented in various and practical data collection scenarios.

References

1. Cao, X., Jia, J., Zhenqiang Gong, N.: Data poisoning attacks to local differential privacy protocols. arXiv preprint arXiv:1911.02046 (2019)
2. Cheu, A., Smith, A., Ullman, J.: Manipulation attacks in local differential privacy, In: 2021 IEEE Symposium on Security and Privacy (SP), pp. 1–18. San Francisco, CA, USA (2021)
3. Narayan, A., et al.: Verifiable differential privacy. In: Proceedings of the Tenth European Conference on Computer Systems (2015)
4. Ambainis, A., Jakobsson, M., Lipmaa, H.: Cryptographic randomized response techniques. In: Bao, F., Deng, R., Zhou, J. (eds.) PKC 2004. LNCS, vol. 2947, pp. 425–438. Springer, Heidelberg (2004). https://doi.org/10.1007/978-3-540-24632-9_31
5. Evfimievski, A., Gehrke, J., Srikant,R.: Imiting privacy breaches in privacy preserving data mining. In: Proceedings of the Twenty-Second ACM SIGMOD-SIGACT-SIGART Symposium on Principles of Database Systems (2003)
6. Dwork, C., McSherry, F., Nissim, K., Smith, A.: Calibrating noise to sensitivity in private data analysis. In: Halevi, S., Rabin, T. (eds.) TCC 2006. LNCS, vol. 3876, pp. 265–284. Springer, Heidelberg (2006). https://doi.org/10.1007/11681878_14
7. Erlingsson, Ú, Pihur, V., Korolova, A.: RAPPOR: randomized aggregatable privacy-preserving ordinal response. In: Proceedings of the 2014 ACM SIGSAC Conference on Computer and Communications Security (2014)
8. Apple differential privacy team. learning with privacy at scale. Mach. Learn. J. (2017)
9. Ding, B., Kulkarni, J., Yekhanin, S.: Collecting telemetry data privately. In: NeurIPS (2017)
10. Kairouz, P., Oh, S., Viswanath, P.: Extremal mechanisms for local differential privacy. In: NeurIPS (2014)
11. Wang, T., Blocki, T., Li, N., Jha, S.: Locally differentially private protocols for frequency estimation. In: USENIX Security (2017)
12. EU GDPR: https://www.eugdpr.institute/. Accessed 21 Mar 2021
13. Brazil's General Data Protection Law: https://iapp.org/media/pdf/resource_center/Brazilian_General_Data_Protection_Law.pdf. Accessed 21 Mar 2021
14. Facebook Cambridge Analytica Data Scandal (wikipedia). https://en.wikipedia.org/wiki/Facebook-CambridgeAnalyticadatascandal
15. Top10 data breaches of 2020. https://www.securitymagazine.com/articles/94076-the-top-10-data-breaches-of-2020. Accessed 21 Mar 2021
16. Kairouz, P., Bonawitz, K., Ramage, D.: Discrete distribution estimation under local privacy. In: ICML (2016)
17. Wang, T., et al.: Answering multi-dimensional analytical queries under local differential privacy. In: SIGMOD (2019)
18. Wang, T., Li, N., Jha, S.: Locally differentially private frequent itemset mining. In S&P (2018)
19. Wang, T., Lopuhaä-Zwakenberg, M,, Li, Z., Skoric, B,, Li. N.: Locally differentially private frequency estimation with consistency. In: NDSS (2020)

20. Gennaro, Rosario, Gentry, Craig, Parno, Bryan, Raykova, Mariana: Quadratic span programs and succinct NIZKs without PCPs. In: Johansson, Thomas, Nguyen, Phong Q.. (eds.) EUROCRYPT 2013. LNCS, vol. 7881, pp. 626–645. Springer, Heidelberg (2013). https://doi.org/10.1007/978-3-642-38348-9_37

21. Pedersen, T.P.: Non-interactive and information-theoretic secure verifiable secret sharing. In: Feigenbaum, J. (ed.) CRYPTO 1991. LNCS, vol. 576, pp. 129–140. Springer, Heidelberg (1992). https://doi.org/10.1007/3-540-46766-1_9

22. Naor, M., Pinkas, B.: Efficient oblivious transfer protocols. In: Proceedings of the Twelfth Annual ACM-SIAM Symposium on Discrete Algorithms (SODA), vol. 1 (2001)

23. Cramer, R., Damgård, I., Schoenmakers, B.: Proofs of partial knowledge and simplified design of witness hiding protocols. In: Desmedt, Y.G. (ed.) CRYPTO 1994. LNCS, vol. 839, pp. 174–187. Springer, Heidelberg (1994). https://doi.org/10.1007/3-540-48658-5_19

24. Bellare, M., Rogaway, P.: Random oracles are practical: a paradigm for designing efficient protocols. In: Proceedings of the 1st ACM Conference on Computer and Communications Security (1993)

25. Goldreich, O.: Secure Multi-Party Computation. Final (Incomplete) Draft, 27 October 2002

26. Do, C.T., et al.: Game theory for cyber security and privacy. ACM Comput. Surv. 50(2)1–37 (2017)

27. Prelec, D.: A Bayesian truth serum for subjective data. Science 306, 5695, 462–466 (2004)

28. Waguih, D.A., Berti-Equille, L.: Truth discovery algorithms: an experimental evaluation. arXiv preprint arXiv:1409.6428 (2014)

29. Sabt, M., Achemlal, M., Bouabdallah, A.: Trusted execution environment: what it is, and what it is not. In: 2015 IEEE Trustcom/BigDataSE/ISPA, vol. 1. IEEE (2015)

30. Costan, V., Devadas, S.: Intel SGX explained. IACR Cryptol. ePrint Arch. 2016(86), 1–118 (2016)

31. Feige, U., Shamir, A.: Witness indistinguishable and witness hiding protocols. In: Proceedings of the Twenty-Second Annual ACM Symposium on Theory of Computing (1990)

32. Warner, S.L.: Randomized response: a survey technique for eliminating evasive answer bias. J. Am. Stat. Assoc. 60, 309, 63–69(1965)

Cryptology I

Simple Storage-Saving Structure for Volume-Hiding Encrypted Multi-maps
(A Slot in Need is a Slot Indeed)

Jiafan Wang and Sherman S. M. Chow(✉) (iD)

Department of Information Engineering, Chinese University of Hong Kong,
Shatin, N.T., Hong Kong
{wj016,sherman}@ie.cuhk.edu.hk

Abstract. Severe consequences in volume leakage (subject to the conditions required by specific attacks) stimulate a new research direction (Eurocrypt 2019) of volume-hiding structured encryption (STE), particularly encrypted multi-maps (EMM), in which all queries should share the same (as the largest) response size unless the scheme is lossy. Meanwhile, note that the responses are originated from the actual ciphertexts outsourced to the server. Conventional wisdom suggests that the ciphertexts (to be accessed by the server while answering a query) should also contain many dummy results to make a query look uniform with others. Supporting updates is also natural; however, attaching dummy results to a query also complicates the operation and leakage of updates, which excludes many advanced data structures, e.g., cuckoo hashing (CCS 2019). This paper proposes a space-efficient EMM without storing any dummy ciphertext, which is volume hiding against passive adversaries (SP 2021) and compatible with dynamic extensions. Its crux structure is a hash ring, which is famous for load balancing but rarely appears in any STE. Efficiency-wise, our scheme beats the state-of-the-art (Eurocrypt 2019, CCS 2019), maintaining the necessary communication overhead and downsizing the server storage to be linear in the number of values in the EMM, while ruling out any data loss due to truncations or differential privacy.

Keywords: Volume hiding · Encrypted multi-maps · Consistent hashing · Structured encryption · Dynamic symmetric searchable encryption

1 Introduction

Structured encryption (STE), introduced by Chase and Kamara [6], enables a client to outsource an encrypted data structure to an untrusted cloud server for

Sherman S. M. Chow is supported by General Research Fund (Project Numbers: CUHK 14210217 and CUHK 14209918) from Research Grant Council.

© IFIP International Federation for Information Processing 2021
Published by Springer Nature Switzerland AG 2021
K. Barker and K. Ghazinour (Eds.): DBSec 2021, LNCS 12840, pp. 63–83, 2021.
https://doi.org/10.1007/978-3-030-81242-3_4

later private queries. Encrypted multi-map (EMM) is an important instance of STE. A multi-map (MM) usually contains a set of keys, each associated with a tuple of values. Treating each key as a keyword and the values as (identifiers of the) documents containing the keyword, EMM realizes searchable symmetric encryption (SSE) for keyword searches. It serves as the core of SSE for more complex queries (e.g., range [7,8,30], graph [24,25], skyline [31], and SQL [13]) and improved security for dynamism (e.g., forward and backward privacy [3,4, 25]). Many EMMs are well-known to be lightweight and efficient for millions of records.

An STE is deemed secure with respect to a given leakage profile if its operations reveal nothing beyond well-defined functions capturing the leakage of the private input data. Leakages allow STE to be efficient, but whether they are "benign" enough is still under study. Though most leakage-abuse attacks can be challenging to realize in practice, e.g., require knowing a large portion of data and some queried keys [5,12] or typical distributions of client queries [10,18,22], they stimulate the research of minimizing the leakage of STE in the first place.

This work focuses on the design of volume-hiding EMMs, a notion recently proposed by Kamara and Moataz [14] and further formalized by Patel et al. [27]. The volume refers to *the number of associated values* of any key. A very recent work [21] demonstrates that a passive attacker solely observing EMM accesses can still exploit the volume leakage of range queries to reconstruct the private plaintext. A volume-hiding EMM can hide the *response length* of a query, which effectively mitigates the damage caused by volume-abusing attacks [2,10,18,21, 22]. In this work, we propose volume-hiding EMMs against such adversaries, which compare favorably with state-of-the-art [14,27] in efficiency.

1.1   Related Work – Many Dummies in Storage and Communication

For a multi-map MM, let ℓ be *the maximum volume*, m be *the number of distinct keys*, and n be *the total number of values over all keys*. Usually, $m \cdot \ell \gg n$. For volume hiding, the response volume is maintained to be at least $\Theta(\ell)$.

The naïve padding approach pads up to $(\ell-1)$ dummy values to any key with a volume less than ℓ, which increases the server-side storage to $m \cdot \ell$. Another solution is oblivious RAMs (ORAM) [9]. Among the oblivious accesses to each associated value, up to ℓ extra fake accesses are made. While ORAM has been made practically efficient, it inherently incurs large communication overheads. Both approaches are undesirable in practice, motivating cleverer constructions.

The first construction VLH [14] reduces the storage overhead of the naïve approach roughly by half under specific parameters, yet still far from the ideal goal of $\Theta(n)$. For each key, it determines its volume via a pseudorandom function (PRF). It pads dummies when the PRF-derived volume is larger than the real one and truncates values when it is smaller. It is thus lossy, which is often undesirable. Although the number of truncated keys is relatively small under the Zipf's distribution, no data loss guarantee is given for a general MM.

The second construction AVLH [14] achieves $\Theta(n)$ storage. For each key, it chooses a set of ℓ related bins uniformly at random. Each value is placed into a distinct bin related to its associating key. All bins are padded with dummies to the maximum bin size after all the n values have been arranged. According to the balls-into-bins analysis, the bin size would be $\Omega(\log n)$, resulting in $\Theta(\ell \cdot \log n)$ communication and computation overheads for any query. AVLH further optimizes the server storage from $\Theta(n)$ to $\Theta(n - \sqrt{m} \cdot \text{polylog}(m))$ for *concentrated* MM with many values associated with a large number of keys. It ensures that these values only appear once among all bins. However, this variant requires the hardness of the densest subgraph problem [20], which is not thoroughly studied in literature and hard to determine related parameters for concrete security.

Patel *et al.* [27] criticized (A)VLH [14] for lossiness, large storage and query overheads, or reliance on a less-studied assumption. Observed that minimizing storage overhead is a typical hashing problem: placing n items to $\Theta(n)$ locations that can be looked up by probing a small number of locations, they proposed two schemes exploiting "cuckoo hashing with a stash" [20], which store each value in one of two hash slots or a client stash, and abort if the stash overflows.

The first scheme dprfMM outsources a hash table of length $(2 + \alpha)n$, which contains those key-value pairs survived from cuckoo hashing evictions and some dummies. A query token consists of 2ℓ hash slots possibly related to the queried key. The server returns 2ℓ ciphertexts retrieved from these slots, each containing a desirable value, a value of other keys, or a dummy. The client combines the value tuple of the queried key from both the local stash and the server response. The token size could be reduced by delegatable PRF [19] with extra computation.

The second construction dpMM [27] outsources a cuckoo hash table storing the volume of each key, in addition to the one used in dprfMM. It leads to the $(2+\alpha)(n+m)$ storage overhead. To query, it first takes the client one roundtrip to fetch the volume of the queried key. With the noise from the Laplacian distribution, the client issues a token consisting of hash slots, whose size is the volume of the queried key plus the (possibly negative) noise and an extra adjustable parameter. The following query steps remain the same as dprfMM. With a small amount of perturbation over the response length, the query communication of dpMM is close to the optimal case, which outperforms any other schemes, but only under the relaxed (differential) privacy guarantee with potential data loss.

Our construction, achieving the known best communication overhead (*i.e.*, ℓ) under the consideration of volume hiding against passive adversaries [21], avoids any locally stored MM (*cf.*, the stash from cuckoo hashing in dprfMM and dpMM [27]) or any loss of values (*cf.*, due to differential privacy in dpMM [27] or PRF-based transform in VLH [14]), while further improving the server storage by wiping out dummies (*cf.*, dummy padding required in the state of the art [14,27]).

1.2 Volume-Hiding EMM with "Volume-Affecting" Updates

To make an EMM scheme dynamic, existing dynamic structures [1,3,16,23–25] may not be applicable to volume-hiding EMMs given the more rigorous leakage control. As discussed, the security and efficiency of volume-hiding EMMs are impacted by the maximum volume of the input MM, which may be changed by even a single update. To hide the key with the (real) maximum volume, the client must expensively introduce a dummy for each distinct key.

Indeed, the only two dynamic volume-hiding EMMs [14], VLH^d and AVLH^d, only support a limited form of updates (*cf.*, [25]). They are extended from VLH and AVLH, respectively. AVLH^d only supports edition, which modifies the value of a key-value tuple, but not the key. VLH^d further supports "addition" and "deletion." Addition can only add a key that never exists in the MM. In other words, adding one more value to an existing key, which affects ℓ, is not possible. Similarly, VLH^d can only delete all values associated with a key. All these updates are relatively straightforward, mostly without affecting the maximum volume. Furthermore, both constructions require the client to get back the tuple to be edited by querying before (re-)uploading the updated tuple. Finally, they lack forward privacy and backward privacy [3,4], which have become the *de facto* standard of dynamic STE due to the adaptive injection attack [33].

One may attempt to upgrade dpMM and dprfMM [27] with dynamic updates. It will involve insertions to the underlying cuckoo hashing table, and the chain of evictions it may incur becomes more complicated in the EMM setting. Firstly, the server cannot determine whether the slot to be inserted is empty unless the client decides to help by an extra round of communication. Also, the eviction chain will affect multiple slots. To avoid extra leakage, such as the occupancy information, one needs to make sure insertions for all updates result in the same eviction time, which significantly burdens the computation complexity. This showcases that designing a volume-hiding EMM that is compatible with updates in the first place appears to be a better approach, which also steers our EMM design.

1.3   Our Contributions and a Technical Overview

We observe that most volume-hiding EMMs hide the volume of a key using the values of other keys and dummies. Dummies carrying no meaningful information are the culprits of the significantly larger storage overhead. Our idea is to wipe out these dummies and realize volume hiding *solely* with meaningful value tuples. Beyond the storage of tuples themselves, we can further free the server from storing a global data structure filled with many dummies here and there (*cf.*, the entire hash table outsourced by dprfMM/dpMM [27] of size $(2+\alpha)n$ or $(2+\alpha)(n+m)$ for a positive constant α). Our volume-hiding EMM thus beats the best-known server storage (from $(2+\alpha)n$ to $2n$) with an even lower communication overhead (from 2ℓ to ℓ). The query overhead remains asymptotically the same as AVLH for a general (*i.e.*, not necessarily concentrated) MM [14]. Notably, our scheme neither stores any local (partial) MMs nor suffers from any data loss due to truncations or differential privacy. Table 1 summarizes the comparison[1].

---

[1] Even powerful cloud/server-side security enclaves would not hide the communication volume. A highlight of our scheme is its simplicity (fits in a single page) and its usage of lightweight tools (*e.g.*, PRF). No significant constant is hidden in the asymptotic evaluations. There are no (non-colluding) servers. These emphases might be random; however, we found the need to respond to a very negative review (not from DBSec) of an imagined paper, if not related subfields (volume-hiding SSE/ORAM/*etc.*).

Table 1. Comparison of volume-hiding encrypted multi-maps

Scheme	Communication (response size)	Computation (for query)	Server storage (encrypted multi-maps)	No stash	No loss
Naïve padding	ℓ	ℓ	$m \cdot \ell$	✓	✓
VLH [14]	$\Theta(\ell)$	$\Theta(\ell)$	$\Theta(m \cdot \ell)$	✗	✓
AVLH [14] (General MM)	$\Theta(\ell \cdot \log n)$	$\Theta(\ell \cdot \log n)$	$\Theta(n)$	✓	✓
AVLH [14] (Concentrated)	$\Theta(\ell \cdot \frac{n}{\text{polylog}(m)})$	$\Theta(\ell \cdot \frac{n}{\text{polylog}(m)})$	$\Theta(n - \sqrt{m} \cdot \text{polylog}(m))$	✓	✓
dprfMM [27]	2ℓ	2ℓ	$(2 + \alpha)n$	✓	✗
dpMM [27]	$2\ell(\text{key}) + \omega(\log \lambda')$	$2\ell(\text{key}) + \omega(\log \lambda')$	$(2 + \alpha)(n + m)$	✗	✗
Our $\mathcal{S}^4$	ℓ	$\Theta(\ell \cdot \log n)$	$2n$	✓	✓

(Legends) m: the number of distinct keys, n: the total number of values across all keys
ℓ: the maximum volume across all keys, $\ell(\text{key})$: the volume of key
λ': a security parameter for differential privacy, α: a positive constant
"No stash" means no need for the client to locally store any partial multi-map.
"No loss" means no loss of data, *e.g.*, due to differential privacy or truncations.

As the trade-off for the above benefits, we consider passive adversaries who can observe EMM accesses of the data owner but cannot force arbitrary accesses of their wishes, capturing realistic scenarios of cloud storage applications. Nevertheless, solely storing meaningful tuples for supporting volume-hiding queries is still non-trivial. To avoid leakages of the underlying MM incurred by merely outsourcing the EMM, the responses for each query should be randomly distributed across the storage space. However, when the responses of a query can only be padded with those of another query, it appears that the client needs a sophisticated mapping, say, by solving a combinatorics and optimization problem, to carefully "plan ahead" what are the responses to be "borrowed" as dummies, which is essentially a rephrasing of the EMM problem we aim to solve.

As a simple storage-saving structure, our EMM is named $\mathcal{S}^4$. It is based on a variant of hash ring [17], a technique that we first bring to the context of STE (SSE included) to our knowledge. The ring stores the encrypted contents of the original MM (notably without dummies). The query for a key specifies ring slots associated with it. Intuitively, any query should contain the same amount of ring slots, or metaphorically speaking, shelters for the key-value pairs. We still need to ensure that, when a shelter is not in real need, *i.e.*, when a key associates with less than ℓ values, the shelter pointed to by the key is storing the real value of another key in need. Our strategy of using the hash ring makes this possible.

We also explore how to update volume-hiding EMMs efficiently. We borrow an approach from the database community [7] of setting up a new instance for updates and periodically merging them, which matches the feature of some real-world systems (*e.g.*, commercial database Vertica [26]) that handles updates efficiently in batch. We further consider forward and backward privacy, resulting in the first dynamic volume-hiding EMM with such privacy guarantees. A concurrent work of Gui *et al.* [11] proposed a scheme that is forward and backward private while *mitigating* the volume leakage. Nevertheless, its bucketization technique exposes an approximate volume of each query even to a passive adversary.

2 Preliminary

2.1 Notation, Basic Structure and Primitive

Notation. Let λ be the security parameter and all algorithms take λ implicitly as input. $\mathsf{negl}(\lambda)$ is a negligible function in λ. PPT stands for probabilistic polynomial-time. For a set X, $x \leftarrow_\$ X$ samples an element x uniformly from X. For an algorithm A, $x \leftarrow A$ means x is an output of A. $\parallel$ denotes string concatenation. $[n]$ denotes the set $\{1, \ldots, n\}$. $[a..b]$ denotes $\{a, a+1, \ldots, b\}$.

Basic Data Structures. We recall operations of arrays, dictionaries, and multi-maps used in this work. An array A of the capacity n is a sequence of n values indexed by integer identifiers from 0 to $n-1$. For an index i and a value v, we set the i-th value via $\mathsf{A}[i] \leftarrow v$, and get the i-th value via $v \leftarrow \mathsf{A}[i]$.

A dictionary DX is a collection of key-value pairs. For key in the key space $\mathcal{K}$ and a value v, $v \leftarrow \mathsf{DX}[\mathsf{key}]$ (resp. $\mathsf{DX}[\mathsf{key}] \leftarrow v$) represents getting (resp. putting) v from (resp. into) $\mathsf{DX}[\mathsf{key}]$. A multi-map MM is a collection of key-tuple pairs $\{(\mathsf{key}, \vec{v})\}$. One could get/put tuples associated with a specific key in MM with similar notions as DX. Specifically, we denote the number of values in the tuple $\vec{v}$ by $[|\vec{v}|]$. For $i \in [|\vec{v}|]$, one could get/put the i-th value of $\vec{v}$ with handle $\vec{v}[i]$.

Basic Cryptographic Tools. Pseudorandom functions (PRF) are polynomial-time computable functions that cannot be distinguished from a truly random function by any PPT adversary. The symmetric-key encryption scheme (SKE) used in this work is *random-ciphertext-secure against chosen-plaintext attacks* (RCPA), which requires ciphertexts to be computationally indistinguishable from random even if the PPT adversary adaptively accesses the encryption oracle. It can be obtained from the standard PRF-based SKE or AES in counter mode.

2.2 Structured Encryption for Multi-maps

A structured encryption scheme STE encrypts a data structure such that the client could privately query its content. It can be either response-revealing (revealing the query response in plaintext) or response-hiding (keeping it private). An STE is non-interactive if the response can be fetched in a single roundtrip.

We focus on non-interactive and response-hiding STE for MMs, *i.e.*, EMMs.

Definition 1 (Non-Interactive Structured Encryption for MMs). *An* STE *scheme for* MM $\Sigma = (\mathsf{Setup}, \mathsf{Query}, \mathsf{Reply}, \mathsf{Result})$ *is defined as follows.*

- $(K, \mathsf{st}, \mathsf{EMM}) \leftarrow \mathsf{Setup}(1^\lambda, \mathsf{MM})$ *is a probabilistic algorithm executed by the client that takes as input the security parameter* λ *and the input multi-map* MM. *It outputs a private key* K, *a state* st, *and the encrypted multi-map* EMM. *The client stores* (K, st) *while sending* EMM *to the server.*
- $\mathsf{tk} \leftarrow \mathsf{Query}(K, \mathsf{st}, \mathsf{key})$ *is a (possibly) probabilistic algorithm executed by the client that takes as input the private key* K, *the state* st, *and the* key *from the key universe* $\mathcal{K}$ *of* MM. *It outputs a query token* tk *to be sent to the server.*
- $\mathsf{res} \leftarrow \mathsf{Reply}(\mathsf{tk}, \mathsf{EMM})$ *is a (possibly) probabilistic algorithm executed by the server that takes as input the query token* tk *and the encrypted multi-map* EMM. *It outputs the response* res *to the client.*
- $\vec{v} \leftarrow \mathsf{Result}(K, \mathsf{key}, \mathsf{res})$ *is a deterministic algorithm executed by the client that takes as input the private key* K *and the queried* key *which produced* tk, *and the response* res *due to* tk. *It outputs a value tuple* $\vec{v}$ *associated with* key.

Security is captured under the real/ideal simulation paradigm with leakage functions $\mathcal{L} = (\mathcal{L}_\mathsf{S}, \mathcal{L}_\mathsf{Q})$. It guarantees that the encrypted multi-maps structure EMM from Setup reveals nothing beyond the setup leakage $\mathcal{L}_\mathsf{S}$, and the query procedure over EMM reveals nothing beyond the query leakage $\mathcal{L}_\mathsf{Q}$. More precisely, Reply is the only algorithm during the query procedure (Query, Reply, Result) that is executed by the server and possibly leaks information to it.

Definition 2 (Adaptive Security of Structured Encryption for MMs).
Let $\Sigma = (\mathsf{Setup}, \mathsf{Query}, \mathsf{Reply}, \mathsf{Result})$ *be a non-interactive response-hiding structured encryption scheme for multi-maps. We say that* Σ *is* $(\mathcal{L}_\mathsf{S}, \mathcal{L}_\mathsf{Q})$-*adaptively-secure, if for all* PPT *adversary* $\mathcal{A}$, *there exists a* PPT *simulator* $\mathcal{S}$ *such that*

$$| \Pr[\mathbf{Real}_\mathcal{A}(1^\lambda) = 1] - \Pr[\mathbf{Ideal}_{\mathcal{A},\mathcal{S},\mathcal{L}}(1^\lambda) = 1]| \leq \mathsf{negl}(\lambda)$$

where **Real** *and* **Ideal** *are probabilistic experiments defined below.*

$\mathbf{Real}_{\Sigma,\mathcal{A}}(1^\lambda)$: *A generates an input multi-map* MM *and receives* EMM *from the challenger, where* $(K, \mathsf{st}, \mathsf{EMM}) \leftarrow \mathsf{Setup}(1^\lambda, \mathsf{MM})$. *Then* $\mathcal{A}$ *adaptively makes a polynomial number of queries. For each queried* key, *the challenger executes* (Query, Reply) *with* $\mathcal{A}$, *and* $\mathcal{A}$ *receives* $\mathsf{tk} \leftarrow \mathsf{Query}(K, \mathsf{st}, \mathsf{key})$.

$\mathbf{Ideal}_{\Sigma,\mathcal{A},\mathcal{S}}(1^\lambda)$ *A generates an input multi-map* MM *and receives* EMM *from the simulator* $\mathcal{S}$, *where* EMM *is generated using only* $\mathcal{L}_\mathsf{S}$. *Then* $\mathcal{A}$ *adaptively makes a polynomial number of queries. For each queried* key, *the simulator* $\mathcal{S}$ *executes* (Query, Reply) *with* $\mathcal{A}$ *by generating* tk *using only* $\mathcal{L}_\mathsf{Q}$.

In both experiments, $\mathcal{A}$ returns a bit $b \in \{0, 1\}$ eventually. Looking ahead, our construction can be proven secure under the adaptive definition, but an active adversary, in principle, can manipulate the leakages to be defined below.

2.3 Typical Leakage Functions

We recall typical leakage functions summarized for encrypted multi-maps [15].

- *Query equality pattern* qeq, usually referred to as the search pattern of SSE, reports whether two queries are issued for the same key or not. Formally, for a sequence of t non-cryptographic keys $qeq(key_1, \ldots, key_t) = M$, where M is a $t \times t$ matrix such that $M[i][j] = 1$ if $key_i = key_j$ for $i, j \in [t]$.
- *Response length pattern* rlen reports the number of values associated with an input key. Formally, for a multi-map $MM = \{(key_i, \vec{v}_i)\}_{key_i \in \mathcal{K}}$ with key space $\mathcal{K}$, $rlen(MM, key_i) = \ell(key_i) = |\vec{v}_i|$. This is the leakage we aim to avoid.
- *Maximum response length pattern* mrlen reports the maximum number of values associated with any key. Formally, for a multi-map MM with key space $\mathcal{K}$, $mrlen(MM) = \ell = \max_{key \in \mathcal{K}} \ell(key)$, where $\ell(key)$ is the number of values associated with key.
- *Data size pattern* dsize reports the total number of values over all keys. Formally, for a multi-map MM with key space $\mathcal{K}$, $dsize(MM) = n = \sum_{key \in \mathcal{K}} \ell(key)$.

We further introduce the access intersection pattern aintx as follows.

- *Access intersection pattern* aintx reports the intersection of EMM slots accessed by two queries. Formally, for a sequence of t keys, $aintx(EMM, key_1, \ldots, key_t) = M'$, where M' is a $t \times t$ matrix such that $M'[i][j]$ contains the common slots of EMM accessed when querying key_i and key_j for $i, j \in [t]$.

2.4 Volume Hiding Against Passive Adversaries

Volume hiding requires that the number of values associated with any single key remains private. Patel *et al.* [27] defined it as a property of the leakage functions in a game where the active adversary receives the setup and query leakage of a chosen multi-map. We instead consider passive adversaries who can observe all EMM accesses but cannot actively perform adversarially chosen accesses. Furthermore, they have no knowledge of the plaintext multi-map used to generate EMM. An EMM that is volume-hiding against passive adversaries means they cannot determine the actual volume of any key in the multi-map.

An active adversary has a high level of "control" of the data owner who keeps explicitly authorizing many queries, which an actively malicious cloud can exploit toward exposing the entire multi-map. Our model is weaker; however, it remains meaningful in a realistic setting that the cloud server cannot actively sabotage the data owner or would not risk being caught. This also matches the required strength of the adversary in the very recent volume-abusing attacks [21].

2.5 Hash Ring or Consistent Hashing

Hash ring [17] is best known as a solution to the load-balancing problem (*e.g.*, Chord [29] for peer-to-peer protocols). To assign a set of objects to some servers, the hash function, taking either (the identifier of) an object or a server as input, outputs s bits as a slot-identifier in an identifier ring modulo 2^s (*i.e.*, 0 to $2^s - 1$). For large s, the probability of the collision of identifiers is negligible. Each object will be assigned to the server that is placed into the first slot clockwise from it.

3   Space-Efficient Volume-Hiding Multi-map Encryption

3.1   Overview

A classical hash-ring application (Fig. 1a) assigns a set of objects to some servers by placing both objects and servers to some ring slots. The object is then assigned to the next server that appears on the ring in clockwise order, also dubbed as a *successor* of the object slot. *Successor discovery* thus refers to finding the next occupied slot. In our scenario (Fig. 1b), we pick a slot (red node) for each *value index of a key* via hashing. For each interval bounded by red nodes, a slot for storing *the key-value tuple* (blue node) is *randomly* picked.

For any key with a volume less than the maximum volume ℓ, we hash each *non-existent (fake) value index* (up to ℓ) to a slot (red ring) that lies in a distinct interval bounded by the slots of two key-value tuples (blue nodes). It means no other fake value index (red ring) or real value index of the same key (red node) is allowed in the interval (but value indexes for any other key are fine). We may need to re-sample these slots to place them accordingly, yet our analysis shows it is unlikely to happen. To finish the EMM setup, the client outsources the ciphertext of each key-value tuple and its ring slot location (blue node). Notably, the server does not need to store the hash ring (*cf.*, the entire hash table is outsourced by dprfMM/dpMM [27]) or any dummies! Our construction thus requires less storage than any existing solutions.

To query for a key, the client issues a token of ℓ ring slots related to it. The server then carries out successor discovery for each slot from the token and returns the ciphertext stored in the successor. Due to the slot arrangement in the setup, the ciphertexts surely contain all the real values associated with the queried key, and there will be one and only one ciphertext (either associated with the queried key or another key) for each of the ℓ slots of any key, thus maintaining the same response volume (*i.e.*, ℓ) for any key.

Our communication overhead is ℓ, which is as good as all prior schemes (except dpMM [27] with possible data loss). Successor discovery takes $\Theta(\log n)$ with the binary search. It can be accelerated to $\Theta(\log \log n)$ with advanced structures [32] at the cost of extra $\Theta(n)$ storage, providing a trade-off.

Figure 1b depicts a sample EMM of $\{(\mathsf{key}_A, \vec{v}_A), (\mathsf{key}_B, \vec{v}_B)\}$ with $|\vec{v}_A| = 3$, $|\vec{v}_B| = 1$, and the maximum volume $\ell = |\vec{v}_A| = 3$. Each key and its value index, *i.e.*, $(\mathsf{key}_A, 1), (\mathsf{key}_A, 2), (\mathsf{key}_A, 3)$, and $(\mathsf{key}_B, 1)$ is mapped to a ring slot (red node). As the volume of key_B is less than ℓ, two extra slots (red ring) are derived by hashing $(\mathsf{key}_B, 2)$ and $(\mathsf{key}_B, 3)$ in a way that the slots of key_B are ensured to have distinct successors ($\mathsf{ct}_2 = \mathsf{Enc}(\mathsf{key}_A || \vec{v}_A[1])$ and $\mathsf{ct}_3 = \mathsf{Enc}(\mathsf{key}_A || \vec{v}_A[3])$) in our example). To query for key_B, the token consists of $\ell = 3$ slots related to key_B in a randomly permuted order. The server fetches the successor of each

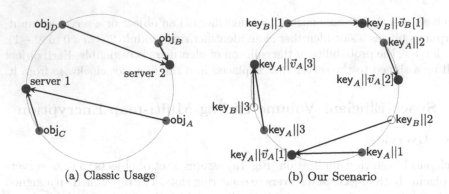

Fig. 1. Consistent hashing and its application in load balancing and our EMM (Color figure online)

slot and returns the corresponding ciphertext, *i.e.*, $\{\mathsf{Enc}(\mathsf{key}_B \| \vec{v}_B[1]), \mathsf{ct}_2, \mathsf{ct}_3\}$ (an unordered set). After client decryption, $\vec{v}_B[1]$ will be identified as the result.

3.2 Description

Figure 2 describes the details of our construction. Let $F : \{0,1\}^\lambda \times \{0,1\}^* \to \{0,1\}^s$ be a pseudorandom function, and $\mathsf{SKE} = (\mathsf{Enc}, \mathsf{Dec})$ be an RCPA-secure encryption scheme. FindSuccessor is a simple function for successor discovery in the hash ring. It takes as input an integer and a sorted array, and outputs the smallest value in the array that is larger than the input integer. If the integer is larger than any value in the array, it outputs the minimum of the array.

Setup. *Cryptographic Keys.* The client randomly picks a PRF seed K_{PRF} and an SKE key K_{SKE} as the secret (cryptographic) key K.

Data Structure. Let n be the number of values over all (non-cryptographic) keys of the MM. The client initializes arrays $\mathsf{U}[0..n-1], \mathsf{V}[0..n-1], \mathsf{HR}[0..2^s-1]$ for the hash ring, and dictionaries $\mathsf{DX}_C, \mathsf{DX}_S$, which are all initially empty.

Client State DX_C. For each key_i with $i \in [m]$, the client samples a λ-bit randomness r_i. $\mathsf{DX}_C[\mathsf{key}_i]$ stores the volume $|\vec{v}_i|$ and the randomness r_i of key_i.

Hash Ring HR. For each value $\vec{v}_i[j]$ associated with key_i, the client determines a hash ring slot by $F_{K_{\mathsf{PRF}}}(r_i \| j)$, puts it in some unoccupied slot of U, and puts the key-value tuple $(\mathsf{key}_i, \vec{v}_i[j])$ in $\mathsf{HR}[F_{K_{\mathsf{PRF}}}(r_i \| j)]$.

Red Nodes U as Boundaries for Ciphertexts. Eventually, n-out-of-2^s slots of HR are occupied. These locations stored in U are then sorted in ascending order.

Blue Nodes V for Ciphertexts. For each $i \in [n]$, the client randomly samples an integer t from $[\mathsf{U}[i-1]+1 .. \mathsf{U}[i \bmod n]]$, with the abused notation for $i = n$ where $[\mathsf{U}[n-1]+1 .. \mathsf{U}[0]] = [\mathsf{U}[n-1]+1 .. 2^s-1] \cup [0 .. \mathsf{U}[0]]$. The client sets $\mathsf{V}[i-1]$ as t, and stores in $\mathsf{DX}_S[t]$ an encryption of the key-value tuple at $\mathsf{HR}[\mathsf{U}[i-1]]$.

- $(K, \mathsf{st}, \mathsf{EMM}) \leftarrow \mathsf{Setup}(1^\lambda, \mathsf{MM} = \{\mathsf{key}_i, \vec{v}_i\}_{i \in [m]})$
  1. Sample PRF seed $K_{\mathsf{PRF}} \leftarrow\$ \{0,1\}^\lambda$ and encryption key $K_{\mathsf{SKE}} \leftarrow\$ \{0,1\}^\lambda$.
  2. Initialize array HR of size 2^s and arrays U, V of size n.
  3. Initialize two dictionaries $\mathsf{DX}_C, \mathsf{DX}_S$.
  4. For each $i \in [m]$:
     (a) Sample $r_i \leftarrow\$ \{0,1\}^\lambda$.
     (b) For each $j \in [|\vec{v}_i|]$,
        i. Insert $F_{K_{\mathsf{PRF}}}(r_i||j)$ into U.
        ii. Set $\mathsf{HR}[F_{K_{\mathsf{PRF}}}(r_i||j)] \leftarrow (\mathsf{key}_i, \vec{v}_i[j])$.
     (c) Set $\mathsf{DX}_C[\mathsf{key}_i] \leftarrow (|\vec{v}_i|, r_i)$.
  5. Sort elements in U in ascending order.
  6. For each $i \in [n]$:
     (a) Sample t from $[\mathsf{U}[i-1]+1 .. \mathsf{U}[i \bmod n]]$.
     (b) Set $\mathsf{V}[i-1] \leftarrow t$.
     (c) Set $\mathsf{DX}_S[\mathsf{V}[i-1]] \leftarrow \mathsf{Enc}(K_{\mathsf{SKE}}, \mathsf{HR}[\mathsf{U}[i-1]])$.
  7. Sort elements in V in ascending order.
  8. For each $i \in [m]$:
     (a) Sample $r'_i \leftarrow\$ \{0,1\}^\lambda$, and set $(t_1, \ldots, t_\ell) =$

     $$\{F_{K_{\mathsf{PRF}}}(r_i||1), \ldots, F_{K_{\mathsf{PRF}}}(r_i||(|\vec{v}_i|)), F_{K_{\mathsf{PRF}}}(r'_i||(1 + |\vec{v}_i|)), \ldots, F_{K_{\mathsf{PRF}}}(r'_i||\ell)\}.$$

     (b) If there exist distinct indices $j, k \in [\ell]$ such that $\mathsf{FindSuccessor}(t_j, \mathsf{V}) = \mathsf{FindSuccessor}(t_k, \mathsf{V})$, redo Step 8a, which includes the sampling of r'_i.
     (c) Set $\mathsf{DX}_C[\mathsf{key}_i] \leftarrow \mathsf{DX}_C[\mathsf{key}_i]||r'_i$.
  9. Delete HR, U.
  10. Return $K \leftarrow (K_{\mathsf{PRF}}, K_{\mathsf{Enc}})$, $\mathsf{st} \leftarrow \mathsf{DX}_C$, and $\mathsf{EMM} \leftarrow (\mathsf{DX}_S, \mathsf{V})$.
- $\mathsf{tk} \leftarrow \mathsf{Query}(K, \mathsf{st}, \mathsf{key})$
  1. Parse K as $(K_{\mathsf{PRF}}, K_{\mathsf{Enc}})$ and st as DX_C.
  2. Parse $\mathsf{DX}_C[\mathsf{key}]$ as $(|\vec{v}|, r)||r'$.
  3. Set $\mathsf{tk} \leftarrow \{F_{K_{\mathsf{PRF}}}(r||j)\}_{j \in [|\vec{v}|]} \cup \{F_{K_{\mathsf{PRF}}}(r'||(j + |\vec{v}|))\}_{j \in [\ell - |\vec{v}|]}$.
  4. Permute elements in tk.
  5. Return tk.
- $\mathsf{res} \leftarrow \mathsf{Reply}(\mathsf{tk}, \mathsf{EMM})$
  1. Parse EMM as $(\mathsf{DX}_S, \mathsf{V})$.
  2. Initialize an empty set res.
  3. For each $t \in \mathsf{tk}$, set $\mathsf{res} \leftarrow \mathsf{res} \cup \mathsf{DX}_S[\mathsf{FindSuccessor}(t, \mathsf{V})]$.
  4. Return res.
- $\vec{v} \leftarrow \mathsf{Result}(K, \mathsf{key}, \mathsf{res})$
  1. Parse K as $(K_{\mathsf{PRF}}, K_{\mathsf{Enc}})$ and res as $\{\mathsf{ct}_i\}_{i \in [\ell]}$.
  2. Initialize an empty array $\vec{v}$.
  3. For $i \in [\ell]$:
     (a) Set $(\mathsf{key}', \mathsf{value}') \leftarrow \mathsf{Dec}(K_{\mathsf{Enc}}, \mathsf{ct}_i)$.
     (b) If $\mathsf{key}' = \mathsf{key}$, insert value' into $\vec{v}$.
  4. Return $\vec{v}$.

Fig. 2. Efficient volume-hiding encryption for multi-maps

Red Rings for Generating Token tk. Corresponding to each key_i for $i \in [m]$, the client samples another λ-bit randomness r'_i, and outputs a tuple $(t_1, \ldots, t_\ell)$:

$$\{F_{K_{\mathsf{PRF}}}(\mathsf{r}_i \| 1), \ldots, F_{K_{\mathsf{PRF}}}(\mathsf{r}_i \| (|\vec{v}_i|)), F_{K_{\mathsf{PRF}}}(\mathsf{r}'_i \| (1 + |\vec{v}_i|)), \ldots, F_{K_{\mathsf{PRF}}}(\mathsf{r}'_i \| \ell)\}$$

where the first $|\vec{v}_i|$ elements are key_i-related slots in HR, and the rest are "fake."

Note that r'_i needs to be chosen in the way that any distinct $j, k \in [\ell]$ satisfy $\mathsf{FindSuccessor}(t_j, \mathsf{V}) \neq \mathsf{FindSuccessor}(t_k, \mathsf{V})$. Otherwise, we re-sample a new r'_i to fulfill the condition[2]. We show that the client does not need to re-sample with the probability tending to 1 in Sect. 3.3. The client concatenates r'_i to $\mathsf{DX}_C[\mathsf{key}_i]$.

The client deletes $(\mathsf{HR}, \mathsf{U})$, sets DX_C as st, and outsources $(\mathsf{DX}_S, \mathsf{V})$ as EMM.

Query. To query key, the client retrieves $(|\vec{v}|, \mathsf{r}) \| \mathsf{r}'$ from $\mathsf{DX}_C[\mathsf{key}]$, and computes:

$$\mathsf{tk} = \{\{F_{K_{\mathsf{PRF}}}(\mathsf{r} \| j)\}_{j \in [|\vec{v}|]}, \{F_{K_{\mathsf{PRF}}}(\mathsf{r}' \| (j + |\vec{v}|))\}_{j \in [\ell - |\vec{v}|]}\}.$$

The client permutes elements in tk before sending it to the server.

For each element $t \in \mathsf{tk}$, the server figures out its successor of the hash ring by $t' = \mathsf{FindSuccessor}(t, \mathsf{V})$ and retrieves the ciphertext stored in $\mathsf{DX}_S[t']$. The collection of these ciphertexts is returned as the query response res.

The client decrypts the ciphertexts in res with K_{SKE}, and collects all decrypted values associated with key as the query result $\vec{v}$.

3.3 Analysis

Security. For Setup, the server receives the EMM, which contains the ciphertexts of all key-value tuples and randomly sampled hash-ring slots. Thus, the setup leakage $\mathcal{L}_S$ only contains the data size dsize.

For Query, the server will know ℓ, the maximum volume of the input multimap by the size of any query token. It could determine the repetition of the same query and the common slots accessed by different queries. Thus, the query leakage $\mathcal{L}_Q$ contains the query-equality pattern qeq, the access intersection pattern aintx, and the *maximum* response length $\mathsf{mrlen} = \ell$. Formally, Theorem 1 (proven in Appendix A) asserts the adaptive security of our construction with the leakage function $(\mathcal{L}_S, \mathcal{L}_Q) = (\mathsf{dsize}, (\mathsf{qeq}, \mathsf{aintx}, \mathsf{mrlen}))$.

Theorem 1. *If* $\mathsf{SKE} = (\mathsf{Enc}, \mathsf{Dec})$ *is an* RCPA-*secure encryption scheme and F is a pseudorandom function, our construction is an* $(\mathcal{L}_S, \mathcal{L}_Q)$-*adaptive-secure encryption scheme for multi-maps, where* $(\mathcal{L}_S, \mathcal{L}_Q) = (\mathsf{dsize}, (\mathsf{qeq}, \mathsf{aintx}, \mathsf{mrlen}))$.

Nevertheless, an active adversary with the knowledge of EMM could adaptively issue queries and try to match the corresponding access intersection aintx with any prior background information about the actual volume. This seems to be inherent when we "reuse" real slots for realizing volume-hiding EMM efficiently. Passive adversaries cannot exploit such kinds of strategies. Intuitively, the response length rlen for any key remains hidden since the slots for different keys are pseudorandom as our scheme decides them via pseudorandom functions.

---

[2] A similar re-sampling is required in AVLH [14], while dprfMM and dpMM [27] abort when the stash of cuckoo hashing overflows due to collisions and evictions.

No Re-sampling Probability. Our construction requires each of the last $\ell - |\vec{v}_i|$ elements of $(t_1, \ldots, t_\ell)$ for key_i, i.e., $F_{K_{\mathsf{PRF}}}(r_i' \| (1 + |\vec{v}_i|)), \ldots, F_{K_{\mathsf{PRF}}}(r_i' \| \ell)$, has a unique successor. For ease of analysis, we model F as a random function. Obviously, the elements within the same interval will share the same successor, and the probability that an element is assigned to an interval by the random function depends on the interval length. Suppose that we divide the hash ring of length 2^s into n intervals with n values among $\{0,1\}^s$ in V. Let c_i be the length of the interval where t_i is placed for $i \in [\ell]$. Consider the case that the first $k-1$ elements of $(t_{|\vec{v}_i|+1}, \ldots, t_\ell)$ belong to distinct intervals. The probability that the k-th element will be assigned to an interval different from any previous elements is $(1 - \frac{c_1 + \cdots + c_{|\vec{v}_i|+k-1}}{2^s})$. Then, the probability that the remaining $\ell - |\vec{v}_i|$ elements are all placed in distinct intervals, i.e., they have distinct outputs of FindSuccessor and no re-sampling is needed for key_i, is:

$$p = \prod_{k=1}^{\ell - |\vec{v}_i|} \left(1 - \frac{c_1 + \cdots + c_{|\vec{v}_i|+k-1}}{2^s}\right) \geq \left(1 - \frac{(\ell - |\vec{v}_i|) \times c_{\mathsf{avg}}}{2^s}\right)^{\ell - |\vec{v}_i|}$$

where c_{avg} is the average length of $(c_1, \ldots, c_{\ell-1})$. We take the expected interval length of the hash ring for c_{avg}, i.e., $\frac{2^s}{n}$. The probability becomes $(1 - \frac{\ell - |\vec{v}_i|}{n})^{\ell - |\vec{v}_i|}$, which is approximate to $e^{-\frac{(\ell - |\vec{v}_i|)^2}{n}}$. It tends to 1 assuming[3] $\ell = o(n)$.

Efficiency. Our construction avoids padding of dummies or storing the entire hash ring. Consider an input multi-map MM with maximum volume ℓ, m distinct keys, and n key-value tuples. The server stores a dictionary DX_S of size n for all key-value tuples of MM and an auxiliary array V of size n to quickly determine the successor of any slot, outperforming the state-of-the-art dprfMM and dpMM [27]. The client stores the state of size m rather than a stash for any parts of the input MM. It could be outsourced as m is independent of ℓ.

To query for any key, the client only needs to generate ℓ PRF values as slots in the token. The server determines exactly ℓ successors correspondingly for these ℓ slots. With binary search, the server can figure out a successor from V with FindSuccessor in $\Theta(\log n)$ time. The overall query complexity is thus $\Theta(\ell \cdot \log n)$, which is asymptotically the same as AVLH for general MMs. ℓ ciphertexts are returned to the client as a response. The communication overhead only falls behind the differentially private scheme [27] with possible data loss.

The prior arts we improved upon are practically efficient [27]. Moreover, unlike a particular existing scheme [14], the efficiency of our scheme is not affected by the data distribution. As stressed in Footnote 3, we only made one reasonable assumption of $\ell = o(n)$, which appears to be necessary for any practically-efficient volume-hiding EMM. (If the volume is in the order of magnitude of the database size, any volume-hiding scheme ought to be inefficient.) Our underlying cryptographic operations are mostly PRF as typically used in many

---

[3] This is our only assumption on the input multi-map. AVLH [14] made more constraints on the multi-map for security or efficiency concerns, including this one.

practically-efficient STEs. It means that our asymptotic improvements over the prior arts can be directly translated to improvements in concrete efficiency, following the doctrine in computer science that asymptotic analysis of complexity remains meaningful, regardless of the ever-changing computational power of the time.

Different Trade-Offs on Server. For EMM = (DX_S, V) outsourced to the server, the dictionary keys in DX_S are essentially the elements stored in the auxiliary array V. Our construction can choose to further outsource V or directly process over DX_S, which improves the storage overhead to n at the cost of either communication overhead or computation overhead for the server.

With more advanced data structures, it is possible to reduce the computation overheads of the binary search for successors, which accelerates the query, while the server storage remains asymptotically the same, *i.e.*, $\Theta(n)$. For example, we could arrange V in a y-fast trie data structure [32], which efficiently supports successor queries for integers using extra storage. Concretely, this trade-off offers a computation overhead of $\Theta(\log \log n)$ for successor discovery with $\Theta(n)$ server-side overhead, while the performance of other criteria remains the same.

4   Volume-Hiding Dynamic EMM via Batching Updates

A dynamic EMM enables the client to insert or remove one or multiple value(s) associated with a key. Dynamic updates can be challenging to realize when we simultaneously consider security and efficiency (*e.g.*, minimizing leakage while maintaining a parallelly-traversable encrypted structure [23–25]). Similarly, we face a dilemma in designing a volume-hiding EMM. The root cause is that even a single update for a key-value pair may change the maximum volume of the multi-map. On one hand, the client should do dummy updates for *distinct* keys to hide whether the real one is associated with the maximum volume. On the other, it might leak more as those dummy updates for *distinct* keys it triggers probably cause extra leakage. For dynamic update, it sounds better to start from a static EMM that is free from dummies (or troubles), which guided our design.

Another way round is to let the client query the tuple regarding the key to be updated in advance, and re-upload the modified tuple, which is confined to the queried key but not any other keys. VLH^d and AVLH^d [14] follow this approach. However, they are still subject to the maximum-volume limit (unless those extra ones are truncated by the pseudorandom transform in VLH^d). They also fail to consider either forward or backward privacy emphasized in recent dynamic STE.

When saving space is one of our design goals, it is natural that the resulting EMM is "highly" optimized with respect to the maximum volume, which makes "volume-affecting" updates tricky. We thus consider dynamic EMM for batch updates [7]. Namely, we set up a new EMM instance for batches and periodically merge instances for efficiency. Setting up new instances for later merging may sound conceptually straightforward; however, it is not an approach practically applicable to all volume-hiding EMMs. As in Table 1, our static construction has decent communication overhead and server storage, which will be scaled up by the number of batches in such an approach (*e.g.*, for storing many EMM instances), making ours more applicable than others. This also aligns with the update feature of some commercial databases (*e.g.*, bulk-loading in Vertica [26]).

4.1 Definition

Syntax. We consider the batch of updates in a similar way as the multi-maps. The μ-th batch of updates MM_μ is defined as $\{\mu, \mathsf{key}_i, \vec{v}_i\}_{\mathsf{key}_i \in \mathcal{K}}$, where $\vec{v}_i$ now is a tuple of operation-value pairs regarding key_i. For each $j \in \ell_\mu(\mathsf{key})$, $\vec{v}[j] = (\vec{o}[j], \vec{u}[j])$, where $\vec{o}[j] \in \{\mathsf{add}, \mathsf{del}\}$ refers to insertion or deletion of value $\vec{u}[j]$ for key, and $\ell_\mu(\mathsf{key})$ is the number of updates associated with key in MM_μ. We treat the initial input multi-map for Setup as MM_0 with an insertion to be done.

We introduce the following update algorithm for non-interactive response-hiding structured encryption scheme for multi-maps $\varSigma$:

- $(K_\mu, \mathsf{st}', \mathsf{EMM}_\mu) \leftarrow \mathsf{Update}(\mathsf{MM}_\mu, \mathsf{st})$ is a probabilistic algorithm executed by the client that takes as input a batch of updates MM_μ and a state st. It outputs a private key K_μ, an updated state st', and the encrypted multi-map EMM_μ. The client stores (K_μ, st') while sending EMM_μ to the server.

We extend the leakage function to $\mathcal{L} = (\mathcal{L}_\mathsf{S}, \mathcal{L}_\mathsf{Q}, \mathcal{L}_\mathsf{U})$, where $\mathcal{L}_\mathsf{U}$ is the update leakage. STE security guarantees that Update reveals no information beyond $\mathcal{L}_\mathsf{U}$.

The definition of adaptive security follows Definition 2, except that the simulator answers the update request on MM_μ from $\mathcal{A}$ with EMM_μ generated using $\mathcal{L}_\mathsf{U}$. Volume hiding is still considered with respect to passive adversaries.

Dynamic STE schemes [1,3,4,25] attach great importance to forward privacy, which requires updates to reveal nothing on the updated key, and backward privacy, which ensures the deleted value remains inaccessible to the server. While a recent work [28] pointed out that forward-private STE still suffers from subtle attacks, forward privacy could mitigate the severe adaptive file injection attacks [33], thus becoming the *de facto* standard of dynamic STE nowadays. We formalize forward and backward privacy for batch updates in Appendix B.

4.2 Description

For the μ-th update batch $\mathsf{MM}_\mu = \{\mu, \mathsf{key}_i, \vec{v}_i = (\vec{o}_i, \vec{u}_i)\}_{\mathsf{key}_i \in \mathcal{K}}$, the client generates $(K_\mu, \mathsf{st}', \mathsf{EMM}_\mu)$ similar to Setup in Fig. 2. The new state is concatenated to the old one according to keys. The only difference is that, for each $\mathsf{key} \in \mathcal{K}$ and each $j \in \ell_\mu(\mathsf{key})$, $(\mathsf{key}, \vec{o}[j], \vec{u}[j])$ is encrypted instead of a key-value pair.

Suppose the client has issued b batches of updates, and the server stores $b+1$ encrypted multi-maps, including the initial one in Setup. Upon a query, the client issues $b+1$ tokens with Query, and the server answers with Reply proposed in our static construction, respectively. The client removes those values marked with del and returns the remaining values associated with the queried key.

4.3 Analysis

Efficiency. For an update batch MM_μ, let n_μ be the total number of updates over all keys, m_μ be the number of distinct keys, and ℓ_μ be the maximum number of updates for any key. To update with MM_μ, the server stores EMM_μ of size $2n_\mu$. It takes the client $\Theta(m_\mu \cdot \ell_\mu)$ to generate this EMM_μ.

For a query over our EMM with b batches of updates, the communication overhead is $\sum_{\mu=0}^{b} \ell_\mu$, while the computation complexity is $\Theta(\sum_{\mu=0}^{b} \ell_\mu \cdot \log n_\mu)$.

Note that our dynamic extension inherits the nice properties of the underlying static EMM, with neither client stash for overflowed multi-maps nor data loss.

For amortized efficiency, we could exploit the periodic-merge approach of Vertica [26] as also adopted by a prior STE [7]. The client decides a consolidation step s and the EMM for each update batch is organized as leaves of an s-ary tree, created bottom-up. When s nodes are occupied in the same level, the client downloads these s EMMs, consolidates existing values associated with the same keys, and re-generates an encrypted multi-map, which will be assigned as the parent of these s nodes. We refer to [7, 26] for a more detailed description.

Despite the periodic cost for consolidation, this optimization can save the server-side storage by reclaiming the space for deleted values, and reduce the query overhead when the consolidation does not increase the maximum volume. The actual consolidation strategy can be different for different scenarios.

Security. We first introduce two new leakage functions for batch updates.

- *Update-batch pattern* updb indicates the batches with any updates on key happened. Formally, $\mathsf{updb}(\mathsf{MM}, \mathsf{key}) = \{\mu \mid \ell_\mu(\mathsf{key}) > 0\}$ with $\mathsf{MM} = \cup \mathsf{MM}_\mu$.
- *Update size pattern* usize reports the total number of updates over all keys in an update batch. Formally, for a batch of updates MM_μ with (non-cryptographic) key space $\mathcal{K}$, $\mathsf{usize}(\mathsf{MM}_\mu) = \sum_{\mathsf{key} \in \mathcal{K}} \ell_\mu(\mathsf{key})$.

Theorem 2. *If* $\mathsf{SKE} = (\mathsf{Enc}, \mathsf{Dec})$ *is an RCPA-secure encryption scheme and* F *is a* PRF*, our dynamic construction is an* $(\mathcal{L}_\mathsf{S}, \mathcal{L}_\mathsf{Q}, \mathcal{L}_\mathsf{U})$*-adaptive-secure dynamic encryption scheme for multi-maps with forward privacy, level-II backward privacy, and volume hiding against passive adversaries where* $(\mathcal{L}_\mathsf{S}, \mathcal{L}_\mathsf{Q}, \mathcal{L}_\mathsf{U}) = (\mathsf{dsize},$ $(\mathsf{qeq}, \mathsf{aintx}, \mathsf{mrlen}, \mathsf{updb}), \mathsf{usize})$.

The proof is omitted due to the page limit and since it can be done in a standard way. Roughly, given usize, $\mathsf{Update}()$ can be simulated similar to simulating $\mathsf{Setup}()$ as in Theorem 1. The update leakage contains usize of the update batch, not revealing the volume associated with any key against passive adversaries.

The construction is forward-private as the token generated for previous encrypted multi-maps could not be applied to any subsequent updates and no key information is revealed during updates. It also satisfies our level-II backward privacy for batch updates (Appendix B) as the construction is response-hiding and nothing except the batches is revealed when updates over the queried key.

5   Conclusion

Structured encryption is volume hiding when each query has the same volume as the largest one. It was originally proposed as a stringent requirement and might be perceived as a theoretical notion with no practical realization.

Motivated by passive attacks abusing volume leakage in plaintext reconstruction [21], this work considers a practically relevant setting in which the cloud is honest-but-curious and does not launch active attacks, which could alert the data owner. Achieving the volume-hiding property remains non-trivial given our primary efficiency goal is to save the actual volume of encrypted data. The novelty of our scheme is that the slots needed in hiding the volume across the queries are all real slots storing some meaningful ciphertexts, in contrast to the known paradigm of volume-hiding structured encryption. Thanks to its nice performance and compatibility with batch updates, we extend it to the first dynamic volume-hiding encrypted multi-maps with forward and backward privacy.

Conceptually, this work brings new ideas to the (volume-hiding) structured encryption realm and hopefully inspires more research. We leave as an open problem to provide volume-hiding against stronger adversaries and handle sporadic volume-affecting updates while maintaining a similar efficiency level.

A   Security Proof for Theorem 1

Proof. We construct the simulator S as follows.

To simulate EMM with $\mathcal{L}_S$ = dsize, the simulator S initializes two arrays V, U of size n = dsize(MM), and a dictionary DX_S. S fills each slot of U with a uniformly random value sampling from $\{0,1\}^s$, and then sorts U. For $i \in [n]$, S randomly selects a uniformly random integer t as $V[i-1]$ from the interval $[U[i-1]+1 .. U[i \bmod n]]$ and sets $DX_S[V[i-1]]$ as a uniformly random value sampling from $\{0,1\}^\lambda$. S sorts and keeps V. (DX_S, V) is returned as EMM.

To simulate tk for the i-th query of key_i with $\mathcal{L}_Q$ = (qeq, aintx, mrlen), the simulator S checks whether key_i has been queried before using qeq($key_1, \ldots, key_i$). If so, S returns the same tk as that of the previous queries of key_i. Otherwise, S checks aintx(EMM, $key_1, \ldots, key_i$) and get the common slots of key_i with previous queries. S picks ℓ = mrlen(MM) intervals in V, with parts of them determined by the common slots from aintx and the rest picked randomly without repeating, and samples a uniformly random value from each selected interval. These ℓ values are returned as tk.

We show that for all PPT adversary $\mathcal{A}$, the outputs of the real-world game and ideal-world game are indistinguishable. We derive a standard game sequence from the real-world game $\mathbf{Real}_{\Sigma,\mathcal{A}}(1^\lambda)$ to the ideal-world game $\mathbf{Ideal}_{\Sigma,\mathcal{A},S}(1^\lambda)$.

- $\mathbf{Game}_0$ is the same as $\mathbf{Real}_\mathcal{A}(1^\lambda)$.
- $\mathbf{Game}_1$ replaces the pseudorandom function F in $\mathbf{Game}_0$ with a random function (and recalled when needed).
- $\mathbf{Game}_2$ replaces the RCPA-secure encryption scheme in $\mathbf{Game}_1$ with a random function.

- **Game$_3$** replaces the outputs of random functions in **Game$_2$** with values chosen uniformly at random.
- **Game$_4$**, for any query, randomly picks $\ell = \mathsf{mrlen}(\mathsf{MM})$ intervals in V and samples a uniformly random value from each selected interval. **Game$_4$** is the same as $\mathbf{Ideal}_{\Sigma,\mathcal{A},\mathcal{S}}(1^\lambda)$.

Game$_0$ and **Game$_1$** are indistinguishable; otherwise, it violates the security of the pseudorandom function. **Game$_1$** and **Game$_2$** are indistinguishable; otherwise, it violates the RCPA security of the encryption scheme. By the definition of random functions, winning **Game$_2$** or **Game$_3$** shares an equal probability. The probabilities of winning **Game$_3$** and **Game$_4$** are also equal since $\mathsf{DX}_\mathcal{S}$, V, tk follow the same distributions in both games. By combining these (in)equalities, we have $|\Pr[\mathbf{Real}_\mathcal{A}(1^\lambda) = 1] - \Pr[\mathbf{Ideal}_{\mathcal{A},\mathcal{S},\mathcal{L}}(1^\lambda) = 1]| \leq \mathsf{negl}(\lambda)$.

B   Forward and Backward Privacy for Batch Updates

Forward privacy for batch updates requires that any batch of updates reveals nothing about the keys to be updated. Thus, the adversary cannot figure out the relation between newly updated multi-maps and any previous query. We extend the definition from that designed for a single key-value update [3].

Definition 3 (Forward Privacy). *We say that an $\mathcal{L}$-adaptively-secure structured encryption scheme for multi-maps Σ over key space $\mathcal{K}$ is forward private, if the update leakage function $\mathcal{L}_\mathsf{U}$ can be written as $\mathcal{L}_\mathsf{U}(\mathsf{MM}_\mu) = \mathcal{L}'(\{\vec{v}_i\}_{\mathsf{key}_i \in \mathcal{K}})$, where $\mathsf{MM}_\mu = \{\mu, \mathsf{key}_i, \vec{v}_i\}_{\mathsf{key}_i \in \mathcal{K}}$, and $\mathcal{L}'$ is stateless.*

Backward privacy hides deleted values during subsequent queries. Like the previous definition [4], we formalize it for *batch updates* by introducing leakage functions constructed from the union of historical update batches $\mathsf{MM} = \cup \mathsf{MM}_\mu$:

- *Value-batch pattern* valb reports the values currently associated with key and in which batches they are inserted. Formally, $\mathsf{valb}(\mathsf{MM}, \mathsf{key}) = \{(\mu, u) \mid (\mu, \mathsf{key}, (\mathsf{add}, u)) \in \mathsf{MM} \land \forall \mu', (\mu', \mathsf{key}, (\mathsf{del}, u)) \notin \mathsf{MM}\}$.
- *Delete-batch pattern* delb lists the batch-pairs of deletions and corresponding insertions on key. Formally, $\mathsf{delb}(\mathsf{MM}, \mathsf{key}) = \{(\mu, \mu') \mid \exists u \text{ s.t. } (\mu, \mathsf{key}, (\mathsf{del}, u)) \in \mathsf{MM} \land (\mu', \mathsf{key}, (\mathsf{add}, u)) \in \mathsf{MM}\}$.

Definition 4 (Backward Privacy). *We say that an $\mathcal{L}$-adaptively-secure structured encryption scheme for multi-maps Σ over key space $\mathcal{K}$ is*

- *insertion-pattern revealing backward-private (Level-I) if*

$$\mathcal{L}_\mathsf{U}(\mathsf{MM}_\mu) = \mathcal{L}'(\{\vec{o}_i\}_{\mathsf{key}_i \in \mathcal{K}}),$$
$$\mathcal{L}_\mathsf{S}(\mathsf{MM}, \mathsf{key}) = \mathcal{L}''(\mathsf{valb}(\mathsf{MM}, \mathsf{key}), \ell(\mathsf{key})),$$

- *update-pattern revealing backward-private (Level-II) if*

$$\mathcal{L}_\mathsf{U}(\mathsf{MM}_u) = \mathcal{L}'(\{\mathsf{key}_i, \vec{o}_i\}_{\mathsf{key}_i \in \mathcal{K}}),$$
$$\mathcal{L}_\mathsf{S}(\mathsf{MM}, \mathsf{key}) = \mathcal{L}''(\mathsf{valb}(\mathsf{MM}, \mathsf{key}), \mathsf{updb}(\mathsf{MM}, \mathsf{key})),$$

– *weakly backward-private (Level-III) if*

$$\mathcal{L}_U(\mathsf{MM}_u) = \mathcal{L}'(\{\mathsf{key}_i, \vec{o}_i\}_{\mathsf{key}_i \in \mathcal{K}}),$$
$$\mathcal{L}_S(\mathsf{MM}, \mathsf{key}) = \mathcal{L}''(\mathsf{valb}(\mathsf{MM}, \mathsf{key}), \mathsf{delb}(\mathsf{MM}, \mathsf{key})),$$

where $\mathsf{MM} = \cup\mathsf{MM}_\mu$ *with* $\mathsf{MM}_\mu = \{\mu, \mathsf{key}_i, \vec{v}_i = (\vec{o}_i, \vec{u}_i)\}_{\mathsf{key}_i \in \mathcal{K}}$. $\ell(\mathsf{key})$ *is the volume of* key *in* MM. $\mathcal{L}'$ *and* $\mathcal{L}''$ *are stateless.*

References

1. Amjad, G., Kamara, S., Moataz, T.: Breach-resistant structured encryption. Proc. Privacy Enhanc. Technol. (PoPETs) **2019**(1), 245–265 (2019)
2. Blackstone, L., Kamara, S., Moataz, T.: Revisiting leakage abuse attacks. In: Network and Distributed System Security Symposium (NDSS) (2020)
3. Bost, R.: $\sum o\varphi o\varsigma$: forward secure searchable encryption. In: ACM Conference on Computer and Communications Security (CCS), pp. 1143–1154 (2016)
4. Bost, R., Minaud, B., Ohrimenko, O.: Forward and backward private searchable encryption from constrained cryptographic primitives. In: ACM Conference on Computer and Communications Security (CCS), pp. 1465–1482 (2017)
5. Cash, D., Grubbs, P., Perry, J., Ristenpart, T.: Leakage-abuse attacks against searchable encryption. In: ACM Conference on Computer and Communications Security (CCS), pp. 668–679 (2015)
6. Chase, M., Kamara, S.: Structured encryption and controlled disclosure. In: Abe, M. (ed.) ASIACRYPT 2010. LNCS, vol. 6477, pp. 577–594. Springer, Heidelberg (2010). https://doi.org/10.1007/978-3-642-17373-8_33
7. Demertzis, I., Papadopoulos, S., Papapetrou, O., Deligiannakis, A., Garofalakis, M.N.: Practical private range search revisited. In: ACM International Conference on Management of Data (SIGMOD), pp. 185–198 (2016)
8. Faber, S., Jarecki, S., Krawczyk, H., Nguyen, Q., Rosu, M., Steiner, M.: Rich queries on encrypted data: beyond exact matches. In: Pernul, G., Ryan, P.Y.A., Weippl, E. (eds.) ESORICS 2015, Part II. LNCS, vol. 9327, pp. 123–145. Springer, Cham (2015). https://doi.org/10.1007/978-3-319-24177-7_7
9. Goldreich, O., Ostrovsky, R.: Software protection and simulation on oblivious RAMs. J. ACM **43**(3), 431–473 (1996)
10. Grubbs, P., Lacharité, M., Minaud, B., Paterson, K.G.: Pump up the volume: practical database reconstruction from volume leakage on range queries. In: ACM Conference on Computer and Communications Security (CCS), pp. 315–331 (2018)
11. Gui, Z., Paterson, K.G., Patranabis, S., Warinschi, B.: SWiSSSE: system-wide security for searchable symmetric encryption. IACR Cryptology ePrint Archive: 2020/1328 (2020)
12. Islam, M.S., Kuzu, M., Kantarcioglu, M.: Access pattern disclosure on searchable encryption: ramification, attack and mitigation. In: Network and Distributed System Security Symposium (NDSS) (2012)
13. Kamara, S., Moataz, T.: SQL on structurally-encrypted databases. In: Peyrin, T., Galbraith, S. (eds.) ASIACRYPT 2018, Part I. LNCS, vol. 11272, pp. 149–180. Springer, Cham (2018). https://doi.org/10.1007/978-3-030-03326-2_6
14. Kamara, S., Moataz, T.: Computationally volume-hiding structured encryption. In: Ishai, Y., Rijmen, V. (eds.) EUROCRYPT 2019, Part II. LNCS, vol. 11477, pp. 183–213. Springer, Cham (2019). https://doi.org/10.1007/978-3-030-17656-3_7

15. Kamara, S., Moataz, T., Ohrimenko, O.: Structured encryption and leakage suppression. In: Shacham, H., Boldyreva, A. (eds.) CRYPTO 2018, Part I. LNCS, vol. 10991, pp. 339–370. Springer, Cham (2018). https://doi.org/10.1007/978-3-319-96884-1_12

16. Kamara, S., Papamanthou, C., Roeder, T.: Dynamic searchable symmetric encryption. In: ACM Conference on Computer and Communications Security (CCS), pp. 965–976 (2012)

17. Karger, D.R., Lehman, E., Leighton, F.T., Panigrahy, R., Levine, M.S., Lewin, D.: Consistent hashing and random trees: distributed caching protocols for relieving hot spots on the world wide web. In: Annual ACM Symposium on the Theory of Computing (STOC), pp. 654–663 (1997)

18. Kellaris, G., Kollios, G., Nissim, K., O'Neill, A.: Generic attacks on secure outsourced databases. In: ACM Conference on Computer and Communications Security (CCS), pp. 1329–1340 (2016)

19. Kiayias, A., Papadopoulos, S., Triandopoulos, N., Zacharias, T.: Delegatable pseudorandom functions and applications. In: ACM Conference on Computer and Communications Security (CCS), pp. 669–684 (2013)

20. Kirsch, A., Mitzenmacher, M., Wieder, U.: More robust hashing: Cuckoo hashing with a stash. SIAM J. Comput. **39**(4), 1543–1561 (2009)

21. Kornaropoulos, E.M., Papamanthou, C., Tamassia, R.: Response-hiding encrypted ranges: revisiting security via parametrized leakage-abuse attacks. In: IEEE Symposium on Security and Privacy (S&P), pp. 750–767 (2021)

22. Lacharité, M., Minaud, B., Paterson, K.G.: Improved reconstruction attacks on encrypted data using range query leakage. In: IEEE Symposium on Security and Privacy (S&P), pp. 297–314 (2018)

23. Lai, R.W.F., Chow, S.S.M.: Structured encryption with non-interactive updates and parallel traversal. In: International Conference on Distributed Computing Systems (ICDCS), pp. 776–777 (2015)

24. Lai, R.W.F., Chow, S.S.M.: Parallel and dynamic structured encryption. In: Deng, R., Weng, J., Ren, K., Yegneswaran, V. (eds.) SecureComm 2016. LNICST, vol. 198, pp. 219–238. Springer, Cham (2017). https://doi.org/10.1007/978-3-319-59608-2_12

25. Lai, R.W.F., Chow, S.S.M.: Forward-secure searchable encryption on labeled bipartite graphs. In: Gollmann, D., Miyaji, A., Kikuchi, H. (eds.) ACNS 2017. LNCS, vol. 10355, pp. 478–497. Springer, Cham (2017). https://doi.org/10.1007/978-3-319-61204-1_24

26. Lamb, A., et al.: The vertica analytic database: C-store 7 years later. Proc. VLDB Endow. **5**(12), 1790–1801 (2012)

27. Patel, S., Persiano, G., Yeo, K., Yung, M.: Mitigating leakage in secure cloud-hosted data structures: volume-hiding for multi-maps via hashing. In: ACM Conference on Computer and Communications Security (CCS), pp. 79–93 (2019)

28. Salmani, K., Barker, K.: Don't fool yourself with forward privacy, your queries still belong to us! In: ACM Conference on Data and Application Security and Privacy (CODASPY), pp. 131–142 (2021)

29. Stoica, I., Morris, R.T., Karger, D.R., Kaashoek, M.F., Balakrishnan, H.: Chord: a scalable peer-to-peer lookup service for internet applications. In: ACM Conference on Applications, Technologies, Architectures, and Protocols for Computer Communication (SIGCOMM), pp. 149–160 (2001)

30. Wang, J., Chow, S.S.M.: Forward and backward-secure range-searchable symmetric encryption. IACR Cryptology ePrint Archive: 2019/497 (2019)

31. Wang, J., Du, M., Chow, S.S.M.: Stargazing in the dark: secure skyline queries with SGX. In: Nah, Y., Cui, B., Lee, S.-W., Yu, J.X., Moon, Y.-S., Whang, S.E. (eds.) DASFAA 2020, Part III. LNCS, vol. 12114, pp. 322–338. Springer, Cham (2020). https://doi.org/10.1007/978-3-030-59419-0_20
32. Willard, D.E.: Log-logarithmic worst-case range queries are possible in space theta(n). Inf. Process. Lett. (IPL) **17**(2), 81–84 (1983)
33. Zhang, Y., Katz, J., Papamanthou, C.: All your queries are belong to us: the power of file-injection attacks on searchable encryption. In: USENIX Security Symposium, pp. 707–720 (2016)

Nowhere to Leak: A Multi-client Forward and Backward Private Symmetric Searchable Encryption Scheme

Alexandros Bakas[✉] and Antonis Michalas

Tampere University, Tampere, Finland
{alexandros.bakas,antonios.michalas}@tuni.fi

Abstract. Symmetric Searchable Encryption (SSE) allows users to outsource encrypted data to a possibly untrusted remote location while simultaneously being able to perform keyword search directly through the stored ciphertexts. An ideal SSE scheme should reveal no information about the content of the encrypted information nor about the searched keywords and their mapping to the stored files. However, most of the existing SSE schemes fail to fulfil this property since in every search query, some information potentially valuable to a malicious adversary is leaked. The leakage becomes even bigger if the underlying SSE scheme is dynamic. In this paper, we *minimize the leaked information* by proposing a *forward and backward private* SSE scheme in a multi-client setting. Our construction achieves optimal search and update costs. In contrast to many recent works, each search query only requires one round of interaction between a user and the cloud service provider. In order to guarantee the security and privacy of the scheme and support the multi-client model (i.e. synchronization between users), we exploit the functionality offered by AMD's Secure Encrypted Virtualization (SEV).

Keywords: Backward privacy · Cloud security · Forward privacy · Multi-client · Symmetric Searchable Encryption

1 Introduction

Symmetric Searchable Encryption (SSE) is a promising encryption technique that squarely fits the cloud paradigm and can pave the way for the development of cloud services that will respect users' privacy even in the case of a compromised Cloud Service Provider (CSP) [6]. Additionally, SSE schemes can be seen as a first, fundamental step for protecting users' data from both external and *internal* attacks (e.g. a malicious administrator). This is because since in an SSE scheme, users generate all the secret information (encryption key) locally and encrypt all

This work was funded by the ASCLEPIOS: Advanced Secure Cloud Encrypted Platform for Internationally Orchestrated Solutions in Healthcare Project No. 826093 EU research project.

© IFIP International Federation for Information Processing 2021
Published by Springer Nature Switzerland AG 2021
K. Barker and K. Ghazinour (Eds.): DBSec 2021, LNCS 12840, pp. 84–95, 2021.
https://doi.org/10.1007/978-3-030-81242-3_5

of their data on the client-side (i.e. the encryption key is *never* revealed to the CSP). The service offered by the CSP is only used for storing and retrieving the generated ciphertexts. In contrast to traditional encryption schemes, SSE offers a remarkable functionality – it allows users to search for specific keywords directly through the stored ciphertexts. This is done by asking the CSP to execute search queries in a privacy-preserving way. In other words, the CSP can find all the ciphertexts containing a specific keyword but without knowing the underlying keyword or anything about the content of the corresponding files.

An ideal SSE scheme should reveal no information about the content of the encrypted information nor about the searched keywords and their mapping to the stored files. However, most SSE schemes fail to fulfill this property since in every search or update query, some information potentially valuable to a malicious adversary is leaked. In the early years of SSE, researchers utilized techniques such as oblivious RAM (ORAM). However, according to [20], adopting such a technique is even less efficient than downloading and decrypting the entire database locally. As a result, researchers have come to a silent agreement that *"nothing should be revealed beyond some well defined and "reasonable" leakage"* [3].

Leaked information in SSE schemes has become a problem of paramount importance since it is the main factor in defining the overall level of security. In works such as [11] and [16] it is pointed out that even a small leakage can lead to several privacy attacks. These works were further extended in [24] where the authors assumed that an active adversary can perform file-injection attacks and record the output. This "new" ability allowed the adversary to recover information about past queries only after ten file insertions. This result led researchers to design *forward private* SSE schemes [8,13]. Forward privacy is a notion introduced in [22] and guarantees that that newly added files cannot be related to past search queries. While forward privacy is a very important property, unfortunately it has been shown to also be vulnerable to certain file-injection attacks [24].

While forward privacy secures the content of a past query, its binary property, *backward privacy*, ensures the privacy of future queries. Backward privacy was formalized in [9] where three different flavors were defined. A backward private SSE scheme ensures that queries do not reveal their association with deleted documents. To the best of our knowledge, there are only a handful of backward private schemes per flavor where *none* of them supports the multi-client model.

Our Contribution: We extend the work proposed in [13] by constructing a forward and backward private Dynamic SSE scheme that supports a multi-client model. We deal with the problem of synchronization between multiple clients by utilizing the functionality offered by AMD's SEV [2]. In particular, our construction: *(1) Provides Forward Privacy. (2) Provides Backward Privacy. (3) Is asymptotically optimal. (4) Is Parallelizable.*

2   Related Work

Our work is based on [13] where the authors presented a Symmetric Searchable Encryption scheme with Forward Privacy, a notion first introduced in [22]. Their construction however is only forward private and limited to a single client model. In this work, we make use of the functionality offered by AMD's SEV VMs to both extend the original scheme to be backward private and to support the multi-client setting. Another single-client forward private SSE scheme is presented in [8], where the authors designed *Sophos*. While *Sophos* achieves asymptotically optimal search and update costs, $O(\ell)$ and $O(m)$ respectively, a file addition requires $O(m)$ asymmetric cryptographic operations on the user's side. An improvement in the search time of *Sophos* is presented in [17]. Authors of [8] extended their work in [9] by designing a number of SSE schemes that are both forward and backward private. Out of those schemes, $Diana_{del}$ and *Janus* are the most efficient but at the same time they satisfy the weakest notion of backward privacy. In particular, *Janus* achieves its security by using public puncturable encryption. *Fides* is among the first efficient backward private SSE schemes with stronger security guarantees. However, it only satisfies the single-client model. Moreover, the search operation requires two rounds of interaction while our scheme only requires one. An improvement of *Janus* is presented in [23] where authors design *Janus++*. While *Janus++* is more efficient than Janus as it is based on symmetric puncturable encryption, Janus++ can only achieve the same security level as Janus. An SGX-based forward/backward private scheme called *Bunker-B* is presented in [3]. Our construction is similar to that in the sense that we use a trusted execution environment (TEE) to reduce the number of required rounds to one. However, as in the case of *Sophos* and *Fides*, *Bunker-B* only supports the single-client model. Moreover, we believe that SGX in not a suitable TEE for a cloud-based service due to its limitations.

Table 1. N: number of (w, id) pairs, n: total number of files, m: total number of keywords, p: number of processors, k: number of keys, a_w: number of updates matching w, d_w: number of deleted entries matching w, ℓ: result size ($\ell = a_w - d_w$), $(t_{PE.Enc}, t_{PE.Dec})$: encryption and decryption times for a public pancturable encryption scheme, $(t_{SPE.Enc}, t_{SPE.Dec})$: encryption and decryption times for a symmetric pancturable encryption scheme MC: Multi-Client, FP: Forward Privacy, BP: Backward Privacy. $\widetilde{O}$ notation hides polylogarithmic factors.

Scheme	MC	FP	BP	Search time	Update time	Client storage
Comparison						
Bunker-B	✗	✓	Type-II	$O(\ell)$	$O(1)$	$O(m log n)$
Fides	✗	✓	Type-II	$O(\ell)$	$O(1)$	$O(m log n)$
$Diana_{del}$	✗	✓	Type-III	$O(\ell)$	$O(log N)$	$O(m log n)$
Janus	✗	✓	Type-III	$O(\ell d) t_{PE.Dec}$	$O(\ell)(t_{PE.Enc} \vee t_{PE.Dec})$	$O(m log n)$
Janus++	✗	✓	Type-III	$O(\ell d) t_{SPE.Dec}$	$O(\ell)(t_{SPE.Enc} \vee t_{SPE.Dec})$	$O(m log n)$
Moneta	✗	✓	Type-I	$\widetilde{O}(a_w \log N + \log^3 N)$	$\widetilde{O}(\log^2 N)$	$O(1)$
Orion	✗	✓	Type-I	$O(\ell \log N^2)$	$O(log N^2)$	$O(1)$
Ours	✓	✓	Type-II	$O(\ell/p)$	$O(m/p)$	None

In [10] authors presented *HardIDX*, a scheme that also supports range queries with the use of SGX [12] based on B^+ trees. *HardIDX* minimizes the leakage by hiding the search pattern but at the same time, their construction is *static*. As a result, it does *not* support file insertions after the generation of the initial index. Therefore, even though the scheme achieves logarithmic search cost, a direct comparison to our scheme is *not* possible. *ORAM-based approaches:* The first forward private SSE scheme was proposed in [22], where the authors presented an ORAM-based construction. More recently, in [9], authors proposed *Moneta*, an SSE scheme that achieves the strongest level of backward privacy, but at the cost of efficiency. *Moneta* is based on the TWORAM construction presented in [14]. However, as argued in [15], the use of TWORAM renders *Moneta* impractical for realistic scenarios and the scheme can serve mostly as a theoretical result for the feasibility of stronger backward private schemes schemes. Finally, in [15], another ORAM-bsed scheme, *Orion*, is proposed. While *Orion* outperforms *Moneta*, the number of interactions between the user and the CSP depends on the size of the encrypted database. It needs to be noted that in this work, we do not deal with the revocation of the users as for example in [4,5]. In Table 1 we see a comparison of the aforementioned schemes to our construction.

3 Background

Notation. Let $\mathcal{X}$ be a set. We use $x \leftarrow \mathcal{X}$ if x is sampled uniformly from $\mathcal{X}$ and $x \xleftarrow{\$} \mathcal{X}$, if x is chosen uniformly at random. If $\mathcal{X}$ and $\mathcal{Y}$ are two sets, we denote by $[\mathcal{X}, \mathcal{Y}]$ all the functions from $\mathcal{X}$ to $\mathcal{Y}$ and by $\overline{[\mathcal{X}, \mathcal{Y}]}$ all the injective functions from $\mathcal{X}$ to $\mathcal{Y}$. $R(\cdot)$ is a truly random function and $R^{-1}(\cdot)$ its inverse. A function $negl(\cdot)$ is called negligible, iff $\forall c \in \mathbb{N}, \exists n_0 \in \mathbb{N} : \forall n \geq n_0, negl(n) < n^{-c}$. If $s(n)$ is a string of length n, we denote by $\overline{s}(l)$ its prefix of length l and by $\underline{s}(l)$, its suffix of length l, where $l < n$. A file collection is represented as $\mathbf{f} = (f_1, \ldots, f_z)$ while the corresponding collection of ciphertexts is $\mathbf{c} = (c_{f_1}, \ldots, c_{f_z})$. The universe of keywords is $\mathcal{W} = (w_1, \ldots, w_k)$ and the distinct keywords in a file f_i are $w_i = (w_{i_1}, \ldots, w_{i_\ell})$.

Definition 1 (DSSE Scheme). *A Dynamic Symmetric Searchable Encryption (DSSE) scheme consists of the following PPT algorithms:*

- $(\mathsf{In_{CSP}}, \mathbf{c})(\mathsf{In_{TA}})(\mathsf{K}) \leftarrow \mathsf{Setup}(\lambda, \mathbf{f})$: *The data owners runs this algorithm to generate the key K as well as the CSP index $\mathsf{In_{CSP}}$ and a collection of ciphertexts $\mathbf{c}$ that will be sent to the CSP. Additionally, the index $\mathsf{In_{TA}}$ that is stored on a remote location is generated.*
- $(\mathsf{In'_{CSP}}, R_{w_{i_j}})(\mathsf{In'_{TA}}) \leftarrow \mathsf{Search}(\mathsf{K}, w_{i_j}, \mathsf{In_{TA}})(\mathsf{In_{CSP}}, \mathbf{c})$. *This algorithm is executed by a user in order to search for all files f_i containing a specific keyword w_{i_j}. The indexes are updated and the CSP also returns to the user the ciphertexts of the files that contain w_{i_j}.*
- $(\mathsf{In'_{CSP}}, \mathbf{c}')(\mathsf{In'_{TA}}) \leftarrow \mathsf{Update}(\mathsf{K}, f_i, \mathsf{In_{TA}})(\mathsf{In_{CSP}}, \mathbf{c}, op)$, *where $op \in \{\mathrm{add}, \mathrm{delete}\}$: A user is running this algorithm to update the collection of ciphertexts $\mathbf{c}$. Based on the value of op, a new file is either added to the collection or an existing one is deleted.*

Definition 2 *(L-Adaptive Security of DSSE).   Let DSSE = (Setup, Search, Update) be a dynamic symmetric searchable encryption scheme and $\mathcal{L} = (\mathcal{L}_{stp}, \mathcal{L}_{search}, \mathcal{L}_{update})$ be the leakage function of the DSSE scheme. We consider the following experiments between an adversary $\mathcal{ADV}$ and a challenger $\mathcal{C}$:*

Real$_{\mathcal{ADV}}(\lambda)$

$\mathcal{ADV}$ outputs a set of files $\mathbf{f}$. $\mathcal{C}$ generates a key K, and runs Setup. $\mathcal{ADV}$ then makes a polynomial number of adaptive queries $q = \{w, f_1, f_2\}$ such that $f_1 \notin \mathbf{f}$ and $f_2 \in \mathbf{f}$. For each q, she receives back either a search token for w, $\tau_s(w)$, an add token, τ_α, and a ciphertext for f_1 or a delete token τ_d for $\{w, f_2\}$. $\mathcal{ADV}$ outputs a bit b.

Ideal$_{\mathcal{ADV},\mathcal{S}}(\lambda)$

$\mathcal{ADV}$ outputs a set of files $\mathbf{f}$. $\mathcal{S}$ gets $\mathcal{L}_{\mathsf{setup}}(\mathbf{f})$ to simulate Setup. $\mathcal{ADV}$ then makes a polynomial number of adaptive queries $q = \{w, f_1, f_2\}$ such that $f_1 \notin \mathbf{f}$ and $f_2 \in \mathbf{f}$. For each q, $\mathcal{S}$ is given either $\mathcal{L}_{\mathsf{Search}}(w)$ or $\mathcal{L}_{\mathsf{update}}(f_i)$, $i \in \{1,2\}$. $\mathcal{S}$ then simulates the tokens and, in the case of addition, a ciphertext. Finally, $\mathcal{ADV}$ outputs a bit b.

We say that the DSSE scheme is secure if $\forall$ PPT adversary $\mathcal{ADV}$, $\exists \mathcal{S}$ such that:

$$|Pr[(\mathsf{Real}_{\mathcal{ADV}}) = 1] - Pr[(\mathsf{Ideal}_{\mathcal{ADV},\mathcal{S}}) = 1]| \leq negl(\lambda) \qquad (1)$$

A DSSE scheme is said to be *forward private* if for all file insertions that take place after the successful execution of the Setup algorithm, the leakage is limited to the size of the file, and the number of unique keywords contained in it. On the other hand, a DSSE scheme is said to be *backward private* if whenever a keyword/document pair $(w, id(f))$ is added into the database and then deleted, subsequent search queries for w do not reveal $id(f)$. More formally:

Definition 3 (Forward Privacy). *An L-adaptively SSE scheme is forward private iff the leakage function $\mathcal{L}_{Update}$ can be written as:*

$$\mathcal{L}_{\mathsf{update}}(op, id(f)) = \mathcal{L}'(op, \#w \in f) \qquad (2)$$

where $\mathcal{L}'$ is a stateless function.

Definition 4 (Backward Privacy). *There are three different flavors of backward privacy (listed in decreasing strength):*

- *Type-I: Backward Privacy with insertion pattern leaks the documents currently matching w and when they were inserted i.e. their timestamps $TimeDB(w)$.*
- *Type-II: Backward Privacy with update pattern leaks the documents currently matching w, $TimeDB(w)$ and a list of timestamps $Updates(w)$ denoting when the updates on w happened.*
- *Type-III: Weak Backward Privacy leaks the documents currently matching they keyword w, $TimeDB(w)$ and $DelHist(w)$, where $DelHist(w)$ reveals the timestamps of the delete updates on w together with the corresponding entries that they remove.*

Our scheme satisfies Type-II backward privacy.

Definition 5. *An $\mathcal{L}$-adaptively SSE scheme is update pattern revealing backward private iff the search and update leakage functions can be written as:*

$$\mathcal{L}_{\text{search}}(w) = \mathcal{L}'(\text{TimeDB(w)}, \text{Updates(w)})$$
$$\mathcal{L}_{\text{update}}(op, w, id) = \mathcal{L}''(op, w_i) \tag{3}$$

Where the functions $\mathcal{L}'$ and $\mathcal{L}''$ are stateless.

Finally, the leakage function $\mathcal{L}_{stp}$ associated with the setup operation is formalised as follows:

$$\mathcal{L}_{stp} = (N, n, c_{id(f_i)}), \forall f_i \in \mathbf{f} \tag{4}$$

Where N is the total size of all the (keyword/filename) pairs, and n is the total number of files in the collection $\mathbf{f}$.

3.1 Secure Encrypted Virtualisation

The main advantages of SEV in comparison to its main competitor -Intel SGX- are *(1)* memory size, *(2)* efficiency and *(3)* No SDK or code refactoring are required. In particular, SGX allocates only 128 MB of memory for software and applications and thus, making it a good candidate for microtranscations and login services. However, SEV's memory is up to the available RAM and hence, making it a perfect fit for securing complex applications. Moreover, in situations where many calls are required, like in the case of a multi-client cloud service, SEV is known to be much faster and efficient than SGX. The above are summarized in Table 2. More information can be found in [19].

Table 2. SEV-SGX comparison

TEE	Memory Size	SDK	Code Refactoring
SEV	Up to Available Ram	Not Required	Not Required
SGX	Up to 128MB	Required	Major Refactoring

4   Architecture

In this section, we introduce the system model by describing the entities participating in our construction.

Users. We denote with $\mathcal{U} = \{u_1, \ldots, u_n\}$ the set of all users that have been already registered in a cloud service that allows them to store, retrieve, update, delete and share encrypted files while at the same time being able to search over encrypted data by using our DSSE scheme. The users in our system model are mainly classified into two categories: data owners and simple registered users that they have not yet upload any data to the CSP. A data owner first needs to locally parse all the data that wishes to upload to the CSP. During this process, she generates three different indexes:

1. No.Files[w] which contains a hash of each keyword w along with the number of files that w can be found at
2. No.Search[w], which contains the number of times a keyword w has been searched by a user.
3. Dict a dictionary that maintains a mapping between keywords and encrypted filenames.

Both No.Files[w] and No.Search[w] are of size $O(m)$, where m is the total number of keywords while the size of Dict is $O(N) = O(nm)$, where n is the total number of files. To achieve the multi-client model, the data owner outsources No.Files[w] and No.Search[w] to a trusted authority (TA). These indexes will allow registered users to create consistent search tokens. Dict is finally sent to the CSP.

Cloud Service Provider (CSP). We consider a cloud computing environment similar to the one described in [21]. The CSP must support SEV-enabled since core entities will be running in the trusted execution environment offered by SEV. The CSP storage will consist of the ciphertexts as well as of the dictionary Dict. Each entry of Dict is encrypted under a different symmetric key K_w. Thus, given K_w and the number of files containing a keyword w, the CSP can locate the files containing w.

Trusted Authority (TA). TA is an index storage that stores the No.Files and No.Search indexes that have been generated by the data owner. All registered users can contact the TA to access the No.Files[w] and No.Search[w] values for a keyword w. These values are needed to create the search tokens that will allow users to search directly on the encrypted database. Similarly to the CSP, the TA is also SEV-enabled.

Deletion Authority (DelAuth). DelAuth is responsible for the deletion of files. Every time a user performs a search operation, the CSP forwards the result R to DelAuth. DelAuth decrypts the result, removes the Dict entries to be deleted and then re-encrypts the remaining filenames and sends them back to the CSP. Like the CSP and TA, DelAuth is also SEV enabled.

TA and DelAuth can be individual entities. For simplicity, we will assume that they are part of the same host.

5   Nowhere to Leak

Formal Construction: Our construction constitutes of three different algorithms, namely Setup, Search and Update. Let $G : \{0,1\}^\lambda \times \{0,1\}^* \to \{0,1\}^*$ be an invertible pseudorandom function (IPRF) [7]. Moreover, let $\mathsf{SKE} = (\mathsf{Gen}, \mathsf{Enc}, \mathsf{Dec})$ be an IND-CPA secure symmetric key cryptosystem and $h = \{0,1\}^* \to \{0,1\}^\lambda$ be a cryptographic hash function. Due to space constrains, we present the Setup, Search and Update Algorithms, in Figs. 1 and 2 respectively.

```
KG ← GenIPRF(1^λ)
KSKE ← SKE.Gen(1^λ)
return K = (KG, KSKE)
c = PKE.Enc(pkTA, KG)
Send KG to the TA and K to DelAuth
c = {}
AllMap = {}
    For all fi
    Run the Update algorithm with op = add to generate cfi and Mapi
c = c ∪ cfi
AllMap = [{AllMap ∪ Mapi}, cid(fi)]
Send (AllMap, c) to the CSP
Send InTA = {No.Files, No.Search} to the TA
CSP stores AllMap in InCSP = Dict
```

Fig. 1. Setup algorithm

Theorem 1. *Let* SKE = (Gen, Enc, Dec) *be a CPA-secure symmetric key encryption scheme. Moreover, let G be an IPRF and h a cryptographic hash function. Then our construction is $\mathcal{L}$-adaptively secure according to Definition 2.*

Proof Sketch. To prove the security of our construction against the threat model defined in Sect. 3, we construct a simulator $\mathcal{S}$ that simulates a perfect view of the real world for a PPT adversary $\mathcal{ADV}$. To do, we make use of a hybrid argument. More precisely, we will design five hybrids $\mathcal{H}_0, \mathcal{H}_1, \mathcal{H}_2, \mathcal{H}_3$ and $\mathcal{H}_4$ such that $\mathcal{H}_0$ is the real experiment and $\mathcal{H}_4$ the ideal one. Each hybrid $\mathcal{H}_i$, will be constructed by replacing a real functionality with a simulated one given the corresponding leakage function $\mathcal{L}_i$. Our goal is to prove that no PPT adversary $\mathcal{ADV}$ will be able to distinguish between two consecutive hybrids $\mathcal{H}_i, \mathcal{H}_{i+1}$. The hybrids are illustrated in Table 3. The complete formal proof can be found in the full version of the paper along with a more detailed presentation of our scheme.

Table 3. Hybrid description

Hybrids	Description
$\mathcal{H}_0$	This is the real experiment
$\mathcal{H}_1$	Simulate Setup given $\mathcal{L}_{stp}$
$\mathcal{H}_2$	Simulate Search given $\mathcal{L}_{search}$
$\mathcal{H}_3$	Simulate Update given $\mathcal{L}_{update}$
$\mathcal{H}_4$	This is the ideal experiment

SEARCH:
User sends $h(w_j)$ to the TA
TA:
$K_{w_j} = G(K_G, h(w_j)||\text{No.Search}[w_j])$
$L_{up} = \{\}$
$\text{No.Search}[w_j] + +$
$K'_{w_j} = G(K_G, h(w_j)||\text{No.Search}[w_j])$
for $i = 1$ to $i = \text{No.Files}[w_j]$
 $\text{addr}'_{w_j} = h(K_{w}', i)$
 $L_{up} = L_{up} \cup \{\text{addr}_{w_j}\}$
Send $\tau_s(w_j) = (K_{w_j}, \text{No.Files}[w_j], L_{up})$ to the CSP
CSP:
$R_{w_j} = \{\}$
for $i = 1$ to $i = \text{No.Files}[w_j]$
 $\text{val}_{w_j} = \text{Dict}[h(K_{w_j}), i]$
 $R_{w_j} = R_{w_j} \cup \{\text{val}_{w_j}\}$
 Forward R_{w_j} and K_{w_j}' to the DelAuth
DelAuth:
Send R_{w_j} to the user and an acknowledgement to the CSP.
CSP:
Delete all Dict entries associated with w_j and insert the addresses from L_{up}

UPDATE:
if $op = add$
 $\text{Map} = \{\}$
 for all $w_{i_j} \in f_i$
  $\text{No.Files}[w_{i_j}] + +$
  $K_{w_{i_j}} = G(K_G, h(w_{i_j})||\text{No.Search}[w_{i_j}])$
  $\text{addr}_{w_{i_j}} = h(K_{w_{i_j}}, \text{No.Files}[w_{i_j}])$
  $\text{val}_{w_{i_j}} = \text{Enc}(K_{SKE}, id(f_i)||\text{No.Files}[w_{i_j}])$
  $\text{Map} = \{\text{addr}_{w_{i_j}}, \text{val}_{w_{i_j}}\}$
  $c_{f_i} \leftarrow \text{Enc}(K_{SKE}, f_i)$
  Send $\tau_a(f_i) = (c_{f_i}, \text{Map})$ to the CSP
else Initiate the Search protocol for a keyword w_j
After DelAuth forwards R to the user:
 for all $c_{id(f_j)} \in R$
  $\text{Dec}(K_{SKE}, c_{id(f_j)}) \rightarrow id(f_j)$
 if $\exists \ell : id(f_\ell) \in L_{TBD}$
  $L_{del} = \{\}$
  Compute $\text{No.Search}[w_j]$ from K_{w_j}
  $\text{No.Files}[w_j] - -$
  $\text{No.Search}[w_j] + +$

  $K_{w_j} = G(K_G, h(w_j)||\text{No.Search}[w_j])$

  for $i = 1$ to $i = \text{No.Files}[w_j] - 1$ $\text{new\_addr} = h(K_{w_j}||i)$
   $\text{new\_value} = \text{Enc}(K_{SKE}, id(f_i)||\text{No.Files}[w_j])$
   $L_{del} = \{(\text{new\_addr}, \text{new\_value})\}$
 else
  Send $\tau_d(f_\ell) = L_{up}$ to the CSP

CSP:
Delete all Dict entries associated with w_j and insert the addresses contained in L_{del}
User:
Update the local indexes in send an acknowledgement to the TA to update its indexes
as well

Fig. 2. Search and update algorithms

6   Experimental Results

We implemented our scheme in Python 2.7 using the PyCrypto library [1]. To test the overall performance, we used files of different sizes and structures. More precisely, we used a collection of five datasets provided in [18]. Table 4 shows the datasets used in our experiments as well as the total number of unique keywords extracted from each set. Our experiments focused on two main aspects: *(1)* Indexing and *(2)* Searching for a specific keyword. Deletion cannot be realistically measured since to completely delete all entries corresponding to a file, we first need to search for all the keywords contained in the file. Additionally, our dictionaries were implemented as tables in a MySQL database. In contrast to other similar works, we did not rely on the use of data structures such as arrays, maps, sets, lists, trees, graphs, etc. While the use of a database system decreases the overall performance of the scheme it is considered as more durable and close to a production level. Conducting our experiments by solely relying on data structures would give us better results. However, this performance would not give us any valuable insights about how the scheme would perform outside of a lab. Additionally, storing the database in RAM raises several concerns. For example a power loss or system failure could lead to data loss (because RAM is volatile memory). To this end, we ran our experiments in the following two different machines: *(1)*, AMD Ryzen^{TM} 7 PRO 1700 Processor at 3.0 GHz (8 cores), 32 GB of RAM running Windows 10 64-bit with AMD SEV Support and *(2)* Microsoft Surface Book laptop with a 4.2 GHz Intel Core i7 processor (4 cores) and 16 GB RAM running Windows 10 64-bit

The reason for measuring the performance on such machines and not only in a powerful desktop – like other similar works – is that in a practical scenario, the most demanding processes of any SSE scheme (e.g. the creation of the dictionary) would take place on a user's machine.

Indexing and Encryption: In our experiments, we measured the total setup time for each one of the datasets shown in Table 4a. Each process was run ten times on the commodity laptop and the average time for the completion of the entire process was measured. As can be seen from Table 4b, the setup time can be considered as practical and can even run in typical users' devices. We compare the setup times for the commodity laptop and the powerful desktop. Based on the fact that this phase is the most demanding one in an SSE scheme the time needed to index and encrypt such a large number of files is considered as acceptable not only based on the size of the selected dataset but also based on the results of other schemes that do not even offer forward privacy as well as on the fact that we ran our experiments on commodity machines and not on a powerful server. This is an encouraging result and we hope that will motivate researchers to design and implement even better and more efficient SSE schemes but most importantly we hope that will inspire key industrial players in the field of cloud computing to create and launch modern cloud-services based on the promising concept of Symmetric Searchable Encryption.

Table 4. Dataset sizes and setup times

No of TXT Files	Dataset Size	Unique Keywords	(w, id) pairs
425	184MB	1,370,023	5,387,216
815	357MB	1,999,520	10,036,252
1,694	670MB	2,688,552	19,258,625
1,883	1GB	7,453,612	28,781,567
2,808	1.7GB	12,124,904	39,747,904

(a) Size of Datasets and Unique Keywords

Dataset / Testbed	Laptop	Desktop
184MB	22.48m	8.49m
357MB	40.00m	13.51m
670	86.43m	29.51m
1GB	141.60m	48.99m
1.7GB	203.28m	68.44m

(b) Setup time (in minutes)

Search: In this part of the experiments we measured the time needed to complete a search over encrypted data. In our implementation, the search time is calculated as the sum of the time needed to generate a search token and the time required to find the corresponding matches at the database. It is worth mentioning that the main part of this process will be running on the CSP (i.e. a machine with a large pool of resources and computational power). To this end, the time to generate the search token was measured on the laptop while the actual search time was measured using the desktop machine described earlier. On average, the time needed to generate the search token on the Surface Book laptop was $9\mu s$. Regarding the actual search time, searching for a specific keyword over a set of 12,124,904 distinct keywords and 39,747,904 addresses required 1.328 s on average while searching for a specific keyword over a set of 1,999,520 distinct keywords and 10,036,252 addresses took 0.131 s.

References

1. PyCrypto - the Python cryptography toolkit (2013). https://pypi.org/project/pycrypto/
2. AMD: Secure Encrypted Virtualization API Version 0.22. Technical Report (07 2019)
3. Amjad, G., Kamara, S., Moataz, T.: Forward and backward private searchable encryption with sgx. In: Proceedings of the 12th European Workshop on Systems Security (EuroSec 2019), Association for Computing Machinery, New York, NY, USA (2019). https://doi.org/10.1145/3301417.3312496, https://doi.org/10.1145/3301417.3312496
4. Bakas, A., Dang, H.V., Michalas, A., Zalitko, A.: The cloud we share: access control on symmetrically encrypted data in untrusted clouds. IEEE Access **8**, 210462–210477 (2020)
5. Bakas, A., Michalas, A.: Modern family: a revocable hybrid encryption scheme based on attribute-based encryption, symmetric searchable Encryption and SGX. In: Chen, S., Choo, K.-K.R., Fu, X., Lou, W., Mohaisen, A. (eds.) SecureComm 2019. LNICST, vol. 305, pp. 472–486. Springer, Cham (2019). https://doi.org/10.1007/978-3-030-37231-6_28
6. Bakas, A., Michalas, A.: Power range: forward private multi-client symmetric searchable encryption with range queries support. In: 2020 IEEE Symposium on Computers and Communications (ISCC). pp. 1–7. IEEE (2020)
7. Boneh, D., Kim, S., Wu, D.J.: Constrained Keys for invertible pseudorandom functions. In: Kalai, Y., Reyzin, L. (eds.) TCC 2017. LNCS, vol. 10677, pp. 237–263. Springer, Cham (2017). https://doi.org/10.1007/978-3-319-70500-2_9

8. Bost, R.: ∑οφος: forward secure searchable encryption. In: Proceedings of the 2016 ACM SIGSAC Conference on Computer and Communications Security, Vienna, Austria, October 24–28, 2016 (2016)
9. Bost, R., Minaud, B., Ohrimenko, O.: Forward and backward private searchable encryption from constrained cryptographic primitives. In: Proceedings of the 2017 ACM SIGSAC Conference on Computer and Communications Security (CCS 2017), pp. 1465–1482. ACM, New York, NY, USA (2017)
10. Brasser, F., Hahn, F., Kerschbaum, F., Sadeghi, A.R., Fuhry, B., Bahmani, R.: HardiDX: practical and secure index with SGX (2017)
11. Cash, D., Grubbs, P., Perry, J., Ristenpart, T.: Leakage-abuse attacks against searchable encryption. In: Proceedings of the 22nd ACM SIGSAC Conference on Computer and Communications Security. ACM (2015)
12. Costan, V., Devadas, S.: Intel SUX explained. IACR Cryptol. ePrint Arch. **2016**(086), 1–118 (2016)
13. Etemad, M., Küpçü, A., Papamanthou, C., Evans, D.: Efficient dynamic searchable encryption with forward privacy. Proc. Privacy Enhan. Technol. **2918** (1), 5–20 (2018)
14. Garg, S., Mohassel, P., Papamanthou, C.: Efficient oblivious ram in two rounds with applications to searchable encryption. In: Robshaw, M., Katz, J. (eds.) CRYPTO 2016. LNCS, vol. 9816, pp. 563–592. Springer, Heidelberg (2016). https://doi.org/10.1007/978-3-662-53015-3_20
15. Ghareh Chamani, J., Papadopoulos, D., Papamanthou, C., Jalili, R.: New constructions for forward and backward private symmetric searchable encryption. In: Proceedings of the 2018 ACM SIGSAC Conference on Computer and Communications Security, pp. 1038–1055 (2018)
16. Islam, M.S., Kuzu, M., Kantarcioglu, M.: Access pattern disclosure on searchable encryption: ramification, attack and mitigation. In: NDSS Symposium, vol. 20, p. 12. Citeseer (2012)
17. Kim, K.S., Kim, M., Lee, D., Park, J.H., Kim, W.H.: Forward secure dynamic searchable symmetric encryption with efficient updates. In: Proceedings of the 2017 ACM SIGSAC Conference on Computer and Communications Security, pp. 1449–1463 (2017)
18. Michalas, A.: Text files from Gutenberg database, August 2019
19. Mofrad, S., Zhang, F., Lu, S., Shi, W.: A comparison study of intel SUX and AMD memory encryption technology. In: Proceedings of the 7th International Workshop on Hardware and Architectural Support for Security and Privacy, pp. 1–8 (2018)
20. Naveed, M.: The fallacy of composition of oblivious ram and searchable encryption. IACR Cryptol. ePrint Arch. **2015**, 668 (2015)
21. Paladi, N., Gehrmann, C., Michalas, A.: Providing user security guarantees in public infrastructure clouds. IEEE Trans. Cloud Comput. **5**(3), 405–419 (2017). https://doi.org/10.1109/TCC.2016.2525991
22. Stefanov, E., Papamanthou, C., Shi, E.: Practical dynamic searchable encryption with small leakage. NDSS **71**, 72–75 (2014)
23. Sun, S.F., et al..: Practical backward-secure searchable encryption from symmetric puncturable encryption. In: Proceedings of the 2018 ACM SIGSAC Conference on Computer and Communications Security, pp. 763–780 (2018)
24. Zhang, Y., Katz, J., Papamanthou, C.: All your queries are belong to us: the power of file-injection attacks on searchable encryption. In: 25th USENIX Security Symposium, pp. 707–720 (2016)

Distributed Query Evaluation over Encrypted Data

Sabrina De Capitani di Vimercati[1], Sara Foresti[1], Sushil Jajodia[2],
Giovanni Livraga[1(✉)], Stefano Paraboschi[3], and Pierangela Samarati[1]

[1] Università degli Studi di Milano, Milan, Italy
{sabrina.decapitani,sara.foresti,giovanni.livraga,
pierangela.samarati}@unimi.it
[2] George Mason University, Fairfax, VA, USA
jajodia@gmu.edu
[3] Università degli Studi di Bergamo, Bergamo, Italy
parabosc@unibg.it

Abstract. The availability of a multitude of data sources has naturally
increased the need for subjects to collaborate for distributed computations, aimed at combining different data collections for their elabora-
tion and analysis. Due to the quick pace at which collected data grow,
often the authorities collecting and owning such datasets resort to external third parties (e.g., cloud providers) for their storage and management. Data under the control of different authorities are autonomously
encrypted (using a different encryption scheme and key) for their external
storage. This makes distributed computations combining these sources
hard. In this paper, we propose an approach enabling collaborative computations over data encrypted in storage, selectively involving also subjects that might not be authorized for accessing the data in plaintext
when it is considered economically convenient.

1 Introduction

Our society and economy more and more rely on the knowledge that can be
generated by analysis and computations combining data that are produced and
owned/controlled by different parties. The cloud, thanks to a variety of storage
and computational providers with different costs and performance guarantees,
represents an accelerator for such needs. Data owners can in fact outsource their
data to storage providers, making them (selectively) available for computations
with reduced management burden at their own side. At the same time, users
requiring analysis can (partially) delegate expensive computations to computational providers, with clear performance and economic benefits [9]. However,
there is no such thing as a free lunch, and the scenario can be complicated by
the fact that some of the data can be sensitive, proprietary, or more in general
subject to access restrictions, all factors that can affect the possibility of relying
on external cloud providers for data management and processing.

© IFIP International Federation for Information Processing 2021
Published by Springer Nature Switzerland AG 2021
K. Barker and K. Ghazinour (Eds.): DBSec 2021, LNCS 12840, pp. 96–114, 2021.
https://doi.org/10.1007/978-3-030-81242-3_6

To solve this issue and ensure data protection while permitting the consideration of a large spectrum of providers for computations, a recent approach proposed a simple, yet flexible, authorization model that enriches the traditional yes/no visibility that a subject can have over data with a third visibility level, granting a subject visibility over encrypted versions of the data [9]. In this way, subjects that are economically convenient, but possibly not fully trusted for accessing data content, may still be involved in computations over encrypted data. To enforce the authorization policy, visibility over data is dynamically adjusted by inserting, before passing a dataset to a subject not trusted for plaintext access, on-the-fly encryption operations. Similarly, the encryption layer can be dynamically removed through on-the-fly decryption when requested for operations that cannot be executed over encrypted data.

The authorization model in [9] operates under the assumption that the datasets involved in the distributed computation are stored in plaintext. This assumption is however viable only when data are either stored at their owners, or outsourced at providers that are trusted to access data in plaintext, hindering the consideration of providers that, while being economically convenient, cannot be considered fully trusted. Intuitively, the spectrum of potential providers that could be adopted for storing datasets could be enlarged if data are encrypted, by their owners, before outsourcing. The joint adoption of the authorization model in [9] and of encrypted storage would benefit both users requiring computations, and owners wishing to make their data selectively available to others. Users might in fact leverage economically convenient providers for the computation, and owners can outsource their datasets to economically convenient providers with the guarantee that their data will be improperly accessed neither in storage, nor in computation. The consideration of encrypted storage in collaborative computations brings however complications, since encryption in storage is not specifically inserted according to the computations to be performed and may not support them, which could hence require additional decryption and re-encryption operations.

In this paper, we build on the authorization model in [9] and propose a solution for collaborative computations over distributed data that can be stored, in encrypted form, at external and possibly not fully trusted providers. The main contributions of this paper can be summarized as follows. First, we re-define the information flows enacted by a computation, necessary for authorization enforcement, based on the possibility of some data being stored in encrypted form (Sect. 2). Second, we identify the need, and propose a solution for, re-encryption operations, to be introduced when the encryption adopted in storage (which is pre-determined by the data owner) does not support operation execution (Sect. 3). Third, we provide an approach for computing an economically convenient assignment of computation operations to subjects in complete obedience of authorizations (Sect. 4). We discuss related works in Sect. 5 and conclude the paper in Sect. 6.

2   Relation Profiles and Authorizations

We consider a scenario characterized by three kinds of subjects: *1) data authorities*, each owning one or more relational tables possibly stored at external *storage*

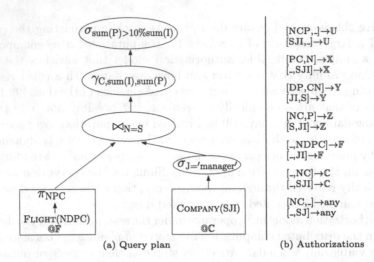

(a) Query plan   (b) Authorizations

Fig. 1. An example of a query plan (a) and of authorizations on relations FLIGHT and COMPANY (b)

providers; *2) users*, submitting queries over relations under the control of different authorities; and *3) computational providers*, which can be involved for query evaluation. Queries can be of the general form "SELECT FROM WHERE GROUP BY HAVING" and can include joins among relations under control of different data authorities. Execution of queries is performed according to a query plan established by the query optimizer, where projections are pushed down to avoid retrieving data that are not necessary for query evaluation. Graphically, we represent query plans as trees whose leaf nodes correspond to base relations, after the projection of the subset of attributes of interest for the query. For simplicity, but without loss of generality, we assume that attributes in the relations have different names.

Example 1. Consider two data authorities, a flight company and a commercial company with one relation each: relation FLIGHT(N, D, P, C) reports the social security Number and Date of birth of passengers, and the Price and Class of their tickets; relation COMPANY(S, I, J) reports the Social security number, Income, and Job of the company employees. These relations are stored in encrypted form at providers $\mathbb{F}$ and $\mathbb{C}$, respectively. The system is characterized by computational providers $\mathbb{X}$, $\mathbb{Y}$, and $\mathbb{Z}$. In our running example, we consider the following query submitted by user $\mathbb{U}$: "SELECT C, SUM(P), SUM(I) FROM FLIGHT JOIN COMPANY ON N = S WHERE J = 'manager' GROUP BY C HAVING SUM(P)>10%SUM(I)", retrieving the classes for which the overall price of tickets is above the 10% of the income of the managers who bought such tickets. Figure 1(a) illustrates a plan for the query.

Relation Profile. Besides the attributes included in its schema, a relation resulting from a computation can convey information on other attributes.

The information content explicitly and implicitly conveyed by a (base or derived, that is, resulting from the evaluation of a sub-query) relation is captured by a *profile* associated with the relation. We extend the definition of relation profile in [9] to model the possible encrypted representation of attributes in storage.

Definition 1 (Relation Profile). *Let R be a relation. The profile of R is a 6-tuple of the form $[R^{vp}, R^{ve}, R^{vE}, R^{ip}, R^{ie}, R^{\simeq}]$ where: R^{vp}, R^{ve}, and R^{vE} are the visible attributes appearing in R's schema in plaintext (R^{vp}), encrypted on-the-fly (R^{ve}), and encrypted in-storage (R^{vE}); R^{ip} and R^{ie} are the implicit attributes conveyed by R, in plaintext (R^{ip}) and encrypted (R^{ie}); $R^{\simeq}$ is a disjoint-set data structure representing the closure of the equivalence relationship implied by attributes connected in R's computation.*

In the definition, R^{vp} corresponds to the set of plaintext attributes visible in the schema of R. We then distinguish between the visible attributes encrypted on-the-fly (R^{ve}) and the visible attributes encrypted in storage (R^{vE}), due to their different nature. In-storage encryption is enforced once, independently from the query to be answered, and uses a scheme and a key (decided by the owning data authority) that do not change over time and are not shared among different data authorities. On-the-fly encryption is enforced at query evaluation time and both the encryption scheme and the encryption key are decided by the user formulating the query and need to be shared among different parties when different attributes need to be compared (e.g., for a join evaluation). Implicit components (R^{ip}, R^{ie}) keep track of the attributes that have been involved in query evaluation for producing relation R. Even if they do not appear in R's schema, query evaluation has left a trace of their values in the query results (e.g., attributes involved in selection or group by operations). Note that we do not distinguish between in-storage and on-the-fly encryption in the implicit component of the profile. Indeed, the information leaked by the evaluation of an operation over an encrypted attribute is not influenced by the time at which encryption has been enforced or the subject enforcing it. The equivalence relationship $(R^{\simeq})$ keeps track of the sets of attributes that have been compared for query evaluation (e.g., for the evaluation of an equi-join). Hence, even if one of the attributes in the equivalence set has been projected out from the relation schema, its values are still conveyed by the presence of other (equivalent) attributes.

The profile of a *base* relation R has all components empty except R^{vp} and R^{vE} that contain the attributes appearing in plaintext and in encrypted form, respectively, in the relation schema. The profile of a *derived* relation resulting from the evaluation of an operation depends on both the operation and the profile of the operand(s). Figure 2 illustrates the profiles resulting from the evaluation of relational algebra operators, and of encryption and decryption operations, which are peculiar of our model. Graphically, we represent the profile of a relation as a tag attached to the relation's node (or the node of the operator producing it in case of a derived relation), with three components: v (visible attributes in R^{vp} and, on a gray background, R^{ve} and R^{vE}), i (implicit attributes in R^{ip} and, on a gray background, R^{ie}), and $\simeq$ (sets of equivalent attributes in $R^{\simeq}$ that have

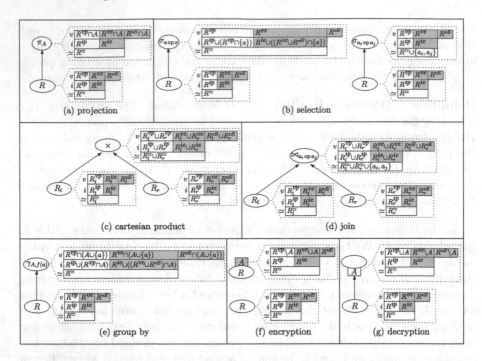

Fig. 2. Profiles resulting from relational, encryption, and decryption operations

been compared for R's computation). We represent encryption and decryption operations as gray and white boxes, respectively, containing the attributes to be encrypted/decrypted, attached to the operand relation or the resulting relation, respectively. Figure 3 illustrates the profiles of the relations resulting from the evaluation of the operations in the query plan in Fig. 1(a), assuming attributes NS and PI are decrypted for enabling computations over them.

Authorizations. Authorizations aim at regulating data flows intended for computations. Authorizations can specify, for each subject, whether she has plaintext visibility, encrypted visibility, or no visibility for performing computations over the attributes in the relations, and are defined as follows.

Definition 2 (Authorization). *Let R be a relation and $\mathcal{S}$ be a set of subjects. An authorization is a rule of the form $[P, E] \rightarrow S$, where $P \subseteq R$ and $E \subseteq R$ are subsets of attributes in R such that $P \cap E = \emptyset$, and $S \in \mathcal{S} \cup \{any\}$.*

Authorization $[P, E] \rightarrow S$ states that subject S can access in plaintext attributes in P, in encrypted form attributes in E, and has no visibility over the attributes in $R \backslash (P \cup E)$. Subject 'any' can be used to specify a default authorization applying to all subjects for which no authorization is defined. Authorizations regulating access for computation over (encrypted) attributes in relation R are defined by the data authority who owns the relation, independently from the provider storing it. Note that the authorizations of storage providers depend on

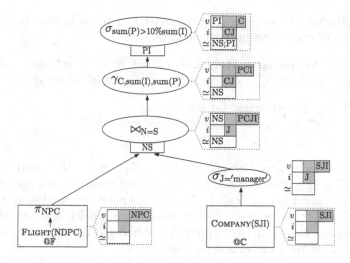

Fig. 3. Query plan with profiles

whether they are to be considered also for computations, independently from the fact that they store a specific relation and its (encrypted or plaintext) form. The user formulating the query is expected to have plaintext visibility over a subset of the attributes in the relational schemas, and we assume that she is authorized for the attributes involved in the query.

Example 2. Figure 1(b) illustrates an example of a set of authorizations regulating access to relations FLIGHT and COMPANY of our running example. User $\mathbb{U}$ has plaintext visibility over a subset of the attributes of the two relations, storage providers $\mathbb{F}$ and $\mathbb{C}$ have encrypted visibility over the attributes in the relation they store, computational providers $\mathbb{X}$, $\mathbb{Y}$, and $\mathbb{Z}$ have plaintext or encrypted visibility over a subset of the attributes in the two relations.

Authorization Verification. To be authorized for a relation, a subject needs the plaintext visibility over plaintext attributes (R^{vp} and R^{ip}) and plaintext or encrypted visibility over encrypted attributes (R^{ve}, R^{vE}, and R^{ie}). Note that there is no need to distinguish between in-storage and on-the-fly encryption for authorization verification, as the information conveyed by encrypted attributes is independent from the time at which it has been applied. The subject also needs to have the same visibility (plaintext or encrypted) over attributes appearing together in an equivalence set. This is required to prevent subjects having plaintext visibility on one attribute in the equivalence set and encrypted visibility on another to be able to exploit knowledge of plaintext values of the former to infer plaintext values of the latter.

In the following, for simplicity, we will denote with $\mathcal{P}_S$ ($\mathcal{E}_S$, respectively) the set of attributes that a subject S can access in plaintext (encrypted, respectively) according to her authorizations. The following definition identifies subjects authorized to access a relation, extending the definition in [9] to take the two kinds of encryption into consideration.

Definition 3 (Authorized Relation). *Let R be a relation with profile $[R^{vp}, R^{ve}, R^{vE}, R^{ip}, R^{ie}, R^{\simeq}]$. A subject $S \in \mathcal{S}$ is authorized for R iff:*

1. $R^{vp} \cup R^{ip} \subseteq \mathcal{P}_S$ *(authorized for plaintext)*;
2. $R^{ve} \cup R^{vE} \cup R^{ie} \subseteq \mathcal{P}_S \cup \mathcal{E}_S$ *(authorized for encrypted)*;
3. $\forall A \in R^{\simeq}$, $A \subseteq \mathcal{P}_S$ *or* $A \subseteq \mathcal{E}_S$ *(uniform visibility)*.

Example 3. Consider a relation R with profile $[P, C, S, \_, \_, \{IP\}]$ and the authorizations in Fig. 1(a). Provider $\mathbb{Z}$ is not authorized for the relation since it cannot access P in plaintext (Condition 1); $\mathbb{C}$ and $\mathbb{F}$ are not authorized since they cannot access P and S, respectively, in any form (Condition 2); $\mathbb{X}$ is not authorized since it does not have uniform visibility on P and I (Condition 3). Provider $\mathbb{Y}$ and user $\mathbb{U}$ are instead authorized for the relation.

For simplicity, in the following we will use notation R_i to denote the relation resulting from the evaluation of node n_i in the query tree plan. When clear from the context, we will use n_i to denote interchangeably the node and the corresponding relation.

3   Extended Minimum Cost Query Plan

Given a query plan T(N) corresponding to a query q formulated by a user $\mathbb{U}$, our goal is to determine, for each node, a subject for its evaluation, possibly extending the query plan with encryption, decryption, and re-encryption operations to guarantee the satisfaction of authorizations and enable the evaluation of operations.

3.1   Candidates

Given a query plan T(N), we first need to identify, for each node, the subjects authorized for evaluating it (i.e., its candidates). Given a node n in a query tree plan, a subject S is authorized for its execution if she is authorized for its operand(s) and for its result. Indeed, S needs to access the operands of the node for its evaluation, and the profile of the result captures all the information directly and indirectly conveyed by the evaluation of the operation. Starting from relations where (a subset of) the attributes are encrypted in storage, it could be necessary to inject decryption and re-encryption (i.e., decryption followed by encryption with a different scheme and/or key) to guarantee that operations can be evaluated when they require plaintext visibility over the involved attributes, or they are not supported by the encryption scheme adopted in storage, respectively. For instance, we cannot expect different data authorities to use the same encryption scheme and key for attributes that will be compared in an equi-join. Hence, even if equality conditions can easily be supported over encrypted data (e.g., using deterministic encryption), the evaluation of equi-joins requires re-encryption of the join attributes.

Besides decryption and re-encryption for enabling query evaluation, also encryption operations could be injected for enforcing authorizations: encryption could enable a subject to perform an operation that she would otherwise not be authorized to evaluate, due to the plaintext representation of some attributes in the operand relation that she can access only in encrypted form.

Example 4. With reference to our running example, $\mathbb{Y}$ can evaluate the join operation if attributes N and S are re-encrypted using a deterministic encryption scheme with the same encryption key. Similarly, attributes P, I, and J must be encrypted for $\mathbb{Z}$ to be authorized for evaluating the group by operation.

We observe that, if all the attributes in the schema of the operand relation(s) appear in encrypted form, the set of subjects who are authorized for evaluating the operation is possibly larger. In fact, encrypted attributes are also accessible by subjects with plaintext visibility. To determine candidates, we therefore assume that all the attributes in the operand relation(s), but those that have to be in plaintext for operation execution, are encrypted. We note that the encryption of the attributes in the operands is always possible, since any attribute can be encrypted by the subject computing the operand (who can see it in plaintext). Similarly, any attribute of the operand(s) can be decrypted by the subject who is in charge for the evaluation of the operation, since otherwise it would not be authorized to evaluate it. Formally, we define candidates for the evaluation of a node as follows.

Definition 4 (Candidate). *Let* T(N) *be a query plan, $n \in N$ be a non-leaf node, and $n_l, n_r \in N$ be its left and right child (if any), $n.A_p$ be the set of attributes that need to be in plaintext for the evaluation of n, and S be a set of subjects. A subject $S \in S$ is a candidate for the execution of a node n iff S is authorized for:*

1) n_l and n_r, assuming the encryption of all the visible attributes (Definition 3);
2) attributes in $n.A_p$ in plaintext;
3) n, assuming the encryption of all the visible attributes in its operand(s) (Definition 3).

The set of candidates for node n is denoted $\Lambda(n)$.

Example 5. Figure 4(a) reports, for each node in the query plan of Example 1, the candidates who can evaluate the operation in the node. In the example, we assume that: *i)* the selection over J and the computation of the sums over I and P can be evaluated over their encrypted in storage representation; *ii)* the evaluation of the join and of the group by require the re-encryption of the involved attributes; and *iii)* the comparison of SUM(P) and SUM(I) can only be done over plaintext values.

The set of candidates along a query plan enjoys a nice *monotonicity* property. In fact, relation profiles never lose attributes, but can only gain new ones (see Fig. 3). Hence, a subject authorized for n is also authorized for its descendants

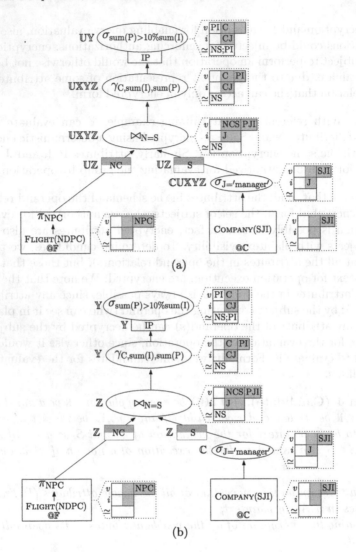

Fig. 4. Extended query plan with candidates (a) and with assignees (b)

in the query plan (the set of candidates monotonically decreases going up in the tree). This is true for all operations that do not require plaintext visibility over attributes, or which leave a trace in the implicit component of the resulting relation profile. A query plan T(N) is extended with encryption, decryption, and re-encryption, generating an *extended query plan*, denoted T'(N'). Figure 4(a) illustrates an example of an extended version of the query plan in Fig. 1(a), where attributes NC and S are re-encrypted (graphically represented with the gray and white rectangles below the join node), and attributes IP are decrypted. Encryption, decryption, and re-encryption are used to adjust visibility and guarantee correct authorization enforcement. Their injection depends on the subjects to which operations are assigned. The injection of encryption and decryption operations does not affect the monotonicity property: the set of candidates of an encryption node corresponds to the one of the node to which encryption applies (i.e., of its child), and the set of candidates of a decryption node corresponds to the one of the node operating on the result of the decryption (i.e., of its parent). For example, candidates for the decryption of IP in Fig. 4(a) are those for the selection to which decryption is connected (and hence are not explicitly reported in the figure). The consideration of re-encryption operations, necessary when the in-storage encryption scheme does not support operation execution, on the other hand, deserves a special treatment. Since the subject in charge of re-encryption must be authorized for the profile of the operand relation, the set of candidates for a re-encryption operation is a subset of the candidates of its operand node n_c. However, the candidates of the parent n_p of the re-encryption operation might not be a subset of the re-encryption candidates. In fact, nothing can be said on the set containment relationship between re-encryption candidates and those of its parent n_p, since a candidate for re-encryption could not be authorized for n_p and vice-versa: while a subject must be authorized for plaintext visibility on the attributes to re-encrypted to be candidate for re-encryption, n_p might not require (and its candidate might not have) plaintext visibility on these attributes. Indeed, the profile of the result of re-encryption is the same as the one of its operand (i.e., it does not move attributes from the encrypted to the plaintext components nor vive-versa). Note that the set of candidates for n_p is a subset of the candidates for n_c, since a candidate for n_p needs to have at least visibility on the relation produced by n_c. Figure 4(a) reports, for the two re-encryption operations, the set of candidates that could re-encrypt the involved attributes.

Given a query plan and the candidates for each of its nodes, it is then necessary to select, for each node, a subject (chosen among its candidates) in charge of the evaluation of the corresponding operation (i.e., the assignee of node n). Given a query plan, there can exist different possible assignments that respect authorizations and permit query execution. In the next section, we discuss how to determine an authorized assignment.

3.2   Authorized Assignment and Minimum Cost Query Plan

Given a query plan $\mathsf{T(N)}$ and the set $\varLambda(n)$ of candidates for each node $n \in \mathsf{N}$, it is possible to determine an assignment of nodes to subjects taken from the corresponding set of candidates by inserting encryption and decryption operations. Such an assignment exists if, for each attribute a that needs to be re-encrypted, there exists a subject who can access a in plaintext, and the other attributes in the schema of the same base relation in encrypted or plaintext form. Encryptions are inserted to enforce authorizations, and decryptions are inserted to adjust attributes visibility for operation evaluation, and are attached to the node requiring each of them. These operations can be performed by the same subject assigned to the nodes to which encryption/decryption are attached. Re-encryption, on the contrary, could be assigned to a different subject, and can be inserted at any point in the query plan, before the node that represents the operation for which re-encryption is needed. We also note that, differently from encryption and decryption operations, the need for re-encryption of an attribute a does not depend on the choice of assignments, but only on: $i)$ the in-storage encryption (scheme and key) of a; and $ii)$ the operations to be evaluated over a for query execution. Hence, independently from the selected assignment, if no subject has plaintext visibility over a and encrypted visibility over all the other attributes in the base relation to which a belongs, there cannot exist any authorized assignment for the query plan. On the contrary, if such a subject exists, there is at least an authorized assignment for the query plan. Indeed, the re-encryption operation can be evaluated as early as when the relation leaves the storage provider.

Example 6. Consider attribute C of our running example, which needs to be re-encrypted for the evaluation of GROUP BY clause. For an authorized assignment, we need a subject who can access attributes N and P in encrypted form and C in plaintext. Since $\mathbb{U}$, $\mathbb{X}$, and $\mathbb{Z}$ can access N and P encrypted and C plaintext, in the worst case scenario, re-encryption of C can be injected as a parent of the leaf node representing base relation FLIGHT and can be assigned to one among $\mathbb{U}$, $\mathbb{X}$, and $\mathbb{Z}$.

The existence of an authorized assignment can be formalized by the following theorem.

Theorem 1 (Existence of an authorized assignment). *Let* T(N) *be a query plan,* $\forall n \in N$, $n.A_e$ *be the set of attributes that need to be re-encrypted for the evaluation of* n, S *be a set of subjects and,* $\forall n \in N$, $\Lambda(n)$ *be the set of candidates for* n. *If* $\forall n \in N$, $\Lambda(n) \neq \emptyset$ *and,* $\forall a \in n.A_e$ *there exists at least a subject* $S \in S$ *s.t.* $a \in \mathcal{P}_S$ *and* $R \subseteq \mathcal{P}_S \cup \mathcal{E}_S$, *with* R *the base relation to which* a *belongs, then there exists at least an extended query plan* T'(N') *of* T(N) *and an assignment* $\lambda : N' \to S$ *of subjects to nodes in* T'(N'), *with* $\lambda(n) \in \Lambda(n)$, *that does not violate any authorization.*

We can then conclude that, if there exists an authorized assignment for the query plan, any combination of subjects chosen from the candidate sets of the nodes in the query plan can be made authorized by injecting encryption, decryption, and re-encryption operations. For instance, Fig. 4(b) illustrates an extended query plan that makes the assignment on the left of each node authorized according to the authorizations in Fig. 1(a).

Among the possible assignments, we expect the user formulating the query to be interested in selecting the one that optimizes performance, economic costs, or both of them. In the considered cloud scenario, we expect the economic cost to be the driving factor in the choice of the candidates. The economic cost for the evaluation of a query includes two main factors: $i)$ *computational cost* for the evaluation of the operations in the query plan; and $ii)$ *data transfer cost* for the relations exchanged between subjects for query evaluation. The cost of query evaluation is obtained by summing these two cost components, taking into consideration also the encryption, decryption, and re-encryption operations. Formally, the problem of computing an assignment that minimizes the cost of query evaluation is formulated as follows.

Problem 1 (Minimum cost query plan). Let $T(N)$ be a query plan and S be a set of subjects. Determine an extended query plan $T'(N')$ of T and an assignment $\lambda : N' \rightarrow S$ such that:

1. $\forall n \in N'$, $\lambda(n) \in \Lambda(n)$, that is, the subject in charge of the evaluation of a node is one of its candidates;
2. $\forall n \in N'$, $\lambda(n)$ is authorized for the profiles of n and of its children;
3. $\nexists T''$, λ' such that T'' is an extended query plan of T and λ' an assignment for T'' such that $\forall n \in N'$, $\lambda'(n) \in \Lambda(n)$ and $cost(T'', \lambda') < cost(T', \lambda)$

The problem of computing a minimum cost query plan is hard. We therefore propose a heuristic approach for its solution.

4   Computing Assignment

The proposed heuristics operates in three phases (see Fig. 5). The first phase identifies the set of candidates associated with the nodes of the query plan given as input. The second phase chooses, for each operation in the query plan, the subject (among the corresponding candidates) in charge of its execution, and inserts the needed re-encryption operations. The third phase inserts the encryption and decryption operations. The procedures corresponding to these phases are presented in Figs. 5, 6, and 7 and illustrated in the following. In the discussion and in the procedures, given a node n, we denote with n_p its parent, and with n_l and n_r its left child and right child, respectively.

Identify Candidates. Recursive procedure **Identify_Candidates** (Fig. 5) performs a post-order visit of the query plan to identify, for each node, the candidates for its evaluation. For each node n, the procedure computes its profile, assuming that all the attributes in the operands are encrypted unless demanded for the evaluation of n (lines 8–12). The procedure then determines the candidates for n, checking among the candidates of n's operands or, for operations operating on plaintext attributes that do not leave a trace in the implicit component, also among the other subjects (lines 15–21). Note that the set of candidates for leaf nodes is set to the complete set of subjects (line 6), even if leaf nodes are assigned to the storing provider, to simplify the computation of the candidate sets in the query plan. For simplicity, but without loss of generality, we assume all the attributes in base relations to be encrypted in storage. Procedure **Identify_Candidates** also sets variables $n.TotA_p$ ($n.TotA_e$, resp.) to the set of attributes that must be plaintext (encrypted on the fly, resp.) for the evaluation of the subtree rooted at n (lines 7, 13–14).

MAIN(T(N), S)

1: **Compute_Cost**(T.*root*)
2: insert a node *client* as parent of T.*root* assigned to the user U formulating the query
3: **Identify_Candidates**(T.*root*) /* Step 1: identify candidates */
4: *to_enc_dec*=∅
5: **Compute_Assignment**(T.*root*) /* Step 2: compute assignment and inject re-encryption */
6: **Extend_Plan**(T.*root*) /* Step 3: inject encryption/decryption */

Identify_Candidates(n)

1: if $n_l \neq$NULL then **Identify_Candidates**(n_l)
2: if $n_r \neq$NULL then **Identify_Candidates**(n_r)
   /* compute the profile of the node over its (encrypted) children */
3: if $n_l = n_r =$NULL /* n is a leaf node */
4: **then** $n.vp = n.ve = n.ip = n.ie = n.eq = \emptyset$
5:         $n.vE = R$ /* all the attributes in the relation schema are encrypted */
6:         $\Lambda(n) = S$ /* any subject */
7:         $n.TotA_p = n.TotA_e = \emptyset$
8: **else** let $n.A_p$ be the set of attributes that need to be plaintext for evaluating n
9:         let $n.A_e$ be the set of attributes that need to be (re)encrypted on-the-fly for evaluating n
10:        $n_l =$**encrypt**($n_l - n.A_p$, **decrypt**($n.A_p \cup n.A_e$, n_l))
11:        $n_r =$**encrypt**($n_r - n.A_p$, **decrypt**($n.A_p \cup n.A_e$, n_r))
12:        **Compute_Profile**(n) /* compute the relation profile according to Figure 2 */
13:        $n.TotA_p = n.A_p \cup n_l.TotA_p \cup n_r.TotA_p$
14:        $n.TotA_e = n.A_e \cup n_l.TotA_e \cup n_r.TotA_e$
15:        $\Lambda(n) = \emptyset$
16:        if $n_l.A_p \cup n_r.A_p \subseteq n.ip$
17:        **then** $Cand = \Lambda(n_l) \cup \Lambda(n_r)$
18:        **else** $Cand = S$
19:        **for** each $S \in Cand$ **do**
20:             if S is authorized for n_l, n_r, n
21:             **then** $\Lambda(n) = \Lambda(n) \cup \{S\}$

Fig. 5. Pseudocode of our heuristic algorithm and of procedure **Identify_Candidates**

Choose Assignment. Recursive procedure **Compute_Assignment** (Fig. 6) performs a pre-order visit of the query plan. Intuitively, for each visited node, the procedure chooses between assigning the evaluation of the node to the same subject as its parent n_p (without paying any transfer cost), or move it to a different subject, if economically convenient. Economic convenience is evaluated comparing the cost of evaluating the whole subtree rooted at n at each subject S being candidate of the node. To estimate the cost of delegating the evaluation of the subtree rooted at n to S, we consider the following cost components.

- *Data transfer cost* (lines 15–16) applies only when n is assigned to a subject S different from its parent and is computed as the product between the estimated size of the relation generated by n and the transfer cost of the subject in charge of evaluating n (in line with cloud market price lists, we consider only outbound traffic).
- *Computational cost* (line 18) is the sum of the costs of evaluating all the nodes in the subtree rooted at n by subject S. Such a cost is pre-computed by recursive procedure **Compute_Cost**, which visits the query plan in pre-order summing the cost of the evaluation of the subtrees rooted at the children of n with the cost of evaluating n, which is obtained by multiplying the estimated computation complexity of evaluating n in $n.TotA_e$ and $n.TotA_p$ by the computation price of S. The costs precomputed by procedure **Compute_Cost**

Compute_Assignment(n)

```
1:   S_min=NULL
2:   min=+∞
3:   if n_l=n_r=NULL /* n is a leaf node */
4:   then λ(n)=n.S /* storage provider for the corresponding relation */
5:       if to_enc_dec∩R≠∅
6:       then insert a re-encrypt node new for to_enc_dec∩n.vE as parent of n
7:           Λ(new)={S∈S: S is authorized for n and to access to_enc_dec∩n.vE in plaintext}
8:           for each S∈Λ(new) do
9:               cost = (dec_cost(to_enc_dec)+enc_cost(to_enc_dec))·S.comp_price+
10:                  + n.size·(S.transf_price+λ(n).transf_price)
11:               if cost<min
12:               then min=cost, S_min=S
13:           λ(new)=S_min
14:  else if n is not a re-encryption operation
15:      then for each S∈Λ(n) do
16:              if S≠λ(n_p) then cost=n.size·S.transf_price /* transfer cost */
17:              else cost=0 /* transfer cost */
18:              cost = cost+comp_cost[n,S] /* computational cost */
19:              for each a∈(n.TotA_p∪n.TotA_e)∩P_S do /* S decrypts the attribute */
20:                  cost=cost+dec_cost(a)·S.comp_price
21:              for each a∈(n.TotA_e\P_S) do /* need to delegate re-encrypt of a */
22:                  cost = cost+(dec_cost(a)+enc_cost(a))·avg_comp_price+
23:                      a.size(avg_transf_price+S.transf_price)
24:              for each a∈(to_enc_dec∩P_S) do /* S can re-encrypt a */
25:                  cost=cost+(dec_cost(a)+enc_cost(a))·S.comp_price
26:              if cost<min
27:              then min=cost
28:                  S_min=S
                 /* select the subject in charge of the evaluation of n */
29:          λ(n)=S_min
30:          if to_enc_dec∩P_λ(n)≠∅
31:          then insert a re-encrypt node new for to_enc_dec∩P_λ(n) as parent of n
32:              λ(new)=λ(n)
33:              to_enc_dec=to_enc_dec\P_λ(n)
34:          to_enc_dec=to_enc_dec∪(n.A_e\P_λ(n)) /* delegated re-encryption */
35:          if n.A_e∩P_λ(n)≠∅
36:          then insert a re-encrypt node new for n.A_e∩P_λ(n) as child of n
37:              λ(new)=λ(n)
38:  if n_l≠NULL then Compute_Assignment(n_l)
39:  if n_r≠NULL then Compute_Assignment(n_r)
```

Compute_Cost(n)

```
1:   if n_l≠NULL then Compute_Cost(n_l)
2:   if n_r≠NULL then Compute_Cost(n_r)
3:   for each S∈S do
4:       comp_cost[n,S] = comp_cost[n_l,S] + comp_cost[n_r,S] + n.comp_cost·S.comp_price
```

Fig. 6. Pseudocode of procedures **Compute_Assignment** and **Compute_Cost**

are stored in a matrix, $comp\_cost[n,S]$, with a row for each node and a column for each subject.

- *Decryption cost* (lines 19–20) is the cost of decrypting the attributes that need to be plaintext (or encrypted on-the-fly) for the evaluation of n or one of its descendants (i.e., any node in the subtree rooted at n that S is in charge of evaluating). The decryption cost is estimated by multiplying the decryption cost of each attribute a by the computation price of S.
- *Re-encryption cost* (lines 21–25) includes the cost of re-encryption operations performed by S as well as of re-encryption operations necessary to S for the evaluation of n but that need to be delegated to a different subject.

To keep track of the attributes that require re-encryption, we use variable
to_enc_dec, which keeps track of the attributes that require re-encryption for
the evaluation of the ancestors of n. If S can access a subset of the attributes in
to_enc_dec in plaintext, the algorithm assumes that S will take care of their re-
encryption (lines 24–25). If S needs to operate on an attribute encrypted on-
the-fly on which she does not have plaintext visibility, the algorithm estimates
the cost of injecting a re-encryption operation into the query plan, performed
by a third party authorized for it. Such a cost is estimated as the sum of the
costs for encrypting and decrypting the attribute of interest (assuming the
average computation price of the subjects in the system), and the transfer
cost for sending the relation to the subject in charge of re-encryption and
then back to S (lines 21–23).

Among the candidates for the node, procedure **Compute_Assignment**
selects the subject S_{min} with minimum estimated cost (line 29). Depending
on the chosen assignee $\lambda(n)$, the procedure injects re-encryption operations and
updates variable to_enc_dec: $\lambda(n)$ is assigned the re-encryption of attributes in
to_enc_dec that she is authorized to access in plaintext (lines 30–33), and these
attributes are removed from to_enc_dec. Attributes in $n.A_e$ that $\lambda(n)$ cannot
access in plaintext are instead inserted into to_enc_dec, to push re-encryption
down in the query plan (line 34). Attributes in $n.A_e$ that $\lambda(n)$ can access in
plaintext are re-encrypted by $\lambda(n)$. To this purpose, the algorithm injects a re-
encryption operation, assigned to $\lambda(n)$, as a child of n (lines 35–37). Note that
$\lambda(n)$ can decide to decrypt the attributes that need to be re-encrypted before
evaluating n, and encrypt them (on the fly) after the evaluation of n. Since
re-encryption operations are assigned to a subject upon injection in the tree,
procedure **Compute_Assignment** does not need to operate over them.

Leaf nodes deserve a special treatment, since they do not represent operations
and can only be assigned to the provider storing the corresponding base relation
(lines 3–4). We note however that, when the visit reaches a leaf node, it is
necessary to verify whether to_enc_dec is empty. If to_enc_dec is not empty, it
is necessary to insert a re-encryption operation for the attributes in to_enc_dec,
which is assigned to the less expensive subject who can access attributes in
to_enc_dec in plaintext (lines 5–13). The need to involve a subject only for re-
encryption operations happens only if no subject assigned to other operations
in the query plan can access the attribute(s) of interest in plaintext.

Extend Query Plan. Recursive procedure **Extend_Plan** (Fig. 7) performs a
post-order visit of the query plan to inject encryption and decryption operations as needed. For the root node, the procedure injects a decryption of the
encrypted attributes in the root (lines 3–5). For each non-root node n, the procedure injects a decryption operation (as child of n and assigned to $\lambda(n)$) for
those attributes that must be in plaintext for the evaluation of n but that are
encrypted in its operands. The procedure also injects an encryption operation
(as parent of n and assigned to $\lambda(n)$) for the attributes appearing in plaintext
in the profile of n and that the assignee of n_p can access only in encrypted form

Extend_Plan(n)

```
1:   if nₗ≠NULL then Extend_Plan(nₗ)
2:   if nᵣ≠NULL then Extend_Plan(nᵣ)
3:   if n=T.root
4:   then insert a decryption node new for n.ve∪n.vE as parent of n
5:        λ(new)=U
6:   else if nₗ≠NULL AND n.Aₚ\nₗ.vp≠∅
7:        then insert a decryption node new for n.Aₚ\nₗ.vp as parent of nₗ
8:        if nᵣ≠NULL AND n.Aₚ\nᵣ.vp≠∅
9:        then insert a decryption node new for n.Aₚ\nᵣ.vp as parent of nᵣ
10:       if Ɛλ(nₚ)∩n.vp≠∅ then insert an encryption node new for Ɛλ(nₚ)∩n.vp as parent of n
11:       λ(new)=λ(n)
12:  Compute_Profile(new); Compute_Profile(n); Compute_Profile(nₚ)
```

Fig. 7. Pseudocode of procedure **Extend_Plan**

(lines 6–11). The procedure finally updates the profiles of the nodes impacted by the encryption/decryption operation (line 12).

Example 7. Considering the query plan and authorizations in Fig. 1, the algorithm first visits the tree in post-order and identifies the candidates for each node (Fig. 4(a)). The algorithm then visits the tree in pre-order and selects, for each node, the candidate that is more promising from an economic point of view (Fig. 4(b)). For instance, assuming that $\mathbb{Y}$ is less expensive, the root node is assigned to $\mathbb{Y}$. Similarly, we assume that evaluating the GROUP BY clause at $\mathbb{Y}$ is more convenient than moving it to $\mathbb{X}$ or $\mathbb{Z}$. However, since $\mathbb{Y}$ cannot access attribute $C \in n.A_e$ in plaintext, C is inserted into *to_enc_dec* and its re-encryption pushed down in the tree. Assuming that the less expensive alternative for join evaluation is $\mathbb{Z}$, since $\mathbb{Z}$ can re-encrypt C, a re-encryption operation for C is inserted in the tree as child of the join node. Also, since both S and N need to be re-encrypted for the evaluation of the join operation and $\mathbb{Z}$ is authorized do so, $\mathbb{Z}$ decrypts and re-encrypts also S and N. We note that $\mathbb{Z}$ can evaluate the join over plaintext values, being authorized for such visibility, and encrypt their values before sending the join result to $\mathbb{Y}$. Finally, we assume that the selection over J can be evaluated over the attribute encrypted in storage and is then evaluated by the provider storing relation COMPANY (i.e., $\mathbb{C}$). The third step of the algorithm injects encryption and decryption operations as needed: in the example, the decryption of P and I by $\mathbb{Y}$ for the evaluation of the root node.

The algorithm illustrated in this section represents a heuristic approach for solving Problem 1 and operates in $O(|\mathbb{N}|\cdot|\mathcal{S}|\cdot|\mathcal{A}|)$ time, with $\mathcal{A}$ the set of attributes involved in the query.

5   Related Work

Traditional solutions aimed at distributed query evaluation and data analytics do not take into consideration access restrictions (e.g., [2,4,15,17,20]). Solutions aimed at enforcing access restrictions in the relational database scenario

112 112   S. De Capitani di Vimercati et al.

(e.g., view based access control [8,13,21], access patterns [3,6], data masking [16]) instead do not consider encryption as a solution for protecting confidentiality.

The use of encryption for protecting data confidentiality, while supporting query evaluation, has been widely studied (e.g., [1,14,19,24]). Alternative solutions studied the adoption of secure multiparty computation (e.g., [5,7]) and of trusted hardware components (e.g., [23]) to support query evaluation. All these solutions are complementary to our work, which can rely on these techniques to partially delegate query evaluation over encrypted data to subjects who are not authorized for plaintext visibility over (a subset of) the attributes.

Recent works have addressed the problem of protecting data confidentiality in distributed computation. The proposed solutions aim at controlling (explicit and implicit) information flows among subjects as a consequence of distributed computations (e.g., [10,18,22,25]). The work closest to ours is represented by the solution in [9], on which our proposal builds. Indeed, the approach proposed in [9] for distributed query evaluation under access restrictions first proposed the idea of distinguishing between plaintext and encrypted visibility over the data, to the aim of enabling the delegation of computations over encrypted data to non-fully trusted subjects. This authorization model has been integrated into a real world query optimizer in [11]. The work in [9] is based on the assumption that base relations are stored on the premises of the authorities owning them. Hence, base relations are available in plaintext and can be selectively encrypted on the fly, based on the needs for query evaluation. Our proposal extends such an approach to consider the more general scenario where base relations might be stored at an external provider, possibly in encrypted form.

In [12] the authors address a complementary problem allowing users to specify confidentiality requirements in query evaluation to protect the objective of their queries to some providers.

6 Conclusions

We proposed an approach for leveraging storage and computational providers to enable distributed query execution, combining data possibly stored in encrypted form at external storage providers. Our solution allows data authorities to delegate the storage of their data to external providers, while still enabling collaborative query evaluation, selectively involving computational providers to limit the costs of query evaluation. The proposed heuristics aims at limiting the economic cost of query evaluation by choosing, for each node, the candidate that is (locally) more economically convenient.

Acknowledgements. This work was supported in part by the EC within the H2020 Program under projects MOSAICrOWN and MARSAL, by the Italian Ministry of Research within the PRIN program under project HOPE, and by JPMorgan Chase & Co under project "k-anonymity for AR/VR and IoT/5G".

References

1. Agrawal, R., Asonov, D., Kantarcioglu, M., Li, Y.: Sovereign joins. In: Proceedings of ICDE, Atlanta, GA, USA, April 2006
2. Alkowaileet, W., et al.: End-to-end machine learning with Apache AsterixDB. In: Proceedings of DEEM. Houston, TX, USA, June 2018
3. Amarilli, A., Benedikt, M.: When can we answer queries using result-bounded data interfaces? In: Proceedings of PODS, Houston, TX, USA, June 2018
4. Armbrust, M., et al.: Spark SQL: relational data processing in Spark. In: Proceedings of SIGMOD, Melbourne, Australia, May–June 2015
5. Bater, J., Elliott, G., Eggen, C., Goel, S., Kho, A., Duggan, J.: SMCQL: secure query processing for private data networks. PVLDB 10(6), 673–684 (2017)
6. Benedikt, M., Leblay, J., Tsamoura, E.: Querying with access patterns and integrity constraints. PVLDB 8(6), 690–701 (2015)
7. Chow, S.S., Lee, J.H., Subramanian, L.: Two-party computation model for privacy-preserving queries over distributed databases. In: Proceedings of NDSS, San Diego, CA, USA, February 2009
8. De Capitani di Vimercati, S., Foresti, S., Jajodia, S., Livraga, G., Paraboschi, S., Samarati, P.: Fragmentation in presence of data dependencies. IEEE TDSC 11(6), 510–523 (2014)
9. De Capitani di Vimercati, S., Foresti, S., Jajodia, S., Livraga, G., Paraboschi, S., Samarati, P.: An authorization model for multi-provider queries. PVLDB 11(3), 256–268 (2017)
10. De Capitani di Vimercati, S., Foresti, S., Jajodia, S., Paraboschi, S., Samarati, P.: Authorization enforcement in distributed query evaluation. JCS 19(4), 751–794 (2011)
11. Dimitrova, E., Chrysanthis, P., Lee, A.: Authorization-aware optimization for multi-provider queries. In: Proceedings of SAC, Limassol, Cyprus, April 2019
12. Farnan, N., Lee, A., Chrysanthis, P., Yu, T.: PAQO: preference-aware query optimization for decentralized database systems. In: Proceedings of ICDE, Chicago, IL, USA, March–April 2014
13. Guarnieri, M., Basin, D.: Optimal security-aware query processing. PVLDB 7(12), 1307–1318 (2014)
14. Hacigümüs, H., Iyer, B., Mehrotra, S., Li, C.: Executing SQL over encrypted data in the database-service-provider model. In: Proceedings of SIGMOD, Madison, WI, USA, June 2002
15. Kossmann, D.: The state of the art in distributed query processing. ACM CSUR 32(4), 422–469 (2000)
16. Kwakye, M.M., Barker, K.: Privacy-preservation in the integration and querying of multidimensional data models. In: Proceedings of PST, Auckland, New Zealand, December 2016
17. Levy, A.Y., Srivastava, D., Kirk, T.: Data model and query evaluation in global information systems. JIIS 5(2), 121–143 (1995)
18. Oktay, K.Y., Kantarcioglu, M., Mehrotra, S.: Secure and efficient query processing over hybrid clouds. In: Proceedings of ICDE, San Diego, CA, USA, April 2017
19. Popa, R., Redfield, C., Zeldovich, N., Balakrishnan, H.: CryptDB: protecting confidentiality with encrypted query processing. In: Proceedings of SOSP, Cascais, Portugal, October 2011
20. Rheinländer, A., Leser, U., Graefe, G.: Optimization of complex dataflows with user-defined functions. ACM CSUR 50(3), 38:1–38:39 (2017)

21. Rizvi, S., Mendelzon, A., Sudarshan, S., Roy, P.: Extending query rewriting techniques for fine-grained access control. In: Proceedings of SIGMOD, Paris, France, June 2004
22. Salvaneschi, G., Köhler, M., Sokolowski, D., Haller, P., Erdweg, S., Mezini, M.: Language-integrated privacy-aware distributed queries. In: Proceedings of the ACM on Programming Languages, vol. 3 (2019)
23. Sharma, S., Burtsev, A., Mehrotra, S.: Advances in cryptography and secure hardware for data outsourcing. In: IEEE ICDE, Dallas, TX, USA, April 2020
24. Tu, S., Kaashoek, M., Madden, S., Zeldovich, N.: Processing analytical queries over encrypted data. PVLDB 6(5), 289–300 (2013)
25. Zeng, Q., Zhao, M., Liu, P., Yadav, P., Calo, S., Lobo, J.: Enforcement of autonomous authorizations in collaborative distributed query evaluation. IEEE TKDE 27(4), 979–992 (2015)

Cryptology II

Multi-party Private Set Operations with an External Decider

Sara Ramezanian$^{(\boxtimes)}$, Tommi Meskanen, and Valtteri Niemi

Helsinki Institute for Information Technology (HIIT),
University of Helsinki, Helsinki, Finland
{sara.ramezanian,tommi.meskanen,valtteri.niemi}@helsinki.fi

Abstract. A Private Set Operation (PSO) protocol involves at least
two parties with their private input sets. The goal of the protocol is to
learn the output of a set operation, e.g., set intersection, on the parties'
input sets, without revealing any information about the items that are
not in the output set. Commonly, the outcome of the set operation is
revealed to parties and no one else. However, in many application areas
of PSO, the result of the set operation should be learned by an external
participant who does not have an input set. We call this participant *the
decider*. In this paper, we present new variants of multi-party PSO, for
the external decider setting. All parties except the decider have a private
set. Parties other than the decider neither learn this result, nor anything
else from this protocol. Moreover, we studied generic solutions to the
problem of PSO in the presence of an external decider.

Keywords: Private Set Operation · Applied cryptography · Secure
multi-party computation · Homomorphic encryption · Privacy
enhancing technologies · Data security

1 Introduction

Private Set Operations (PSO) [1] such as Private Set Intersection (PSI) [2],
Private Set Union (PSU) [3], and Private Membership Test (PMT) [4], are special
cases of Secure Multi-party Computation that have been in the interest of many
researchers. In particular, the problem of PSI has been studied a lot and many
PSI protocols have been proposed [5].

Electronic voting [6], botnet detection between different ISPs [7], and genomic
applications [8] are a few real-life examples of set operations where the sets are
about sensitive data. Also, there is a growing need for privacy-preserving data
analysis, e.g., in the context of 5G networks. Therefore, a comprehensive study of
the factors that affect the feasibility and efficiency of a PSO protocol is needed,
and a general but still feasible solution to any multi-party PSO problem is worth
seeking for.

Multi-party PSO is a protocol with several parties, each holding a private
set, who want to perform a function on their input sets. Usually, at the end

© IFIP International Federation for Information Processing 2021
Published by Springer Nature Switzerland AG 2021
K. Barker and K. Ghazinour (Eds.): DBSec 2021, LNCS 12840, pp. 117–135, 2021.
https://doi.org/10.1007/978-3-030-81242-3_7

of the protocol all parties learn the outcome of this function, and nothing else. However, in many realistic PSO scenarios, instead of parties themselves learning the outcome, there is a special party who does not have an input set but who needs to learn the result of the PSO. We call this special party *an external decider*, or for short, *a decider*. In our setting, the decider is a trusted third party who typically does not have an incentive to get any information about the input sets, except the output of the set operation. In Sect. 2 we present several examples of real life scenarios where the result of PSO is obtained only by the decider, and no one else.

The contributions of this work are as follows:

- We classify the problem of PSO with different criteria.
- We present a comprehensive study of PSO problem with an external decider. To the best of our knowledge, this variant of PSO has been studied only in special cases, such as secure electronic voting, but not in the general case.
- We specifically study the case where the set elements are chosen from a universe of limited size. We present a general solution to any PSO problem with external decider and with limited universe.
- We present another general solution to any PSO problem with external decider, where the universe is not limited. This protocol solves the emptiness and cardinality of the output set.
- We assume that all parties are semi-honest, but we also present a modification that provides protection in the presence of one malicious party.
- Finally, we implement our protocols and compare the efficiency of our protocols against the existing work for PSO problems.

Next we briefly explain the necessary concepts that are required to understand the rest of the paper.

Homomorphic encryption (HE) schemes allow computations to be carried out using encrypted values, without the need to decrypt them first. In our paper, we are interested in *additively* homomorphic encryption such as Paillier cryptosystem [9]. If two ciphertexts, $c_1 = enc(m_1)$ and $c_2 = enc(m_2)$, are generated using an additively homomorphic encryption scheme, then the product (or result of some other operation) of these two ciphertexts is decrypted to the sum of the two plaintexts: $dec(c_1 \cdot c_2) = m_1 + m_2$.

Keyed hash function, such as HMAC [10], is a cryptographic hash function with a secret key that is utilized to create fingerprints of messages. The key is only known to the designated parties. A keyed hash function is a collision-resistant one-way function [11].

2   Motivational Examples of PSO with the Decider

In this section, we present several examples of scenarios where most parties provide only input data while the result of the set operation goes to an external decider. In each scenario preserving the privacy is important.

- **Example-1:** Several proposals are on the table for a board meeting of an organization. Every board member would choose which proposals are acceptable for them. However, they are not willing to reveal their choices to others. Therefore, a PSO protocol is used and the secretary (the decider) computes what proposals are acceptable for everyone.
- **Example-2:** A traffic office has installed surveillance cameras in a city and each of them is collecting plate numbers of the passing cars. The office wants to collect all kinds of statistics from the traffic. For example, how many different cars were observed during a day, or how many of those cars that were seen at either point A or point B in the morning were also seen at one of these points in the afternoon. Of course, technically the cameras could simply send all observed data to the central office but this would be a privacy violation. Instead, the central office is an external decider and all cameras deliver input for various PSOs.
- **Example-3:** A decentralized social networking platform such as HELIOS[1] is inherently more privacy-friendly than centralized solutions. However, people would still like to make searches in larger setting than within their own direct contacts. For example, number of people have formed a group G (for some purpose). In order to extend the group, one of the members of the group would like to identify, in privacy-preserving manner, whether her/his friend is also a friend of at least three other existing group members and could be asked to join. By utilizing PSI between that group member and other group members, the platform (the decider) can learn which user is potentially interested to join the group G.
- **Example-4:** Privacy preserving digital parental control in 5G networks [12] is another use case where PSO could be needed. A parent has a set of unwanted attributes that websites can have and wants to prevent the child to access websites that contain any of these attributes. A child wants to access a website. Each child under digital parental control has an application installed on their device that is called the *kid-client*. The network analyses the website that the child wants to access and stores the attributes of that website in a private set. The kid-client should decide whether to grant or deny access to this child. This is a PSI between the network's and parent's respective private sets. It is enough for the kid-client (the decider) to learn whether the end result is empty or not.
- **Example-5:** The government of a country wants to find out whether the health-care system functions properly. For instance, the government wants to learn the percentage of the people at higher risk from certain disease who are covered by insurance. In this case, the government wants to know the number of people in a set A, which is the union of people ensured by different companies, that are also in a set B, which is the union of all people identified with higher risk by different hospitals. This is an example of cardinality for PSI+PSU, which will be learned by the government (the decider).

---

[1] HELIOS project homepage (2021). Retrieved from https://helios-h2020.eu/.

- **Example-6:** Conversion rate of an advertisement is measured by finding the size of the intersection between people who see the ad and who completed a relevant transaction. The advertisement is shown in different platforms and the transaction is completed with the owner of the product. The names of the individuals who visited different platforms, and also the names of the buyers are considered to be private. Therefore, the marketing company (the decider) has to use a PSO protocol to learn the conversion rate of its ad. In this example, the decider learns the cardinality of the intersection.

3 Classification of PSO Problems

The scenarios of Sect. 2 lead to different types of PSO problems. In general, PSO problems can be classified using different criteria. In the following, we discuss some of these criteria. Later we study some of the many possible PSO problems further and develop solutions. Note that the list of criteria is certainly not complete and there could be more factors that guide the future research directions.

- **Criterion 1: "What information is wanted from the set".** At least the following three questions could be asked about the end result set: "What is the cardinality of the end result set?" "What are the elements of this set?" "Is it empty?"
- **Criterion 2: "Who gets the outcome".** PSO problems vary also in the way the final result is learned. For instance, in some scenarios it is required that all parties learn the outcome, whereas in some cases only one party learns the result. In this paper, we focus on the case where the result of the protocol goes to an external decider.
- **Criterion 3: "Adversary model".** Semi-honest and malicious adversarial models are two common settings in privacy-preserving protocols.
- **Criterion 4: "Size of the universe".** In each PSO problem, elements of the input sets belong to an a priori defined set of potential elements. Hereafter we call this set *the universe*. Whether the total number of elements in the universe is limited or not is a criterion that could be taken into account when defining a PSO problem.
- **Criterion 5: "Number of parties that have input".** Number of participants can vary a lot, depending on the use case for which the PSO problem is solved.
- **Criterion 6: "What is the set operation".** We need to determine what is the set operation to be computed, defined by combining intersections, unions and complements of sets.
- **Criterion 7: "Size of input sets".** Whether the sets are the same (or almost the same) in size, whether the sets are large or small, or whether some sets are actually singletons.
- **Criterion 8: "Is it possible to use a trusted party".** Many security and privacy solutions become simpler if it is possible to get help from an outsider who can be trusted by all stakeholders.

– **Criterion 9: "On-line or off-line sets"**. There are use cases where at least some part of the input data is known well before the output data is needed. Then it may be possible to do some off-line computations before the PSO protocol is run in on-line fashion.

4 Problem Statement

We study many variants of multi-party PSO problems. For the criterion "Who gets the outcome" we restrict ourselves to the case where the decider gets the result. For the Criterion 8, we only cover the case without the trusted third party. For the Criterion 3, we first assume that the parties are semi-honest. Later we show how to modify the protocols to fit in the malicious adversarial model.

We assume there are $n \geq 2$ parties plus a decider in the protocol. Each party P_i has a private set S_i, where $1 \leq i \leq n$. The decider does not have an input set and all the other parties trust the decider to calculate the result according to the protocol. It depends on the use case who the decider is. In the use cases, the decider does not typically have any incentive to learn more information than the final outcome. Moreover, the decider may want to prove that they cannot learn anything else than what is necessary. This can be for the reason of safety, data minimization, to avoid the necessity to destroy the unwanted data, to prevent conflict of interests, to avoid the need of legal documents, or to convince the public opinion that their privacy is preserved by the decider. For instance, in examples 2 and 5 the deciders want to show the citizens that their private lives are not monitored. In example 1, the decider wants to avoid the possibility of blackmailing, signing Non-disclosure Agreement (NDA) and the process of deleting the data. Deciders of examples 3, 4 and 6 want to show their users that their privacy is preserved, and avoid potential blackmailing and conflicts of interests inside their organizations.

Each example of Sect. 2 can be described as one of the following problems:

– **Problem-1:** All parties have their private sets as input. After executing the protocol, the decider wants to learn the answer to one of the following questions: 1) What are the elements in the union of all n sets. 2) What is the cardinality of the union. 3) Whether the union is an empty set.
– **Problem-2:** All parties have a set. After executing the protocol, the decider learns the answer to one of the following questions: 1) What are the elements in the intersection of n sets of all other parties. 2) What is the cardinality of the intersection. 3) Whether the intersection is empty.
– **Problem-3:** The parties all have their private sets. After executing the protocol, the decider learns the result of any given set operation. The general PSO can be written in Conjunctive Normal Form (CNF):

$$S_T = (A_{1,1} \cup ... \cup A_{1,\alpha_1}) \cap ... \cap (A_{\beta,1} \cup ... \cup A_{\beta,\alpha_\beta}) \tag{1}$$

where $A_{i,j} \in \{S_1, ..., S_n, \bar{S}_1, ..., \bar{S}_n\}$, the set $\bar{S}_i$ is the complement of the set S_i, and $1 \leq \alpha \leq n$ and $\beta \in \mathbb{N}$. After the protocol has been executed the decider learns answer to one of the following questions: 1) What are the elements of S_T. 2) What is the cardinality of S_T. 3) Is S_T empty.

Note that Problem-3 covers both Problem-1 and Problem-2 but we consider those separately for two reasons: 1) so many use cases are only about Problem-1 or Problem-2, and 2) our solutions for Problem-1 and Problem-2 are used when building one of the solutions for Problem-3.

5 Related Work

In this section, we first present the state of the art in PSO protocols, then we present some of the related previous works on secure multi-party computation.

As we mentioned before, private set operations are applicable to variety of use cases, such as privacy-preserving genomic similarity [13] and private profile matching in social media [14]. Therefore, different variants of PSO problems have been studied extensively, see [5]. Kolesnikov et al. in [15] proposed a new function that is called Oblivious Programmable Pseudorandom Function, and used it to design a practical multi-party PSI protocol that is secure in a malicious setting. In 2019, Ghosh and Nilges proposed a novel approach to PSI [16], by utilizing Oblivious Linear Function Evaluation (OLE) to evaluate the intersection.

At the time of writing, the protocol of Kolesnikov et al., in [2] is the fastest two-party PSI protocol. In [2], the authors proposed a variant of Oblivious Pseudorandom Function, and utilized it to achieve a light-weight PSI protocol.

Chun et al., generalized the problem of PSO by studying any PSO problem in disjunctive normal form (DNF) [17]. Wang et al., further studied the general PSO problem in DNF for a limited universe [18]. In this paper, we study the general problem of PSO in the setting with an external decider. To the best of our knowledge, this is the first time that this problem is studied comprehensively.

Feige et al. proposed a general solution to compute any function in the secure multi-party computation setting where there is a party without input set who computes the final result and gives it to the other parties [19]. As an example case, they presented an efficient algorithm to compute the logical AND of n bits, each bit belonging to different party. They assumed that all parties follow the protocol honestly. For the limited universe case, computing the logical AND is essentially equivalent to computing set intersection. Although their protocol for computing the logical AND is efficient, they did not show how this can evolve such that it can solve any set operations in DNF form.

There are several variants of secure multi-party set operations that use a "special party" in their setting. We briefly explain these variants and why these are different from our setting.

The work of Feige et al. in [19], lead to introduction of a variant of secure multi-party computation called *Private Simultaneous Messages* (PSM) [20]. A PSM protocol is between n parties each with a private input, and a "referee" who does not have input data. Parties have access to a common secret. Each party computes a single message by utilizing the common secret and the party's private input, and sends this message to the referee. After receiving all the messages from all the n parties, the referee is able to compute the output of the function. The referee should not learn anything else than this outcome. The construction of PSM protocols focus on information theoretical security, rather than on computational efficiency [21]. In [22] the authors present protocols to compute any function with PSM schemes using a key that can be used only once, thus each protocol needs to be initialized with a different secret key every time it is run. Our solutions differ from these unconditionally secure solutions in that we allow usage of the same keys for several different PSO instances, with different input sets and even with different operations. Moreover, the communication complexity of the protocols in [22] is increasing exponentially when the number of parties growing, whereas in our PSO protocols the communication complexity is independent of the number of participants.

Functional encryption (FE) is a variant of public key encryption that allows the holder of the secret key to learn a function of the plaintext [23]. The notion of FE can also be applied in the case of many parties with inputs [24]. In principle, if the decider is the party who holds the secret key, the setting of multi-input FE can fit into the PSO scenario with the decider. While there are accelerators for special cases of FE, no efficient solutions are known that would cover all cases [25].

In the server aided PSO setting the privacy preserving set operation protocol uses an untrusted server that carries out some part of the computations (similar to secure cloud computing). In other words, the server only helps the parties by performing some part of the computations and does not get the final result [26, 27].

The setting of secret-sharing based secure multi-party computation protocols [28] is typically designed in such a way that all parties execute their share of the computations and send their results to a party who will compute the result (the resulting party/the dealer), and later announces the final result. This setting is more complex than ours because resulting party also has a private set that they need to hide from other parties. Moreover, the resulting party has to prove to other parties that they performed the computations correctly, without revealing any extra information on the private sets to the parties. Therefore, the typical secret sharing based protocols perform much less efficiently than our solutions with a decider.

6   Protocols

In this section, we present our privacy preserving set operation protocols. These protocols are our proposed solutions to the problems of Sect. 4.

In each protocol there are $n+1$ participants involved with the set operations: n parties have input sets and the result of the protocol goes to the decider D, who does not have an input set. Other n parties do not get the final outcome of the protocol. We present our protocols with the assumption that the participants are semi-honest. Later in Sect. 9 we present a solution for malicious model as well.

Protocol 0: The off-line phase.

1 The decider sends the ordering of $U = \{a_1, ..., a_u\}$ to all the other parties.
2 The decider creates public and private keys that fulfil requirements for an additively homomorphic encryption scheme.
3 The decider picks one of the parties randomly and informs this party that they are responsible for sending the final result to the decider. For simplicity, let us assume that the decider chooses party P_n.
4 The decider sends the public key of the encryption scheme to all parties.
5 Each party P_i creates two sets of encrypted values by utilizing the decider's public key. One set contains u instances of $enc(0)$, and the other set contains u instances of $enc(r)$ where r is a random number chosen specifically for that instance.
6 Parties create a shared repository.
7 Parties together create a vector $V = (V_1, ..., V_u)$ with u components, where each component is an instance of $enc(r)$, where r is a random number.

6.1   Protocol-0

In this section and Sects. 6.2, 6.3 and 6.4 we assume that the universe is limited, i.e., it is possible to present it as an ordered set $U = \{a_1, ..., a_u\}$. For simplicity, we assume that the decider creates this ordered set. Parties create a shared repository which contains a vector with u components. Each component represents an item in U, and the order of the components is as defined in U. Each component is a bit string of a certain length. All parties can read and write to the repository but two parties cannot write at the same time, to avoid conflicts. The decider or anybody else than parties P_i does not have access to this repository. The details about the required technologies to create this kind of repository are outside the scope of this paper. The shared repository is the centralized part of our PSO protocol. Protocol 0 is the off-line phase used in all protocols of this section.

6.2   Protocol-1

Let us assume that party P_i has a private input set S_i, for $i = 1, ..., n$. The decider wants to learn one of the following cases about the union of these n sets: 1) What are the elements in $\bigcup_{i=1}^{n} S_i$. 2) What is the cardinality of $\bigcup_{i=1}^{n} S_i$. Our solution for these questions is by Protocol 1.

If the application area of the protocol is such that the decider only needs to know whether the union of all input sets is empty, the protocol can be simplified a lot. Instead of vector V we have just one value V that is initially set to $enc(1)$. Then, each P_i multiplies V by $enc(0)$ if S_i is empty and replaces V by $enc(0)$ if S_i is not empty. When all the parties have altered V, party P_n sends it to D. The decider decrypts and if D gets zero it means that the union is non-empty.

After executing the on-line phase, only the decider learns the outcome, and other parties do not learn anything else about this protocol than that it was run.

Protocol 1:

0 Protocol-0 is run.
1 Each party P_i, where $1 \leq i \leq n$, modifies vector V as follows. If $u_j \in S_i$ then P_i replaces V_j by enc(0), which is one of the encryptions of zero that P_i generated in Protocol-0. If $u_j \notin S_i$ then P_i multiplies V_j by enc(0).
2 When all parties have finished their modifications on vector V then one of the following cases will take place.
3 **Case 1:** Party P_n sends V to the decider. The decider decrypts components of this vector, and $a_j \in \bigcup_{i=1}^{n} S_i$ if and only if $dec(V_j) = 0$.
   Case 2: Party P_n permutes the components of V before sending them to D. The decider decrypts V. The number of zeros in the decrypted vector is equal to the cardinality of $\bigcup_{i=1}^{n} S_i$.

6.3   Protocol-2

Let us again assume that the party P_i has the set S_i. The decider wants to learn one of the following cases about the intersection of these n sets: 1) What are the elements in $\bigcap_{i=1}^{n} S_i$. 2) What is the cardinality of $\bigcap_{i=1}^{n} S_i$. 3) Whether $\bigcap_{i=1}^{n} S_i$ is an empty set. Our solution to these questions is Protocol 2.

After executing this protocol, only the decider learns the result. Other parties do not learn anything about each others' sets or about the result of the protocol.

6.4   Protocol-3: A Generic Solution to Perform Any PSO with an External Decider

It can be shown that every set that is obtained from a collection of sets by operations of intersection, union and complement can equivalently be computed by a conjunctive normal form. Thus, the general PSO problem can be written as presented in Eq. 1.

The general PSO problem can be formalized as follows: Party P_i has a private input set S_i, for $i = 1, ..., n$. The decider D wants to learn one or more of the following cases about the set S_T of Eq. 1: 1) What elements are there in the set S_T. 2) What is the cardinality of S_T. 3) Whether S_T is an empty set. Other parties should not learn anything about this protocol except that it is executed. Our solution for these questions is by Protocol 3.

6.5   Protocol-4: Keyed Hash Functions with Dummies

In this section we drop the assumption that the universe is limited, and present protocols for finding answers to the following questions about the set S_T that can also be written in Disjunctive Normal Form as $S_T = (A_{1,1} \cap ... \cap A_{1,\alpha_1}) \cup ... \cup (A_{\beta,1} \cap ... \cap A_{\beta,\alpha_\beta})$: 1) What is the cardinality of S_T? 2) Is this set empty? In other words, our protocols cover all private set operations but the decider cannot get the elements in the result set. In these protocols we use keyed hash function, and assume that the parties are semi-honest.

Protocol 2:

0 Protocol-0 is run, with one difference: in step 7, all components of vector V are initially set to instances of enc(0).

1 Each party P_i where $1 \leq i \leq n$ modifies every component j of the vector V as follows.

$$V_j = \begin{cases} V_j \cdot enc(0) & \text{if } a_j \in S_i \\ V_j \cdot enc(r) & \text{otherwise.} \end{cases} \tag{2}$$

After all parties have modified the vector V, one of the following cases will be executed.

2 **Case 1:** Party P_n sends V to the decider. After decrypting, for entries that do not decrypt to zero, D learns they are not in the intersection. Similarly, if decryption yields zero, D learns that the corresponding element is in every input set. Therefore, the decider learns $\bigcap_{i=1}^{n} S_i$.

Case 2: Party P_n shuffles V and sends it to D. The decider decrypts and the number of zero values in the decrypted vector is the cardinality of $\bigcap_{i=1}^{n} S_i$.

Case 3: To hide the true cardinality of the intersection, the parties need to create "clones" of elements in the universe. For each element that is in the intersection, there would be many zeros after decryption. The parties would also add many encryptions of non-zero numbers to hide further the number of elements in the intersection.

In the off-line phase of this protocol, the parties P_1 to P_n decide on a shared key k to be used for computing a keyed hash function. The decider should not learn this key. We may even assume that the key k is generated by a trusted hardware and anyway we assume that only the authorized parties (e.g. cameras in Example-2) have access to this key. Because the parties $P_1, ..., P_n$ are semi-honest, they do not collude with the decider and reveal the key to D. In other words, the decider cannot access the key k.

The basic idea is simple. All parties replace elements in their input sets with images of the elements under keyed hash function. However, the parties cannot simply send the images to the decider because then the decider would get lots of information about the input sets.

Let us first consider the question 1. The true cardinalities of all input sets are hidden from the decider and from other parties by adding a big number of "dummies" among the true images of elements in input sets. These dummies are just random bit strings that look like results of the keyed hash function.

We present the protocol in the case of a simple example, for better illustration. It is straight-forward to convert the presentation to the general case but we skip it, for sake of compactness. Let us assume that there are three parties A, B and C in the protocol and the decider wants to know the cardinality of set $(A \cap B \cap \bar{C}) \cup (B \cap C)$. The Venn diagram for the three input sets is shown in Fig. 1. Eight disjoint sets are formed by first choosing either the input set or its complement for each party and taking an intersection of the chosen three sets. Because the total universe could be very large, one of these eight sets (the

Protocol 3:

0 The off-line phase of this generic protocol is as explained in Protocol-0 with only one difference: in the step 7, the parties create β vectors W^k, where $1 \leq k \leq \beta$. Each vector W^k is created similarly to the vector V in step 7 of Protocol-0. On-line phase of Protocol-1 is used as a building block of the on-line phase of the generic solution.

1 In order to compute each W^k, parties should compute each set $(A_{k,1} \cup ... \cup A_{k,\alpha_k})$ by utilizing Protocol-1. Note that party P_i either inputs S_i or $\bar{S}_i$ or does not attend the computation for this term.

2 After all the vectors W^k have been computed, party P_n creates a new vector Z where every entry j of the vector is computed as $Z_j = \prod_{k=1}^{\beta} W_j^k$.

3 Now, one of the following cases will be executed.

4 **Case 1:** Party P_n sends vector Z to D. The decider decrypts Z. For every entry Z_j which decrypts to zero, the decider learns that the corresponding element a_j is in the set S_T. On the other hand, if decryption yields a non-zero, then the corresponding element is not on the set S_T.

Case 2: Party P_n shuffles vector Z and sends it to D. The decider then decrypts the vector Z. The cardinality of S_T is equal to the number of zero values in the decrypted vector.

Case 3: Similarly to Case 3 of Protocol-2, the party P_n creates a new vector Z' from the vector Z, by appending the components of Z and their duplicates to vector Z'. Then, P_n shuffles vector Z', and sends it to the decider. The decider decrypts the vector. If at least one of the values in Z' decrypts to zero, then S_T is non-empty. Otherwise, the set of Equ. 1 is empty.

intersection of complements) is assumed to be non-relevant for the end result of the PSO. For each of the other 7 disjoint sets, the parties A, B and C agree on the number of dummy values. Please note that we can only increase the total number of elements in each set, and if the number of dummies is small compared to the total number, the dummies do not change the whole picture. If the number of dummies is about the same as the total number, then the total number can still be found out. Thus, the number of dummies should dominate the actual size. In other words, the number of dummies should be at least one order of magnitude greater than the typical size of an input set. The agreed numbers of dummy values are shown in Fig. 1.

The parties also need to agree on the actual 97 dummy values that every party adds to the set of values they would later send to the decider. Similarly, A and C have to agree values for the 12 dummies that both of them include among keyed hash images of their input sets. Parties A and B agree on 23 joint dummy values, while B and C agree on 53 joint dummies. Finally, A would freely choose 34 random dummy values while B (resp., C) chooses 88 (resp., 145) dummies. The decider receives all the hash-values and dummy values. From the received values, the decider identifies those that appear in every set, those that have been received from A and B but not from C, and those that have been received from

B and C but not from A. Finally the decider is asked to subtract 173 (= 23 + 97 + 53) from the gross number of collisions explained above.

Note that every PSO problem can be presented in CNF or alternatively in DNF. Our example PSO above was in disjunctive normal form. This is mainly just for making our solution easier to explain.

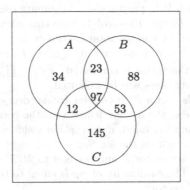

Fig. 1. An example of a PSO for three sets with keyed hash function.

Now, let us now assume that the decider only needs to learn whether the resulting set from the PSO is empty or not. In the off-line phase, each possible collection of parties would agree on several keys that would be used to compute images of elements in sets of parties in the collection. This would be done in addition to the images computed by the common key k. Effectively, for each element that can be found in several input sets there would be many collisions in the data received by the decider. Dummies would be added in addition to this "cloning" of elements. Apart from images by the common key k, parties should skip computing images of a few elements with the other keys. This is to further confuse the true number of collisions from the number of observed collisions.

7   Performance Evaluation

In this section we evaluate the communication and computation complexities of our protocols. We have only implemented the most critical part of our protocols which is the cryptographic part, and we present the results of our experiments. Moreover, we do some comparison between the performance of our protocols with the prior works.

Please note that any set operation can be considered a boolean function, therefore, the number of possible set operations between n different sets is equal to the number of truth tables with n different variables, which is 2^{2^n}. In other words, when the number of input sets increases, the number of possible set operations increases more than exponentially. This means in order to evaluate the performance of Protocol 3 and protocol with keyed hash function, we cannot run the experiment with all the possible set operations (because there are many of those operations). Therefore, we choose one operation that we find interesting, and is compatible in size and the number of operations with the experiments in [18]: In Eq. 1, we assume that $\alpha = \beta = n$. We have compared the complexity of our Protocol 3 against the protocol 3 by Wang et al. in [18], because they already compared their protocol against the state of the art and showed that their protocol 3 is more feasible in practice than other solutions. Moreover, the underlying idea of our protocol is close to that of the protocols in [18]. It is important to note that our setting with an external decider differs from the settings in [18] and other prior art. The difference in setting gives an opportunity to get more efficient solutions.

Wang et al.'s protocol is based on threshold ElGamal while our protocols use Paillier encryption scheme or any additively homomorphic encryption. We have listed the number of operations in Table 1. The entry for number of encryptions includes the number of re-encryptions. Both in ElGamal and in Paillier this is done by multiplying the encrypted value by a random encryption of zero. The speed of multiplication is very fast compared to encryption in both cryptosystems. Thus we have not listed the number of multiplications in the table. The number of operations in a single union or intersection can be calculated from the values of the table by substituting $\beta = 1$.

Table 1. Number of operations of our protocol 3 and the protocol 3 of [18]

Operations	Protocol of Wang et al. in [18]	Our Protocol-3
Encryptions	$O(nu + \alpha\beta u)$	$O(\alpha\beta u)$
Decryptions	$O(nu)$	$O(u)$

We also compare the number of communication rounds in our protocol against the protocol of Wang et al. In [18], the n parties first need to decide together on the key for threshold ElGamal. Then each party needs to individually take part in calculating the union of all sets. Then each party needs to individually take part in calculating the result and finally all parties together decrypt the result.

In our protocol the decider picks the key, then all other parties individually, in any order and maybe even partially simultaneously, modify the values in vectors W^k. Party P_n does the multiplications and sends Z to D that will decrypt it. Thus the number of communication rounds is much smaller in our solution.

The time measurements are obtained by running crypto operations in our protocols on an x86-64 Intel Core i5 processor clocked at 2.7 GHz with a 4 MB L3 cache. In Protocols 1, 2, and 3, any additively homomorphic non-deterministic encryption can be used. For our experiments, we use Paillier cryptosystem. The modulus that Wang et al. used in ElGamal was set to 512 bits, therefore, to be able to do a comparison we set the modulus N for Paillier cryptosystem to non-secure 512 bits as well. In practice, 512 bits is considered to be too short for a public key, because it can be factored. To be in secure side, N should be 2048 bits long. We made experiment when N is 2048 bits long and reported the results in Table 2.

We now compare the execution time of our Protocol 3 and the protocol 3 by Wang et al. in [18]. We first tested the cases where $u \in \{10, 20, 40, 60, 80, 100\}$ and there are 3 parties with input sets in the protocol. The result of our implementation showed that our protocol is 5 times faster than the protocol of Wang et al. For example, when $u = 100$, our protocol 3 needs 0.71 s to compute the outcome set, while protocol 3 of Wang et al. needs 3.7 s. We next compared the execution time of our Protocol 3 and the protocol 3 of Wang et al. in the cases that $n \in \{3, 5, 10, 15, 20\}$ and $u = 20$. In our experiments our Protocol 3 performs significantly faster than the protocol of Wang et al., and the speed difference increases with the number of participants. For example, when $n = 20$, our protocol needs 0.15 s, the protocol of Wang et al. needs 24 s to perform PSO. Therefore, when there are 20 parties with input sets, our protocol performs more that 150 times faster than Wang et al.'s protocol. Please note that we utilized the same computational power as Wang et al. used.

Table 2. Execution time of the on-line phase our Protocol-3 in seconds, when modulo N in Paillier is of length 2048 bits and $u \in \{2^2, 2^5, 2^7, 2^{10}\}$ and $n \in \{3, 5, 10, 15, 20\}$. In Eq. 1, we assume that $\alpha = \beta = n$. The numbers in the table are required time for each party to modify Z with a single thread. When $u = 2^2, 2^5, 2^7, 2^{10}$ the decider needs $0.02, 0.17, 0.68, 5.51$ seconds respectively, to decrypt this vector with 32 threads.

	$n = 3$	$n = 5$	$n = 10$	$n = 15$	$n = 20$
$u = 2^2$	0.001	0.002	0.003	0.005	0.007
$u = 2^5$	0.008	0.013	0.025	0.039	0.05
$u = 2^7$	0.031	0.05	0.1	0.15	0.2
$u = 2^{10}$	0.237	0.391	0.786	1.178	1.56

The evaluation of the keyed hash function is many orders of magnitude faster compare to public key encryptions. For instance, for a set size of one million, the computation of keyed hash values only takes one second.

As we mentioned before, our setting with a decider differs from the typical secret-sharing schemes and the server-aided PSO protocols. In fact in our variant of multi-party computation, the existance of the decider makes the protocol more efficient. For instance, the server-aided PSI protocol in [27] has the computation complexity of $O(u)$ and $O(nu)$ for each party and the server respectively, whereas in our PSI protocol (Protocol-2) the computation complexities are $O(u)$ for each party and also for the decider.

8   Security and Privacy Analysis

In this section, we present the security and privacy analysis of our generic protocols in the semi-honest setting.

In our protocols we assume that the parties communicate through a secure channel. Moreover, we also assume that the repository that the parties $P_1, ..., P_n$ use is accessible by them only. Moreover, this repository has a secure version control system to log the activities of its users [29].

In protocols with homomorphic encryption, if we assume that all parties are semi-honest, all vectors W^k are calculated correctly and the correct vector Z is sent to D. Also the result D gets after decrypting Z is the correct answer.

All values in the vectors in the repository are (very likely to be) different and no party P_i can decrypt them without the help of D and thus parties P_i do not learn anything from them.

The decider does not know the values in the repository and thus does not get any information from the parties in addition to the encrypted vector sent to it in the end of each protocol. Therefore, D does not learn anything else than the decrypted values of the vector that P_n send to D.

Thus in the semi-honest setting no party P_i learns anything and the decider only learns the end result of the protocol.

In the protocols with keyed hash function, D only receives the hash values of the items and dummy values. Because of the one-way property of the hash functions, D cannot derive the items from their hash values. D does not know the key and thus cannot even check if a certain element is in the resulting set or in the input set of some party. Also, hash values are indistinguishable from random dummy values, hence D cannot tell these two apart. Moreover, the dummy values hide the cardinality of the elements that are not in the output set. Addition of "cloned" hash images (for the emptiness protocol) hide the true cardinality of the output set and just reveals whether the cardinality is zero or positive.

9   Modified Protocol with One Malicious Party

We consider now the adversarial model in which we assume that there are no collusions between the parties. Please note that the material of this section is a hint of future work.

If we do not assume that the parties P_i are semi-honest, there are several ways how they can try to cheat: (i) They can use different inputs S_i (or $\bar{S}_i$) in different parts of the protocol (in different unions); (ii) They can use an incorrect complement for their set S_i; (iii) They can calculate the elements in W^k incorrectly: for instance, instead of multiplying the previous value by $\text{enc}(0)$ they replace the element by $\text{enc}(r)$; (iv) Party P_n can send an incorrect vector Z to the decider.

Our protocol can be made secure against these actions in the following way:

1) Each party P_i for each element in V, chooses one encryption of 0 and one encryption of 1, puts these two encrypted values in random order and sends them to all other parties, including D. The decider can confirm that, indeed, the pair is of the form $\{enc(0), enc(1)\}$. The party P_i then, for each pair, chooses a random number a, publishes the number to other parties except

D, and raises both encrypted values to power a. The result is a different pair enc(0) and enc(a) with different a, for each element in the vector V. The party P_i uses only these values later in the protocol. All other parties can later make sure that these are calculated correctly.

2) We repeat our protocol n times and change the order of parties P_i such that everyone must be once in the role of P_1. When a party is in the role of P_1, they must use the values enc(0) and enc(1) they created in step 1) to initialize the vector V. When they are initializing V, they must use the same values for the same element every time a different union is calculated and their input should be the same (S_1 or $\bar{S}_1$). Also, they should use different values for the same element when the inputs are different (one is S_1 and the other is $\bar{S}_1$). Now, the parties cannot cheat by using different inputs for S_i (or $\bar{S}_i$), or miscalculating the set complement.

3) In every repetition every party calculates the vector Z and sends it to the Decider. If every party has acted honestly in every repetition, then (i) everybody should send the same Z in every repetition; (ii) the decryption of Z would give the same result for each repetition. On the other hand, if some party has used different S_i in some repetition than what they used when they were playing the role of P_1, then there is a fair chance that the results from these two repetitions do not match.

4) The previous steps are likely to reveal whether an individual party has been cheating while others have stayed honest. However, it is harder to identify the cheater. The following can be done in order to locate where the cheating has happened. After n runs of the protocol every party independently multiplies together all elements in every vector W^k in every run that they know must be encryptions of zero if computations have been done correctly. Then they further multiply the result with a random encryption of zero before sending it to the Decider. The Decider decrypts everything and verifies that nothing else than zeros come out.

10   Conclusion

Private set operations such as private set intersection and private set union can be used in many privacy sensitive use cases, and therefore they have been studied extensively. In this work, we studied a special variant of PSO where the outcome of the multi-party PSO is learned by an external decider while the parties who have provided an input set to the PSO do not learn the outcome. By providing realistic examples, we showed the importance of the setting with a decider. Furthermore, we showed how our setting is different from what has been presented so far in academia.

We presented two generic solutions (Protocol 3 and protocol with keyed hash) to any PSO problems that the decider seeks the cardinality of the output set or wants to investigate whether this set is empty. Moreover, Protocol 3 is a generic protocol for the decider to learn also elements in the output set, under the assumption that the universe, from where parties choose elements to their sets, is limited in size. In addition to the cardinality and emptiness, the decider can receive the elements in the output set.

We presented the security and privacy analysis of our protocols in the semi-honest setting, and presented a modified protocol for the malicious setting. Lastly, we implemented our protocol. The result of our experiments showed that our solutions for our special setting, i.e., having an external decider, are more efficient than applying state of the art solutions proposed for other settings to our setting. The experiments also show that our solutions are feasible for many real-life use cases.

We present different criteria that can be used to determine which PSO protocol fits a certain setting. The criteria of Sect. 3 can give a future direction to the research in the field of PSO protocols. We only covered the case with one malicious party. The case with more malicious parties is left for future work.

Acknowledgement. This paper is supported by 5GFORCE project funded by Business Finland, and HELIOS H2020 project funded by the European Union's Horizon 2020 research and innovation programme under grant agreement No 825585.

References

1. Kissner, L., Song, D.: Privacy-preserving set operations. In: Shoup, V. (ed.) CRYPTO 2005. LNCS, vol. 3621, pp. 241–257. Springer, Heidelberg (2005). https://doi.org/10.1007/11535218_15
2. Kolesnikov, V., Kumaresan, R., Rosulek, M., Trieu, N.: Efficient batched oblivious PRF with applications to private set intersection. In: Proceedings of the 2016 ACM SIGSAC Conference on Computer and Communications Security, pp. 818–829 (2016)
3. Frikken, K.: Privacy-preserving set union. In: Katz, J., Yung, M. (eds.) ACNS 2007. LNCS, vol. 4521, pp. 237–252. Springer, Heidelberg (2007). https://doi.org/10.1007/978-3-540-72738-5_16
4. Ramezanian, S., Meskanen, T., Naderpour, M., Junnila, V., Niemi, V.: Private membership test protocol with low communication complexity. Digital Commun. Netw. **6**(3), 321–332 (2020)
5. Pinkas, B., Schneider, T., Zohner, M.: Scalable private set intersection based on OT extension. ACM Trans. Privacy Secur. (TOPS) **21**(2), 1–35 (2018)
6. Mursi, M.F.M., Assassa, G.M.R., Abdelhafez, A., Abo Samra, K.M.: On the development of electronic voting: a survey. Int. J. Comput. Appl. **61**(16) (2013)
7. Nagaraja, S., Mittal, P., Hong, C.-Y., Caesar, M., Borisov, N.: BotGrep: finding P2P bots with structured graph analysis. In: USENIX Security Symposium, vol. 10, pp. 95–110 (2010)
8. Erlich, Y., Narayanan, A.: Routes for breaching and protecting genetic privacy. Nat. Rev. Genet. **15**(6), 409–421 (2014)
9. Paillier, P.: Public-key cryptosystems based on composite degree residuosity classes. In: Stern, J. (ed.) EUROCRYPT 1999. LNCS, vol. 1592, pp. 223–238. Springer, Heidelberg (1999). https://doi.org/10.1007/3-540-48910-X_16
10. Krawczyk, H., Bellare, M., Canetti, R.: HMAC: keyed-hashing for message authentication (1997)

11. Turner, J.M.: The keyed-hash message authentication code (HMAC). Federal Information Processing Standards Publication 198:1 (2008)
12. Ramezanian, S., Meskanen, T., Niemi, V.: Parental control with edge computing and 5g networks. In: 2021 29th Conference of Open Innovations Association (FRUCT), pp. 290–300. IEEE (2021)
13. Dugan, T., Zou, X.: A survey of secure multiparty computation protocols for privacy preserving genetic tests. In: 2016 IEEE First International Conference on Connected Health: Applications, Systems and Engineering Technologies (CHASE), pp. 173–182. IEEE (2016)
14. Li, M., Cao, N., Yu, S., Lou, W.: FindU: privacy-preserving personal profile matching in mobile social networks. In: 2011 Proceedings IEEE INFOCOM, pp. 2435–2443. IEEE (2011)
15. Kolesnikov, V., Matania, N., Pinkas, B., Rosulek, M., Trieu, N.: Practical multiparty private set intersection from symmetric-key techniques. In: Proceedings of the 2017 ACM SIGSAC Conference on Computer and Communications Security, pp. 1257–1272 (2017)
16. Ghosh, S., Nilges, T.: An algebraic approach to maliciously secure private set intersection. In: Ishai, Y., Rijmen, V. (eds.) EUROCRYPT 2019. LNCS, vol. 11478, pp. 154–185. Springer, Cham (2019). https://doi.org/10.1007/978-3-030-17659-4_6
17. Chun, J.Y., Hong, D., Jeong, I.R., Lee, D.H.: Privacy-preserving disjunctive normal form operations on distributed sets. Inf. Sci. **231**, 113–122 (2013)
18. Wang, W., Li, S., Dou, J., Runmeng, D.: Privacy-preserving mixed set operations. Inf. Sci. **525**, 67–81 (2020)
19. Feige, U., Killian, J., Naor, M.: A minimal model for secure computation. In: Proceedings of the Twenty-Sixth Annual ACM Symposium on Theory of Computing, pp. 554–563 (1994)
20. Ishai, Y., Kushilevitz, E.: Private simultaneous messages protocols with applications. In: Proceedings of the Fifth Israeli Symposium on Theory of Computing and Systems, pp. 174–183. IEEE (1997)
21. Assouline, L., Liu, T.: Multi-party PSM, revisited. IACR Cryptol. ePrint Arch. **2019**, 657 (2019)
22. Beimel, A., Kushilevitz, E., Nissim, P.: The complexity of multiparty PSM protocols and related models. In: Nielsen, J.B., Rijmen, V. (eds.) EUROCRYPT 2018. LNCS, vol. 10821, pp. 287–318. Springer, Cham (2018). https://doi.org/10.1007/978-3-319-78375-8_10
23. Boneh, D., Sahai, A., Waters, B.: Functional encryption: definitions and challenges. In: Ishai, Y. (ed.) TCC 2011. LNCS, vol. 6597, pp. 253–273. Springer, Heidelberg (2011). https://doi.org/10.1007/978-3-642-19571-6_16
24. Goldwasser, S., et al.: Multi-input functional encryption. In: Nguyen, P.Q., Oswald, E. (eds.) EUROCRYPT 2014. LNCS, vol. 8441, pp. 578–602. Springer, Heidelberg (2014). https://doi.org/10.1007/978-3-642-55220-5_32
25. Bahadori, M., Järvinen, K.: A programmable SOC-based accelerator for privacy-enhancing technologies and functional encryption. IEEE Trans. Very Large Scale Integr. (VLSI) Syst. **28**(10), 2182–2195 (2020)
26. Kamara, S., Mohassel, P., Raykova, M., Sadeghian, S.: Scaling private set intersection to billion-element sets. In: Christin, N., Safavi-Naini, R. (eds.) FC 2014. LNCS, vol. 8437, pp. 195–215. Springer, Heidelberg (2014). https://doi.org/10.1007/978-3-662-45472-5_13
27. Zhang, E., Li, F., Niu, B., Wang, Y.: Server-aided private set intersection based on reputation. Inf. Sci. **387**, 180–194 (2017)

28. Du, W., Atallah, M.J.: Secure multi-party computation problems and their applications: a review and open problems. In: Proceedings of the 2001 Workshop on New Security Paradigms, pp. 13–22 (2001)
29. Guthrie, P., Dale, A., Tolson, M., Buchanan, C.: Distributed secure repository, March 16 2006. US Patent App. 10/943,495

Encrypted-Input Obfuscation of Image Classifiers

Giovanni Di Crescenzo[1]([✉]), Lisa Bahler[1], Brian A. Coan[1], Kurt Rohloff[2], David B. Cousins[3], and Yuriy Polyakov[3]

[1] Peraton Labs, Basking Ridge, NJ, USA
{gdicrescenzo,lbahler,bcoan}@peratonlabs.com
[2] New Jersey Institute of Technology, Newark, NJ, USA
rohloff@njit.edu
[3] Duality Technology Inc., Newark, NJ, USA
{dcousins,ypolyakhov}@dualitytech.com

Abstract. We consider the problem of protecting image classifiers simultaneously from inspection attacks (i.e., attacks that have read access to all details in the program's code) and black-box attacks (i.e., attacks where have input/output access to the program's code). Our starting point is cryptographic program obfuscation, which guarantees some provable security against inspection attacks, in the sense that any such attack is not significantly more successful than a related black-box attack. We actually consider the recent model of encrypted-input cryptographic program obfuscation, which uses a key shared between the obfuscation deployer and the input encryptor to generate the obfuscated program. In this model we design an image classifier program and an encrypted-input obfuscator for it, showing that the classifier program is secure against both inspection and black-box attacks, under the existence of symmetric encryption schemes. We evaluate the accuracy of our classifier and show that it is significantly better than the random classifier and not much worse than more powerful classifiers (e.g., k-nearest neighbor) for which however no efficient obfuscator is known.

Keywords: Inspection attacks · Black-box attacks · Program obfuscation · Image classifiers

1 Introduction

According to web sources, the Internet of Things (IoT) market is expected to grow by $ 421.28 billions during 2021–2025, progressing at a compound annual growth rate of 33%. In many typical IoT applications, servers perform analytics over data received by multiple distributed sensors (see, e.g., [21]). Just like most web or cloud computing services, IoT analytics servers can be subject to a number of attacks. In this paper, we focus on attacks to the server programs, here categorized as inspection attacks (informally defined as attacks that try to

© IFIP International Federation for Information Processing 2021
Published by Springer Nature Switzerland AG 2021
K. Barker and K. Ghazinour (Eds.): DBSec 2021, LNCS 12840, pp. 136–156, 2021.
https://doi.org/10.1007/978-3-030-81242-3_8

access internal data or computation used by the server program), and black-box attacks (informally defined as attacks only use input-output access to the attacked server program, and no access to any internal data or computation).

Our starting point to propose solutions mitigating these attacks is the recent area of cryptographic program obfuscation, which promises a set of solutions with some provable security guarantee in the presence of inspection attacks, but does not address the problem of protecting programs against black-box attacks. Program obfuscation is the problem of modifying a computer program so to hide any sensitive details without changing its input/output behavior. While this problem has been known for several years in computer science, only in the last 20 years or so, researchers have considered the problem of *provable* program obfuscation, where sensitive code details are proved to remain hidden under a widely accepted intractability assumption, such as those often used in cryptography. The most studied security guarantee offered by provable program obfuscation, also called "virtual-black-box" obfuscation [4], says that for any efficient inspection attack to the program (i.e., an attack that has access to all details in the program's code) there exists an efficient black-box attack to the program (i.e., an attack that only has input-output access to the program) that is about as equally successful. Early results in the area implied the likely impossibility of constructing a single program capable of obfuscating any input arbitrary polynomial-time program into a virtual black box [4]. Most recent results show the possibility of constructing, under close to standard intractability assumptions, practically efficient obfuscators for very restricted families of functions, such as point functions and a few extensions of them (see, e.g., [5,6,11,13,17,22]), as well as theoretically feasible obfuscators for large families of functions (e.g., compute-and-compare functions [23]).

On one hand, such provable program obfuscation solutions make inspection attacks to the program's sensitive information essentially useless, in that any inspection attack would not be significantly better than a related black-box attack. On the other hand, the security guarantee does not say anything new about black-box attacks. Recent results (see, e.g., [12,19,20]) show successful black-box attacks to popular programs (e.g., machine learning programs), even undermining the success of the related business model (e.g., MLaaS). Motivated by these results, recent work [10] has considered the problem of augmenting the cryptographic program obfuscation model so to achieve, in at least some class of application scenarios, program confidentiality in the presence of *both* inspection and black-box attacks. A resulting model, called *encrypted-input program obfuscation*, has been proposed as a mixed encryption/obfuscation model with the following security guarantee: for any efficient inspection attack to the program (i.e., an attack that has access to all details in the program's code) there exists an efficient algorithm (note: *not* one that is given black-box attack to the program) that is about as equally successful. Thus, a provable encrypted-input obfuscation solution makes both inspection attacks and black-box attacks to the program's sensitive data essentially useless. Moreover, in this model the parties generating the inputs (e.g., the IoT devices) are assumed to encrypt them by using a key shared with the entity obfuscating the program, which can then work by computing over the obfuscated program and the encrypted (and authenticated) input.

In this paper we continue this effort and specifically focus on posing the problem for machine learning classifiers (instead of arbitrary programs) and on finding a concrete classifier that can be secured against both black-box and inspection attacks in this model. We start by observing that any encrypted-input obfuscator for an arbitrary program can be used to design an encrypted-input obfuscator for a machine learning classifier, but then note that the resulting scheme would be very inefficient (e.g., depending linearly in the dataset) and may likely require conversions to other representations (e.g., circuits). Thus, we consider the problem for a specific task: *image matching classification*, where given a secret image, and an input image, the classifier returns 1 if they belong to the same class, or 0 otherwise. Since no efficient obfuscators are known for complex machine learning classifiers, we opt for designing our own image matching classifiers, using tools like principal component transformations and textbook statistics, for which we know how to produce an encrypted-input obfuscator for the related evaluation program. We study the true positive rate and true negative rate for this classifier and obtain that they are significantly better than the random classifier and not much worse than much more powerful classifiers (e.g., k-nearest neighbor) for which we do not know of any efficient obfuscator.

Table 1. For each computing model (programs or classifiers), 2 security models are considered: cryptographic obfuscation (briefly, obfuscation) and encrypted-input cryptographic obfuscation (briefly, ei-obfuscation). For each computing model and associated security model, the table lists if provable security is guaranteed in the presence of inspection security, and of black-box attack security, which paper first defines the security model, and which appendix in this paper contains the main definition.

Computing model	Security model	Inspection attack security	Black-box attack security	Defined in
Programs	Obfuscation	Yes	No	[3], Appendix B.1
Programs	ei-Obfuscation	Yes	Yes	[10], Appendix B.3
Classifiers	Obfuscation	Yes	No	[9], Appendix B.2
Classifiers	ei-Obfuscation	Yes	Yes	Appendix C

2   Definitions and Models

We present definitions of the computing models of interest (i.e., secret-based programs and matching classifiers) in Sect. 2.1; and an informal discussion of attack classes and resources, previous related obfuscation models, and the model for encrypted-input obfuscation of matching classifiers in Sect. 2.2. Formal definitions of the various obfuscation models from the literature are recalled in Appendices A, B.1, B.2 and B.3, and finally the (new) formal definition for encrypted-input obfuscation of matching classifiers is presented in Appendix C.

2.1   Computing Models: Programs and Classifiers

We consider two computation models: (secret-based) programs and (matching) classifiers. Informally, secret-based programs are programs with both a public and a secret input, and matching classifiers are a pair of programs: a training program and a matching program, with a specific syntax, including inputs of specific data and label types. We now proceed more formally.

(Secret-Based) Programs. We consider *families of (secret-based) functions* as families $F = \{f_{pv,sv}\}$ of maps $f_{pv,sv} : \{0,1\}^n \rightarrow \{0,1\}$ parameterized by some public values $pv \in \{0,1\}^{m_p}$ and secret values $sv \in \{0,1\}^{m_s}$, for some *length parameter* n, and *parameter value lengths* m_p, m_s polynomial in n. We will think of public values pv as being available to all parties, including the adversary, and of secret values sv as encoding the information that has to remain secret from the adversary. In later sections of the paper, we will specifically consider obfuscation of the following classes of (secret-based) programs:

1. the *family of range-membership programs*, where the program computes if an input value belongs to a range, where we keep the two range limits secret but the length of their binary representation public. Formally, we define the family of programs $RM_{pv,sv}$, where $pv = (1^n)$, $sv = (a,b)$, with $[a,b] \subseteq \{0,1\}^n$, and that on input $x \in \{0,1\}^n$, return 1 if $x \in [a,b]$ and 0 otherwise.
2. the *family of conjunctions of range-membership programs*, where the program computes if each input value in a sequence belongs to a (potentially different) range, where we want to keep all the range limits secret but can keep the length of their binary representation public. Formally, we define the family of programs $CRM_{pv,sv}$, where $pv = (1^n, 1^t)$, $sv = (a_1, b_1, \ldots, a_t, b_t)$, with $[a_i, b_i] \subseteq \{0,1\}^n$, and that on input $x_1, \ldots, x_t \in \{0,1\}^n$, return 1 if $x_i \in [a_i, b_i]$ for all $i = 1, \ldots, t$, and 0 otherwise.

Matching Classifiers. By D we denote a *probability distribution*, and by dS we denote a *data space*; that is, the set $dS = \{d_1, \ldots, d_N\}$ of all possible data samples, that can be drawn according to distribution D. For instance, a sample d_i could be an image of an object (e.g., a car) and D could returns images of cars of possibly different brands.

By cS we denote a *class space*; that is, the set $cS = \{c_1, \ldots, c_q\}$ of all possible data classes. For instance, c_i could be the i-th car brand name within a known and pre-specified list.

The *class function* is a function $cF : dS \rightarrow cS$ mapping a data sample in dS to its class in cS. In the given example, function cF would map the data sample $d \in dS$ containing the image of a car to a value $c_i \in cS$ denoting this car's brand name, as from the known list.

We define a *matching classifier* (briefly, *classifier*) *for class function cf* as a pair of algorithms MC = (CTrain, CMatch) such that:

– on input a dataset $ds = (d_1, \ldots, d_n) \in dS^n$ of n data samples, labels $cl = (c_1, \ldots, c_n)$ such that $c_i = cF(d_i)$ for $i = 1, \ldots, n$, and a data sample $d_0 \in dS$, algorithm CTrain returns matching auxiliary input *maux*;

– on input data sample d and matching auxiliary input $maux$, algorithm CMatch returns a bit b. Here, the value $b = 1$ denotes that this classifier predicts that class $c = cF(d)$ of data sample d is equal to the class $c_0 = cF(d_0)$ of data sample d_0; and, naturally, the value $b = 0$ denotes that the classifier predicts that $c \neq c_0$.

Towards defining classifier MC's output accuracy metrics of interest, relatively to a dataset ds with samples independently drawn from distribution D and with labels cl, a data sample d_0, and a random execution of $CTrain(ds, cl, d_0)$ returning matching auxiliary input $maux$, we say that a data sample d can be a

– *true positive*, if $cF(d) = c_0$ and $CMatch(d, maux) = 1$ (predicting $c = c_0$);
– *true negative*, if $cF(d) \neq c_0$ and $CMatch(d, maux) = 0$ (predicting $c \neq c_0$);
– *false positive*, if $cF(d) \neq c_0$ and $CMatch(d, maux) = 1$ (predicting $c = c_0$);
– *false negative*, if $cF(d) = c_0$ and $CMatch(d, maux) = 0$ (predicting $c \neq c_0$).

Let tp (resp., tn, fp, fn) denote an estimate of the number of true positives (resp., true negatives, false positives, false negatives) for matching classifier MC. Based on these definitions, relatively to a dataset ds, a class c, and a random execution of $CTrain(ds, cl, d_0)$ returning matching auxiliary input $maux$, we can then define the following *classification metrics* for MC:

– *true negative rate* (aka specificity): $tn/(tn + fp)$;
– *true positive rate* (aka recall): $tp/(tp + fn)$.

Similarly as for the program computing model, even in the classifier computing model we will think of some inputs as being available to all parties, called *public parameter values* $pv(MC)$, and of values that have to remain secret from the adversary, called *secret parameter values* $sv(MC)$. Specifically, $pv(MC)$ contains a description of CTrain and CMatch, a syntactic description of dataset ds, including the number n of data samples, and a syntactic description of data space dS, including the total number N of possible data samples. Moreover, $pv(MC)$ contains dataset ds, labels cl, data sample d_0 and its class $c_0 = cF(d_0)$.

In later sections of the paper, we will specifically consider obfuscation of the *family of image matching classifiers*, denoted as $iMC = (iCTrain, iCMatch)$. This is defined exactly as a family of matching classifiers, with the (only semantic) difference that data samples are images.

2.2   Modeling Obfuscation of Matching Classifiers

Our goal is to produce a formal model for the encrypted-input obfuscation of image matching classifiers. Since much literature focuses on obfuscation of programs, we first discuss related formal models on program obfuscation from the literature, and then where we extend these models.

Threat Model: Attack Classes and Attacker Resources. We consider attacks trying to infer whether a binary-valued property (possibly involving secret data or program description) is satisfied or not.

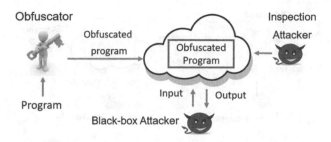

Fig. 1. Usage paradigm for a cryptographic program obfuscator.

An *inspection attack to a program* is defined as a polynomial-time attack, that given access to some or all of the program's code, tries to find out whether a property of the program is satisfied or not. An *inspection attack to a matching classifier* is defined as an inspection attack to the classifier's matching program, where the attacker is additionally given access to some or all of the output *maux* of the classifier's training program.

A *black-box attack to a program* is defined as a polynomial-time attack, that given input-output access to the program (but not given access to any part of the program's code), tries to find out whether a property of the program is satisfied or not. A *black-box attack to a matching classifier* is defined as a black-box attack to the classifier's matching program, whose input *maux* is set as the output of the classifier's training program.

Previous Related Obfuscation Models. Cryptographic program obfuscation, as originally studied in [4,14], is about security of an arbitrary program in the presence of inspection attacks. Here, a virtual black-box obfuscation property is formalized, briefly speaking, as follows: for any efficient inspection attack, there is an efficient black-box attack that is about equally successful. Note that this property does not target black-box attacks. A pictorial description of the usage paradigm for a cryptographic program obfuscator can be found in Fig. 1.

Intrusion-resilient matching classifiers, as recently defined in [9], extend the above cryptographic program obfuscation model to a type of matching classifiers. The extension is necessary due to a different program syntax (specifically, matching classifiers are a pair of programs, instead of a single program), and a different set of public data and information to be kept secret (specifically, in matching classifiers the description of algorithms CTrain and CMatch may be public while the dataset ds, the class labels cl, and a data class c are desired to remain secret). The usage paradigm of a cryptographic obfuscator for a matching classifier is obtained by using, in Fig. 1, a program equal to CMatch($d, maux$), where $maux = $ CTrain(ds, cl, d_0).

Encrypted-input program obfuscation, as recently defined in [10], targets a combined key-based encryption/obfuscation of an arbitrary program, satisfying the following property: for any efficient inspection attack, there is an efficient attack with neither inspection nor black-box access to the program that is equally

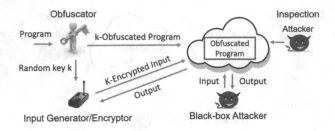

Fig. 2. Usage paradigm for an encrypted-input cryptographic program obfuscator.

successful. The model in [10] can be seen as obtained by applying two modifications to the original model in [4,14]: (1) the obfuscation of the program is performed using a random key that is unknown to the attacker; and (2) the inputs to the program are always generated in encrypted form, using this key. Here, (1) suggests that security against inspection attacks may follow from some form of encryption of the program, and (2) suggests that security against black-box attacks may follow since an attacker, not knowing the key, may not be able to generate valid inputs for a black-box attack. The usage paradigm for an encrypted-input cryptographic program obfuscator is depicted in Fig. 2.

In this paper, we generalize the encrypted-input program obfuscation model of [10] to matching classifiers. The generalization is necessary due to a different program syntax (specifically, matching classifiers are a pair of programs, instead of a single program), and a different set of public data and information to be kept secret (specifically, in matching classifiers the description of algorithms CTrain and CMatch may be public while the dataset ds, the class labels cl, the secret data sample d_0 and its class c_0 are desired to remain secret). Our definition of matching classifiers also slightly generalizes [9], as follows: in our paper the algorithm CMatch of a matching classifier takes as a secret input a data sample d_0, and possibly but not necessarily its data class c_0, while in the previous definition the secret input was just a secret class c. The usage paradigm of an encrypted-input obfuscator for a matching classifier is obtained by using, in Fig. 2, a program equal to CMatch($d, maux$), where $maux = $ CTrain(ds, cl, d_0). It is interesting to compare the original virtual-black-box obfuscation requirement, as rewritten in Appendix B.1 for programs and in Appendix B.2 for classifiers, with the simulated-view obfuscation requirements in Definition 2 for programs and in Definition 3 for classifiers. In the virtual-black-box obfuscation requirements, it is required that the adversary's view can be simulated by an efficient algorithm that is given access to a black-box computing the original program or the original matching algorithm of the classifier, while in the latter the efficient algorithm is *not* given access to any such black box. We derive:

- For an obfuscator satisfying the virtual-black-box obfuscation property, an inspection attack is not significantly more successful than a related black-box attack, but this does not rule out the existence of a black-box attack that learns the (program's or) the classifier's secret parameter values sv.

- For an encrypted-input obfuscator satisfying the simulated-view obfuscation property, if the obfuscation key remains secret, the obfuscated version of the (program or) matching classifier is of no help to inspection or black-box attacks to learn the (program's or) classifier's secret parameters values sv.

3   A General Result on Encrypted-Input Obfuscation

We consider a question naturally arising from our definitions of encrypted-input obfuscation of programs and matching classifiers: is it possible to construct an encrypted-input obfuscator for an arbitrary family of matching classifiers (as formally defined in Definition 3) starting from an encrypted-input obfuscator for any family of programs (as formally defined in Definition 2, recalled in Appendix B.3). We give a positive answer to this question and show the following

Theorem 1. For any family of matching classifiers MC, there exists a family of secret-based programs P such that the following holds. If there is an encrypted-input obfuscator for P then there exists an encrypted-input obfuscator for MC.

To prove Theorem 1, we show a family of secret-based programs for any family of matching classifiers. Specifically, let MC = (CTrain, CMatch) be a family of matching classifiers, as formally defined in Sect. 2.1. We define the family of secret-based programs $P_{pv,sv}$, where

- $pv = (1^n, 1^N, desc(\text{CTrain}), desc(\text{CMatch}), pv(\text{MC}))$,
- $sv = (ds, cl, d_0, sv(\text{MC}))$,

and such that, on input x, it returns $b = \text{CMatch}(x, \text{CTrain}(ds, cl, d_0))$. The computation correctness, low runtime overhead and simulated-view obfuscation properties of the encrypted-input obfuscator for MC directly follow from the analogue properties of the encrypted-input obfuscator for $P_{pv,sv}$.

Remark. In [10] it was observed that Yao's garbling circuit technique [24] can be directly used to construct an encrypted-input obfuscator for any polynomial-time program, hiding all circuit gates of the input circuit equivalent to the program. When combined with Theorem 1, this implies a similar general result for any matching classifier. We caution the reader that a direct use of Theorem 1 would result in an obfuscated matching classifier of size polynomial in the dataset size (which is undesirable as in many practical applications datasets are very large). However, we believe that this result is still encouraging towards finding, in some model, general methods to efficiently and provably secure classifiers against both inspection and black-box attacks.

4   Image Matching

In this section we present our result on image matching classifiers. First, in Sect. 4.1, we recall background definitions of interest, including principal component transformations. Next, in Sect. 4.2, we formally describe our new image matching classifier, and finally in Sect. 4.3 we report our experimental analysis of its accuracy properties over 3 different datasets.

4.1 Background Definitions and Tools

An *attribute* A is a function that maps a data sample $d_i \in dS$ to a numerical *attribute value* v in some *value space* vS. Let $A_1, \ldots, A_m$ denote m distinct attributes. To a data sample $d_i \in dS$, we can then associate a value tuple $v_i = (v_{i,1}, \ldots, v_{i,m})$, where $v_{i,j}$ represents $A_j(d_i)$; that is, the value returned by attribute A_j on input data sample d_i. For instance, if d_i is an image, the value $v_{i,j}$ could represent the numeric value associated with the j-th pixel (or block of pixels) in d_i. We can then define a *data $n \times m$-matrix V* as a matrix where each of the n rows is associated to a data sample d_i, each of the m columns is associated to an attribute A_j, and each entry $v_{i,j}$ contains value $A_j(d_i)$.

For a sequence of values $v_j = (v_{1,j}, \ldots, v_{n,j})$ from value set vS, we use the textbook definitions of *mean*, denoted as $\mu(v_j)$, and *standard deviation*, denoted as $\sigma(v_j)$, for all $j = 1, \ldots, m$.

An important tool used in our classifier construction is a *Principal Components transformation* (briefly, pcT). Informally speaking, pcT is defined as an orthogonal linear transformation converting a data matrix to a new coordinate system where the greatest variance by some scalar data projection lies on the 1st coordinate (also called the 1st principal component), the 2nd greatest variance on the 2nd coordinate, and so on [15]. Thus, truncating this transformation to its first coordinates is sufficient to capture a large part of the data variability of interest in many practical uses. More formally, given a data $(n \times m)$-matrix V, we define the *truncated Principal Components transformation* (briefly, tpcT) of V as the function tpcT that on input V, returns $Y = VW$, where W is an $(m \times \ell)$-matrix, dependent on V, with elements in vS, and the product VW is a matrix product, returning $(n \times \ell)$-matrix Y, where we usually think of ℓ as much shorter than m. Thus, tpcT also serves as a dimensionality reduction method. While one classical way to compute W from V consists of setting W as the matrix whose columns are the first ℓ eigenvectors of matrix $V^{\mathrm{T}}V$, many more efficient variants and generalizations have been studied (see, e.g., [15]).

4.2 Our New Image Matching Classifier

In this subsection we formally describe a new image matching classifier. First, we provide formal definitions for some useful families of classifiers: range membership, and conjunction of range memberships. Then we use these classifiers and the background tools from Sect. 4.1 to present our image classifier IC.

Our Image Matching Classifier IC. Our image matching classifier, denoted as IC = (imCTrain, imCMatch), reduces the problem of image matching (i.e., matching a new test image against a secret image) to the problem of evaluating a conjunction of range memberships (i.e., testing if each of the values in an input test sequence belongs to a prespecified, secret, value range). Informally speaking, this is performed using the following steps: (1) a truncated principal components transformation tpcT maps the dataset images to short vectors capturing a large part of the images' variability; (2) the tpcT output is processed by using basic statistics like mean and standard deviation to capture summary ranges for attribute values relative to each attribute, for each attribute and each image class; and finally (3) such ranges and the input secret image are used to define a conjunction of ranges of attribute values, one range for each attribute, against which any new test image can be later matched.

A formal description follows. (For simplicity, in this description we assume that the class c_0 of the secret data sample d_0 is known to algorithm imCTrain. Later, we show the extension to the more general cases when c_0 is not known.)

Input to Algorithm imCTrain:

- an n-image dataset $ds = (d_1, \ldots, d_n)$
- class labels $cl = (c_1, \ldots, c_n)$ such that $c_i = cF(d_i)$, for $i = 1, \ldots, n$
- a secret data sample d_0 and its class $c_0 = cF(d_0)$
- a parameter 1^m denoting the number of image attributes

Instructions for Algorithm imCTrain:

1. let V be the data $n \times m$-matrix associated with the n-image dataset ds
2. set $W = \text{tpcT}(V)$
3. set $Y = VW$ (i.e., Y is the product of matrices V and W)
4. let A_j denote the attribute that maps the data sample d_i in V's i-th row to the j-th data block in Y's i-th row, for $i = 1, \ldots, n$, and $j = 1, \ldots, m$
5. for all $i = 1, \ldots, n$ such that $cF(d_i) = c_0$,
   set $y_i = (y_{i,1}, \ldots, y_{i,m})$,
     where $y_{i,j} = A_j(d_i)$, for all $j = 1, \ldots, m$
6. let α be a configurable constant (e.g., $\alpha = 1.5$)
7. for all $j = 1, \ldots, m$
   compute $ct_j = \mu(\{y_{i,j} | cF(d_i) = c_0\})$
   compute $std_j = \sigma(\{y_{i,j} | cF(d_i) = c_0\})$
   set $a_j = ct_j - \alpha \cdot std_j$, and $b_j = ct_j + \alpha \cdot std_j$
8. return: $maux = (W, (a_1, b_1), \ldots, (a_m, b_m))$

Input to Algorithm imCMatch:

- a tuple $maux$ returned by imCTrain
- an image data sample d

Instructions for Algorithm imCMatch:

1. write $maux$ as $(W, (a_1, b_1), \ldots, (a_m, b_m))$
2. compute $e = dW$ (i.e., e is the product of vector d and matrix W)
3. let A'_j denote the attribute mapping data sample d to the j-th data block in e, for $j = 1, \ldots, m$
4. set $e = (e_1, \ldots, e_m)$, where $e_j = A'_j(e)$, for $j = 1, \ldots, m$

5. if $e_1 \in [a_1, b_1]$ AND $\cdots$ AND $e_m \in [a_m, b_m]$ then return 1 else return 0.

We assumed, for description simplicity, that the above algorithm imCTrain$_1$ takes as input the class c_0 of the secret data sample d_0, and the algorithm only needed to compute ranges from the data samples d_i such that $cF(d_i) = c_0$. In the case class c_0 is not known, the algorithm can find c_0 as follows. First, it computes ranges from all the data samples d_i such that $cF(d_i) = c$, for all $c \in cS$. Then, the algorithm matches data sample d_0 against the conjunction of range membership statements obtained from all data samples in class c, for all $c \in cS$. Finally, it sets c_0 as the class which maximizes the number of range memberships within the same conjunction.

4.3   Accuracy Properties of Our Image Classifier

To evaluate the accuracy properties of our image matching classifier, we performed experiments with data values from the following 3 datasets (including one often used dataset with well structured image samples, as well as two less structured datasets, one of which including real-life images with 3D objects):

- The MNIST dataset [16], containing images of handwritten digits; specifically, a training set of 60,000 images and a test set of 10,000 images. The digits have been size-centered and normalized into a fixed-size image. Each image is a 28 × 28 pixel array, where the value of each pixel is a positive integer in the range [0; 255]. We used 50,000 images for training ad 10,000 images for testing.
- The 'ETL Character Database' [1], a collection of images of about 1.2 million hand-written and machine-printed numerals, symbols, Latin alphabets and Japanese characters and compiled in 9 datasets (ETL-1 to ETL-9). We have used images from this dataset containing the 26 lower-case letters and the 26 upper-case letters from the English alphabet.
- The 'Statlog (Vehicle Silhouettes) Data Set' [18], where the purpose is to classify a given silhouette as one of four types of vehicle (a double decker bus, Chevrolet van, Saab 9000 and an Opel Manta 400), using a set of features extracted from the silhouette. This particular combination of vehicles was chosen with the expectation that the bus, van and either one of the cars would be readily distinguishable, but it would be more difficult to distinguish between the cars. The vehicles may be viewed from one of many different angles. The original purpose was to find a method of distinguishing 3D objects within a 2D image by application of an ensemble of shape feature extractors to the 2D silhouettes of the objects.

We stress that the design possibilities for our image classifiers were heavily constrained within the small class of functions that have an efficient cryptographic obfuscator in the literature. Thus, our classifiers were limited in that they had to be selected from the family of point functions (see, e.g., [8] and [2] for efficient implementations) or wildcard matching classifier (see, e.g., [7] for an efficient implementation).

With respect to the 3 above datasets (briefly denoted as the 'digits', 'letters' and 'vehicles' datasets), in Table 2 we show the true positive rate and true negative rate of our 2 image matching classifiers, as well as 2 baseline classifiers: a random classifier (i.e., a classifier that, on input an image, returns a random class value as output); and k-nearest neighbor (for this latter classifier, however, we do not know of an efficient cryptographic program obfuscator in the literature).

Table 2. Classification metrics for our image matching classifier and 2 baseline classifiers. The true positive rate and true negative rate are defined in Sect. 2.1. The average rate is simply defined as the average of these two rates.

Datasets	Classifiers	True positive rate	True negative rate	Average rate
Digits	Random	10%	90%	50%
Digits	Conjunction of Range Memberships	84%	85%	84.5%
Digits	k-Nearest Neighbor	89%	90%	89.5%
Letters	Random	98%	2%	50%
Letters	Conjunction of Range Memberships	44%	79%	62%
Vehicles	Random	75%	25%	50%
Vehicles	Conjunction of Range Membership	34%	83%	59%

Main takeaways from this analysis include the following:

1. Even on the 'vehicle' dataset, containing very unstructured and close to real life images, our image classifier IC performs better than the random classifier; moreover, this improvement becomes even larger as we use more structured datasets, like the 'letters' and the 'digits' datasets;
2. On the quite well-structured 'digits' dataset, the average accuracy of our image classifier IC, based on conjunction of range memberships, is no more than 5% less accurate than the much more powerful nearest neighbor classifier (for which however we do not know how to construct an efficient cryptographic program obfuscator); in other words, we showed a classifier for which we can gain the obfuscation property at a very small accuracy loss.

5   Encrypted-Input Obfuscation of Image Matching

In this section we present our encrypted-input obfuscator for the image matching classifier IC from Sect. 4.2. Formally, we obtain the following

Theorem 2. Let $IM_{ipv,isv}$ be the family of image matching classifiers. If there exists a symmetric encryption scheme, then there exists an encrypted-input program obfuscator for $IM_{ipv,isv}$.

In [10] it was proved that the existence of a symmetric encryption scheme suffices to construct an encrypted-input program obfuscator for the family of conjunctions of range-membership programs. Thus, to prove Theorem 2, it suffices to prove the following

Lemma 1. Let $IM_{ipv,isv}$ be the family of image matching classifiers. and let $CRM_{cpv,csv}$ be the family of conjunctions of range-membership programs. If there exists an encrypted-input program obfuscator for $CRM_{cpv,csv}$, then there exists an encrypted-input program obfuscator for $IM_{ipv,isv}$.

The rest of this section is devoted to the proof of Lemma 1. We start with an informal description of the ideas behind our obfuscator, and then present

Obfuscator Description: Denoting as IC = (imCTrain, imCMatch) the image matching classifier to be obfuscated, we have to show an obfuscator icO consisting of 4 algorithms: the key generator $kGen$, the obfuscation generator $oGen$, the input encryptor $iEnc$ and the obfuscation evaluator $oEval$.

The main idea underlying the construction of these 4 algorithms is that icO runs the classifier IC's training algorithm and obfuscates its output using the obfuscator iCO. More specifically, algorithm $oGen$ runs the training algorithm of classifier IC, and obtains the principal component transform matrix W and m ranges $(a_1, b_1), \ldots, (a_m, b_m)$. Then, the matrix W is passed to the input encryptor $iEnc$ and the ranges are passed to the obfuscation evaluator $oEval$, so that the execution of the classifier IC's matching algorithm can be suitably distributed, encrypted and obfuscated by both $iEnc$ and $oEval$, as follows. Algorithm $iEnc$ processes its input by first multiplying it with matrix W and then encrypting it using the input encryptor for obfuscator $crmO$. Algorithm $oEval$ processes its input ranges by obfuscating them using the obfuscation generator for obfuscator $crmO$.

Note that if obfuscator $crmO$ correctly computes a conjunction of range memberships then the execution of classifier IC's matching algorithm is well distributed across $iEnc$ and $oEval$. Moreover, an inspection attacker to algorithm $oEval$ can be turned into an inspection attacker to algorithm $oEval_{crm}$, which is not successful since we assume the existence of an encrypted-input obfuscator for $CRM_{cpv,csv}$. Finally, a black-box attacker to algorithm $oEval$ can be turned into a black-box attacker to algorithm $oEval_{crm}$, but we know from [10] that this attacker is not successful since it cannot produce valid encrypted inputs for $oEval_{crm}$, not having the key used for these encryptions (assuming the existence of symmetric encryption schemes). Now we proceed more formally.

Input to kGen: a security parameter 1^n

Instructions for kGen:

1. run algorithm $kGen_{rm}$ to obtain a random key k
2. Return: key k.

Input to oGen: secret parameter values $sv(\mathrm{IC}) = (ds, cl, d_0, c_0)$, public parameter values $pv(\mathrm{IC}) = (desc(\mathrm{imCTrain}), desc(\mathrm{imCMatch}))$, and key k

Instructions for oGen:

1. Let $(W, (a_1, b_1), \ldots, (a_m, b_m))$ be the output returned by imCTrain on input the n-image dataset ds, the class labels $cl = (c_1, \ldots, c_n)$, the secret image data sample d_0 and its secret class c_0
2. let $t = m$, $pv = (1^n, 1^t)$ and $sv = (a_1, b_1, \ldots, a_t, b_t)$, and run algorithm $oGen_{crm}$ on input k, pv, sv, thus obtaining $gout_{crm}$, the obfuscated version of sv, and $iaux_{crm}$, the auxiliary input for $iEnc_{crm}$
3. Return: obfuscated program $gout = gout_{crm}$ and auxiliary string $iaux = (W, iaux_{crm})$.

Input to iEnc: key k, public parameter values $pv(\mathrm{IC})$, auxiliary string $iaux$, input string d

Instructions for iEnc:

1. Write auxiliary string $iaux$ as $(W, iaux_{crm})$
2. compute $e = dW$ (i.e., $e = $ vector d times matrix W)
3. let A'_j denote the attribute mapping data sample d to the j-th data block in e, for $j = 1, \ldots, m$
4. set $e = (e_1, \ldots, e_m)$, where $e_j = A'_j(e)$, for $j = 1, \ldots, m$
5. run algorithm $iEnc_{crm}$ on input $k, pv, iaux_{crm}, (e_1, \ldots, e_t)$, thus obtaining as output $iout$, an encrypted version of $(e_1, \ldots, e_t)$
6. Return: encrypted input $iout$.

Input to oEval: public parameter values $pv(\mathrm{IC})$, an obfuscated program $gout$ and an encrypted input string $iout$

Instructions for oEval:

1. run algorithm $oEval_{crm}$ on input public parameter values $pv(\mathrm{IC})$, obfuscated program $gout$ and encrypted input $iout$, thus obtaining $eout$, which is intended to be equal to the output of $CRM_{pv,sv}$ on input $(e_1, \ldots, e_t)$.
2. Return: $eout$.

Obfuscator Properties. To complete the proof of Lemma 1, it remains to prove that obfuscator $icO = (kGen, oGen, iEnc, oEval)$ satisfies the properties listed in Definition 3: computation correctness, low runtime overhead, and simulation-based obfuscation.

The computation correctness property of obfuscator icO directly follows from the analogue property of obfuscator $crmO = (kGen_{crm}, oGen_{crm}, iEnc_{crm}, oEval_{crm})$, and by observing that obfuscator icO uses obfuscator $crmO$ to compute the same functionality as the IC classifier. Specifically, the output of algorithm $oEval$ is defined to be equal to the output $eout$ of algorithm $oEval_{crm}$ on input $pv(\mathrm{IC})$, $gout$ and $iout$. By the computation correctness property of obfuscator $crmO$, the output $eout$ is equal to the output of $CRM_{pv,sv}$ on input $(e_1, \ldots, e_t)$. Finally, observe that since $e = dW$, and W is the principal component transformation of the dataset matrix V, the output of $CRM_{pv,sv}$ on input $(e_1, \ldots, e_t)$ is equal to the output of the classifier algorithm imCMatch, by inspection of algorithms imCTrain and imCMatch.

The low runtime overhead property of obfuscator icO follows by algorithm inspection, after observing that the runtime complexity of $oEval$ is essentially the same as that of algorithm $oEval_{crm}$, which, in turn, only requires $O(m \log |D|)$, where by $|D|$ we denote the size of the public domain that is a superset of all secret input ranges $[a_i, b_i]$, for $i = 1, \ldots, m$.

The simulated-view obfuscation property directly of obfuscator icO directly follows from the simulated-view obfuscation property of obfuscator $crmO$ since $oEval$ just runs $oEval_{crm}$ and returns its output.

6 Conclusions

We extend a recent model to consider encrypted-input cryptographic program obfuscation of matching classifiers, and observe that in this model solutions (based on Yao's garbled circuits) are possible for a very general class of matching classifiers. It remains of interest to produce much more efficient constructions for such a general set of classifiers. We then consider the problem of constructing encrypted-input obfuscators for image matching classifiers and show a new classifier for image matching which both has non-trivial accuracy properties and can be obfuscated and protected against both inspection and black-box attacks. It remains of interest to study what other classifiers can be efficiently obfuscated in this model.

Acknowledgements. Part of this work was supported by the Defense Advanced Research Projects Agency (DARPA) via U.S. Army Research Office (ARO), contract number W911NF-15-C-0233. The U.S. Government is authorized to reproduce and distribute reprints for Governmental purposes notwithstanding any copyright annotation hereon. Disclaimer: The views and conclusions contained herein are those of the authors and should not be interpreted as necessarily representing the official policies or endorsements, either expressed or implied, of DARPA, ARO or the U.S. Government.

A   Basic Notations and Definitions

The expression $\{0,1\}^n$ denotes the set of n-bit strings, where n is a positive integer. If S is a set, the expression $x \leftarrow S$ denotes the probabilistic process of uniformly and independently choosing x from S. If A is an algorithm, the expression $y \leftarrow A(x_1, x_2, \ldots)$ denotes the probabilistic process of running algorithm A on input $x_1, x_2, \ldots$ and any random coins, and obtaining y as output.

A function ϵ over the set of natural numbers $\mathbb{N}$ is *negligible* in n if for every polynomial p, there exists an n_0 such that $\epsilon(n) < 1/p(n)$, for all integers $n \geq n_0$.

Two distribution ensembles $\{D_\sigma^0 : \sigma \in \mathbb{N}\}$ and $\{D_\sigma^1 : \sigma \in \mathbb{N}\}$ are *computationally indistinguishable* if for any efficient algorithm A, the quantity

$$\left|\text{Prob}\left[x \leftarrow D_\sigma^0 : A(x) = 1\right] - \text{Prob}\left[x \leftarrow D_\sigma^1 : A(x) = 1\right]\right|$$

is negligible in σ (i.e., no efficient algorithm can distinguish if a random sample came from one distribution or the other).

B Cryptographic Program Obfuscation: Previous Models

B.1 Cryptographic Program Obfuscation: Original Model

Cryptographic program obfuscation schemes are usually defined as a pair of algorithms: an obfuscation generator and an obfuscation evaluator. On input the original program, the obfuscation generator returns an obfuscated version of it, called the obfuscated program. On input the obfuscated program and a program input, the obfuscation evaluator, returns an output, which is intended to be the same as the original program's output for this program input. Program obfuscation schemes are required to satisfy the following requirements: preserving the same computation, adding low runtime overhead and offering the same security as a (virtual) black box. The latter says, informally, that any efficient adversary's output bit on input the obfuscated program can be efficiently simulated given access to a black box computing the program. Now we proceed more formally.

Definition 1. We define a *program obfuscation scheme* for family of functions F as a pair $(oGen, oEval)$ of algorithms satisfying the following syntax

1. on input parameter values pv, sv of function $f_{pv,sv} \in F$, the *obfuscation generator* algorithm $oGen$ returns an output, denoted as $gout$, which is intended to be an obfuscated version of $f_{pv,sv}$.
2. on input parameter values pv and the obfuscation $gout$, the *obfuscation evaluator* algorithm $oEval$ returns an output, which we denote as $eout$, which is intended to be the original program's output, when run on input x;

and the following requirements:

1. *Computation Correctness:* Except with very small probability, it holds that $eout = f_{pv,sv}(x)$.
2. *Low Runtime Overhead:* $oGen$'s runtime is only polynomially slower than the circuit computing $f_{pv,sv}$.
3. *Virtual-Black-Box Obfuscation:* Any efficient adversary's output bit on input the obfuscated program can be efficiently simulated given access to a black box computing the program. A bit more formally: for any efficient adversary Adv given $pv, gout$ as input, there exists an efficient simulator algorithm Sim given pv as input and given oracle access to a black box computing $f_{pv,sv}$, such that the probability that Adv returns 1 and the probability that Sim returns 1 only differ by a negligible amount.

B.2   Cryptographic Obfuscation of Classifiers

We review the definition of cryptographic obfuscation of classifiers, from [9].

In line with the Kerckhoff's principle of modern cryptography, this definition assumes that the description of algorithms CTrain and CMatch is known to an adversary, while the dataset ds and the matching auxiliary input $maux$ returned by CTrain are not. An *intrusion attack* to a classifier is defined as an attack capable of obtaining the matching auxiliary input $maux$ returned by CTrain on input ds and a class number c. Accordingly, classifiers are defined so that an intrusion attack does not leak any true/false property about $maux$, other than what possibly leaked by the capability of performing remote calls to algorithm CMatch$(\cdot, maux)$. In other words, the classifier CMatch$(\cdot, maux)$ would look to an adversary very much like a virtual black box, in the sense of Definition 1. Specifically, this *intrusion-resiliency* notion for classifiers says that an adversary's 1-bit output obtained when given as input the output $maux$ of algorithm CTrain can be simulated, up to a small error, by the output of an efficient algorithm Sim that does not know $maux$ but can query the algorithm CMatch as an oracle. More formally, the classifier (CTrain, CMatch) satisfies *intrusion resiliency* with respect to dataset distribution D if for all probabilistic polynomial-time algorithms A, there exists a probabilistic polynomial-time algorithm Sim with black-box access to Cmatch$(\cdot, maux)$ such that the quantity

$$\left| \mathrm{Prob}\left[\mathrm{AdvExp}_{\mathrm{IR}}(1^k, c) = 1\right] - \mathrm{Prob}\left[\mathrm{SimExp}_{\mathrm{IR}}(1^k) = 1\right] \right|$$

is negligible in the security parameter k, where the probability experiments $\mathrm{AdvExp}_A, \mathrm{SimExp}_{\mathrm{Sim}}$ are detailed below.

$\mathrm{AdvExp}_{\mathrm{IR}}(1^k)$

1. $c \leftarrow cS$
2. $ds \leftarrow D(1^k)$,
3. $maux \leftarrow CTrain(ds, c)$
4. $b \leftarrow A(maux)$
5. return: b

$\mathrm{SimExp}_{\mathrm{IR}}(1^k)$

1. $c \leftarrow cS$
2. $ds \leftarrow D(1^k)$,
3. $maux \leftarrow CTrain(ds, c)$
4. $b \leftarrow Sim^{CMatch(\cdot, maux)}$
5. return: b

B.3   Encrypted-Input Obfuscation: The Model

This model, introduced in [10], extends the original 2-algorithm model of program obfuscation (including an obfuscation generator an an obfuscation evaluator, as reviewed in Appendix B.1) to a 4-algorithm model that additionally includes a key generator and an input encryptor. The key generator returns a random key to be used by the obfuscation generator to generate its obfuscation of the program and by the input encryptor to generate an encrypted input on which the obfuscation evaluator can be run. The changes to the computation correctness requirement are only of syntactic nature. The runtime requirements extend to the input generation algorithm as well. The obfuscation property is strengthened in that the adversary's output on input the obfuscated program can be efficiently simulated by an algorithm given only public information and, in particular, *without* need for black-box access to the original program.

Definition 2. Let F be a family of functions. We define an *encrypted-input program obfuscator* for F as a 4-tuple $(kGen, oGen, iEnc, oEval)$ of algorithms satisfying the following syntax:

1. on input a security parameter, the *key generator* algorithm $kGen$ returns a random key k
2. on input k, and parameter values pv, sv of function $f_{pv,sv} \in F$, the *obfuscation generator* algorithm $oGen$ returns an output, denoted as $gout$, which is intended to be an obfuscated version of $f_{pv,sv}$, and an output, denoted as $iaux$, intended to be an auxiliary input for the input encryptor algorithm.
3. on input k, parameter values pv of function $f_{pv,sv} \in F$, auxiliary input $iaux$, and input x, the *input encryptor* algorithm $iEnc$ returns an output, denoted as $iout$, intended to be an encrypted version of input x to function $f_{pv,sv}$;
4. on input parameter values pv, the obfuscation $gout$ and the encrypted input $iout$, the *obfuscation evaluator* algorithm $oEval$ returns an output, which we denote as $eout$, which is intended to be the original program's output, when run on input x;

and the following requirements:

1. *Computation Correctness:* Except with very small probability, it holds that $eout = f_{pv,sv}(x)$.
2. *Low Runtime Overhead:* the sum of $iEnc$ and $oGen$'s runtime is only polynomially slower than the circuit computing $f_{pv,sv}$.
3. *Simulated-view Obfuscation:* The obfuscated program returned by algorithm $oGen$ can be efficiently simulated given only the program's public parameters. Formally, there exists a polynomial-time algorithm Sim such that for any function $f_{pv,sv}$ from the class of function F, and any distribution hD returning secret parameter values sv with high min-entropy, the distributions D_{view} and D_{sim} are computationally indistinguishable, where

$$D_{view} = \{sv \leftarrow hD; \; k \leftarrow kGen(1^\sigma); gout \leftarrow oGen(k, pv, sv) \; : \; gout\},$$
$$D_{sim} = \{gout \leftarrow Sim(1^\sigma, pv) \; : \; gout\}.$$

C Encrypted-Input Classifier Obfuscation: Formal Model

Let F be a function and let MC = (CTrain,CMatch) a matching classifier for F, as defined in Sect. 2.1. The *public parameter values* of MC, denoted as $pv(MC)$, are available to all parties, including an attacker. The *secret parameter values* of MC, denoted as $sv(MC)$, are only available to the obfuscation generator. In our model, we will consider the description of CTrain and CMatch as being known to the attacker, while the dataset, the secret data sample and its class as unknown to the attacker. Thus, we have that $pv = (desc(\text{CTrain}), desc(\text{CMatch}), n, N)$ and $sv = (ds, cl, d_0, c_0)$, where $c_0 = cF(d_0)$. An encrypted-input obfuscator for MC is defined as a 4-tuple of algorithms: a key generator, an obfuscation generator, an input encryptor, and an obfuscation evaluator; which need to satisfy 3 properties: computation correctness, low runtime overhead and simulated-view obfuscation.

Definition 3. Let F be a function and let MC = (CTrain, CMatch) be a matching classifier for F. We define an *encrypted-input program obfuscator for matching classifier MC* as a 4-tuple icO = (kGen, oGen, iEnc, oEval) of algorithms satisfying the following syntax

1. on input a security parameter, the *key generator* algorithm *kGen* returns a random key k
2. on input key k, public parameter values $pv(\text{MC})$ and secret parameter values $sv(\text{MC})$ for classifier MC, the *obfuscation generator* algorithm *oGen* returns an output, denoted as *gout*, which is intended to be an obfuscated version of MC, and an output, denoted as *iaux*, which is intended to be an encryption auxiliary input for the input encryptor algorithm *iEnc*.
3. on input key k, public parameter values $pv(\text{MC})$ for MC, encryption auxiliary input *iaux*, and data input d, the *input encryptor* algorithm *iEnc* returns an output, denoted as *iout*, which is intended to be an encrypted version of input d to algorithm CMatch;
4. on input public parameter values $pv(\text{MC})$, the obfuscation *gout* and the encrypted input *iout*, the *obfuscation evaluator* algorithm *oEval* returns an output, which we denote as *eout*. The latter is intended to be equal to CMatch's output, when run on input $d, maux$, where *maux* has been returned by algorithm CTrain on input ds, cl, d_0;

and the following requirements:

1. *Computation Correctness:* Except with very small probability, it holds that $eout = \text{CMatch}(d, maux)$, where $maux = \text{CTrain}(ds, cl, d_0)$.
2. *Low Runtime Overhead:* the sum of *iEnc* and *oGen*'s runtime is not significantly slower than the program computing CMatch with public parameter values $pv(\text{MC})$ and secret parameter values $sv(\text{MC})$.
3. *Simulated-view Obfuscation:* The obfuscated program returned by algorithm *oGen* can be efficiently simulated given only the program's public parameters $pv(\text{MC})$. Formally, there exists a polynomial-time algorithm Sim such that for any matching classifier MC for function F, and any distribution hD returning secret parameter values $sv(\text{MC})$ with high min-entropy, the distributions D_{view} and D_{sim} are computationally indistinguishable, where
   - $D_{view} = \{sv(\text{MC}) \leftarrow hD; \ k \leftarrow kGen(1^n);$
     $\qquad\qquad gout \leftarrow oGen(k, pv(\text{MC}), sv(\text{MC})) : gout\}$,
   - $D_{sim} = \{gout \leftarrow Sim(1^n, pv(\text{MC})) : gout\}$.

References

1. National Institute of Advanced Industrial Science and Technology (AIST) and Japan Electronics and Information Technology Industries Association: The ETL Character Database. http://etlcdb.db.aist.go.jp/

2. Bahler, L., DiCrescenzo, G., Polyakov, Y., Rohloff, K., Cousins, D.B.: Practical implementations of lattice-based program obfuscators for point functions. In: International Conference on High Performance Computing & Simulation, HPCS 2017 (2017)

3. Barak, B., et al.: On the (im)possibility of obfuscating programs. In: Kilian, J. (ed.) CRYPTO 2001. LNCS, vol. 2139, pp. 1–18. Springer, Heidelberg (2001). https://doi.org/10.1007/3-540-44647-8_1

4. Barak, B., et al.: On the (im)possibility of obfuscating programs. J. ACM **59**(2), 6 (2012)

5. Bellare, M., Stepanovs, I.: Point-function obfuscation: a framework and generic constructions. In: Kushilevitz, E., Malkin, T. (eds.) TCC 2016. LNCS, vol. 9563, pp. 565–594. Springer, Heidelberg (2016). https://doi.org/10.1007/978-3-662-49099-0_21

6. Canetti, R., Micciancio, D., Reingold, O.: Perfectly one-way probabilistic hash functions (preliminary version). In: Proceedings of 30th ACM STOC, pp. 131–140 (1998)

7. Cousins, D.B., et al.: Implementing conjunction obfuscation under entropic ring LWE. In: IEEE Symposium on Security and Privacy (SP), pp. 68–85 (2018)

8. DiCrescenzo, G., Bahler, L., Coan, B.A., Polyakov, Y., Rohloff, K., Cousins, D.B.: Practical implementations of program obfuscators for point functions. In: Proceedings of HPCS 2016, pp. 460–467 (2016)

9. DiCrescenzo, G., Bahler, L., Coan, B.A., Rohloff, K., Polyakov, Y.: Intrusion-resilient classifier approximation: from wildcard matching to range membership. In: Proceedings of IEEE TrustCom/BigDataSE (2018)

10. DiCrescenzo, G., Bahler, L., McIntosh, A.: Encrypted-input program obfuscation: simultaneous security against white box and black box attacks. In: 8th IEEE Conference on Communications and Network Security, CNS 2020 (2020)

11. Dodis, Y., Smith, A.D.: Correcting errors without leaking partial information. In: Proceedings of 37th ACM STOC, pp. 654–663. ACM (2005)

12. Fredrikson, M., Jha, S., Ristenpart, T.: Model inversion attacks that exploit confidence information and basic countermeasures. In: Proceedings of 22nd ACM SIGSAC Conference on Computer and Communications Security, pp. 1322–1333 (2015)

13. Galbraith, S.D., Zobernig, L.: Obfuscated fuzzy hamming distance and conjunctions from subset product problems. In: Hofheinz, D., Rosen, A. (eds.) TCC 2019. LNCS, vol. 11891, pp. 81–110. Springer, Cham (2019). https://doi.org/10.1007/978-3-030-36030-6_4

14. Hada, S.: Zero-knowledge and code obfuscation. In: Okamoto, T. (ed.) ASIACRYPT 2000. LNCS, vol. 1976, pp. 443–457. Springer, Heidelberg (2000). https://doi.org/10.1007/3-540-44448-3_34

15. Jolliffe, I.T.: Principal Component Analysis. Springer Series in Statistics, 2nd edn. Springer, New York (1986). https://doi.org/10.1007/978-1-4757-1904-8

16. LeCun, Y., Cortes, C., Burges, C.J.: The mnist database of handwritten digits. http://yann.lecun.com/exdb/mnist/

17. Lynn, B., Prabhakaran, M., Sahai, A.: Positive results and techniques for obfuscation. In: Cachin, C., Camenisch, J.L. (eds.) EUROCRYPT 2004. LNCS, vol. 3027, pp. 20–39. Springer, Heidelberg (2004). https://doi.org/10.1007/978-3-540-24676-3_2

18. Mowforth, P., Shepherd, B.: Statlog (vehicle silhouettes) data set. https://archive.ics.uci.edu/ml/datasets/Statlog+%28Vehicle+Silhouettes%29

19. Shokri, R., Stronati, M., Song, C., Shmatikov, V.: Membership inference attacks against machine learning models. In: IEEE Symposium on Security and Privacy, SP 2017, pp. 3–18 (2017)

20. Song, L., Shokri, R., Mittal, P.: Membership inference attacks against adversarially robust deep learning models. In: 2019 IEEE Security and Privacy Workshops, SP Workshops, pp. 50–56 (2019)

21. Souri, H., Dhraief, A., Tlili, S., Drira, K., Belghith, A.: Smart metering privacy-preserving techniques in a nutshell. In: Proceedings of 5th ANT and 4th SEIT. Procedia Comput. Sci. **32** (2014)

22. Wee, H.: On obfuscating point functions. In: 2005 Proceedings of 37th ACM STOC, pp. 523–532 (2005)

23. Wichs, D., Zirdelis, G.: Obfuscating compute-and-compare programs under LWE. In: Proceedings of 58th IEEE FOCS 2017, pp. 600–611 (2017)

24. Yao, A.C.: How to generate and exchange secrets (extended abstract). In: Proceedings of 27th IEEE FOCS 1986, pp. 162–167 (1986)

Preserving Privacy of Co-occurring Keywords over Encrypted Data

D. V. N. Siva Kumar[1]([✉]) and P. Santhi Thilagam[2]

[1] Department of CS&IS, BITS Pilani, Hyderabad Campus, Hyderabad, India
sivakumar@hyderabad.bits-pilani.ac.in
[2] Department of CSE, National Institute of Technology Karnataka, Surathkal,
Mangalore, India
santhi@nitk.edu.in

Abstract. The indexes of ranked searchable encryption contain encrypted keywords and their encrypted relevance scores. The encryption scheme of relevance scores must preserve the plaintext order after encryption so as to enable the cloud server to determine ranks of the documents directly from the encrypted keywords' scores for a given trapdoor. Existing schemes such as Order Preserving Encryption (OPE) and One-to-Many OPE preserve the plaintext order. However, they leak the distribution information, i.e., the frequency of ciphertext values, due to the insufficient randomness employed in these schemes. The cloud server uses frequency analysis attack to infer plaintext keywords of the indexes based on the frequency leakage. In this paper, an Enhanced One-to-Many OPE scheme is proposed to minimize the frequency leakage. The proposed scheme reduces not only the frequency leakage of individual keywords but also the co-occurring keywords of the phrases like "computer network", and "communication network".

Keywords: OPE · Frequency leakage · Index keywords' confidentiality

1 Introduction

In spite of numeric benefits with cloud computing services, privacy, and confidentiality are significant concerns for the data owners who store sensitive documents at the cloud servers. The stored documents are accessible to the snooping administrators, who could misuse the sensitive information for their own benefits. Therefore, storing documents in plaintext form at cloud servers poses a threat to the confidentiality of data owners' sensitive information. Although the encryption guarantees confidentiality, it makes the retrieval process more complicated. A cryptographic paradigm called Searchable Encryption (SE) facilitates searching over encrypted documents by uploading encrypted indexes, i.e., searchable indexes corresponding to the uploaded encrypted documents.

The information in searchable indexes includes unique keywords of the dataset and their corresponding keywords' relevance scores, e.g., Term Frequency

K. Barker and K. Ghazinour (Eds.): DBSec 2021, LNCS 12840, pp. 157–168, 2021.
https://doi.org/10.1007/978-3-030-81242-3_9

(TF), and Term Frequency-Inverse Document Frequency (TF-IDF), which convey keywords' distribution information of the documents and the dataset, respectively. This information should not be directly exposed to the cloud servers. Hence, these relevance scores must be encrypted alongside the index keywords. The relevance scores should be encrypted so that the cloud server could still perform rank-ordering of the documents directly from the keywords' encrypted relevance scores. The encryption scheme must maintain the plaintext order of relevance scores after their encryption to meet this requirement. Various encryption schemes exist such as Order Preservation Encryption (OPE) [2], and One-to-Many OPE [17]. However, they leak frequency information due to the insufficient randomness in mapping plaintext scores to encrypted values. The frequency information is leaked especially when two or more keywords contain the same plaintext score in the same document.

Various attacks such as correlation attack [1] and frequency analysis attack [3, 11] exploit frequency information from the encrypted scores to infer index keywords of the dataset. Hence, it is required to minimize or prevent the frequency information leakage in order to use it for encrypting sensitive information. There are many schemes like Paillier encryption for preventing frequency information leakage, but they do not preserve plaintext order due to which the cloud server cannot perform rank-ordering of the documents. As a result, only OPE schemes should be used but they should not leak frequency information. An Enhanced One-to-Many OPE scheme is proposed in this paper, which reduces the frequency leakage of individual keywords and also the co-occurring keywords of the dataset. The summary of our contributions are provided below.

- An Enhanced One-to-Many OPE scheme is proposed to return a ciphertext value for a given plaintext relevance score.
- A thorough analysis of the proposed approach and its demonstration that it is safer in frequency leakage than the conventional OPE and One-to-Many OPE schemes.

The rest of the paper is presented as follows: The preliminary details are presented in Sect. 2. Section 3 presents the related work, followed by the proposed approach in Sect. 4, results and discussion in Sect. 5, and the conclusion in Sect. 6.

2   Preliminaries

(a) Term Frequency (TF): The TF value of a keyword captures its relevance in the corresponding document. Another measure, TF-IDF, could also be used, but TF is more suitable for demonstrating the frequency leakage. The normalized TF value a of keyword kw of a document d_i can be measured using the below equation.

$$TF(kw, d_i) = \frac{1}{|d_i|}(1 + \log(tf_i)) \tag{1}$$

where, $|d_i|$ denotes the length of the document, i.e., the total number of unique keywords of the document d_i and tf_i denotes the number of a keyword kw occurs in document d_i.

(b) Frequency Analysis Attack: This attack aims to deduce the plaintext index keyword from the encrypted TF values of the encrypted index [7,11]. It is based on the premise that the distribution details of plaintext TF values and the corresponding encrypted TF values will be the same for some specific keywords. The cloud server infers frequently occurring keywords of the dataset from the encrypted index by relating the frequency of the same encrypted TF values to the frequency of plaintext TF values of frequently occurring keywords of the publicly available datasets. Sometimes, index keywords also are guessed by observing the frequency information of some encrypted TF values. This attack is more likely to succeed with the help of background knowledge of the dataset, which could be about what data the data owners stored at the cloud server, and what probable keywords that are most likely to occur. For example, assume that the data owners store Request for Comments [13] dataset in encrypted form at the cloud server. This dataset is all about how information can be transmitted from one host to another over the internet using different "computer network" protocols. It can be assumed that some of the specific keywords like "computer", "network" and "communication" may appear in most of the documents of this dataset. The cloud server then infers one or all of these specific keywords by generating frequency histograms (i.e., showing distribution information) for all encrypted TF values of each encrypted index keyword. The cloud server then examines the histograms and distinguishes those of a much higher frequency (i.e., same encrypted TF values) than the others. The corresponding encrypted keywords of those histograms are noted. The encrypted keywords of histograms may be related to one or all of those specific keywords of the dataset. The cloud server thus infers plaintext index keywords by observing the frequency information of encrypted TF values.

(c) Coordinate Matching: It is a similarity metric that uses the inclusion of query keywords in the document to assess the document's relevance score [18]. The score of a document for a trapdoor can be determined by adding the encrypted TF values of each trapdoor's keywords in that document.

3   Related Work

Order Preserving Encryption (OPE) [2] and Order Revealing Encryption (ORE) [4,8] support ranked search as they preserve plaintext order after encryption. The problem with ORE schemes is the output of these schemes is not numerical, due to which an additional public function is required that will let the cloud server know the order of the given ciphertexts. Swaminathan et al. [16] proposed an OPE scheme based approach, but it leaks frequency information, i.e., returns the same ciphertext value for the same plaintext value. Wang et al. [17] proposed using the One-to-Many OPE scheme, an extension of the OPE scheme, but leaks frequency information when two or more keywords contain the same TF values in the same document's index.

Kerschbaum proposed a frequency hiding OPE scheme [6] to conceal the frequency information. However, this approach still leaks frequency information in the form of the tree's linear depth, where each new ciphertext value is stored in a new node for the same plaintext value. This depth precisely reflects the frequency of plaintext scores. Several works like Grubbs et al. [5] and Maffei et al. [9] demonstrated the inference of plaintext information from the encrypted values of frequency-hiding OPE scheme.

Orencik et al. [12] proposed a multi-keyword ranked SE scheme using TF-IDF and Forward indexing techniques. They used the Paillier Encryption (PE) scheme to encrypt TF-IDF values. A fully homomorphic Encryption (FHE) and a partially homomorphic encryption (PHE) [14] also can be used in ranked search approaches and for encrypting TF-IDF values. The PE, FHE and PHE prevent the frequency leakage, but they do not preserve the plaintext order after the encryption. Roche et al. [15] proposed a partial order preserving encryption (POPE) approach for encrypting keywords' relevance scores. However, all the ciphertext values in this approach do not preserve the plaintext order, and it is maintained only for some relevance scores. Hence, this approach is also not completely suitable for ranked search approaches.

4 Proposed Approach

4.1 Objectives

It is aimed to meet the following two objectives:

1. *Relevance Score Encryption Scheme:* A novel approach is designed based on the existing OPE schemes to minimize frequency leakage.
2. *Efficiency:* The proposed approach should encrypt the keyword's relevance scores efficiently as the existing approaches.

4.2 Proposed Methodology

It consists of the Initialization Phase and the Retrieval Phase.

Initialization Phase: It includes the following activities to be done by data owner, who owns the dataset:

1. *Build the Encrypted Index:* The steps required to generate the encrypted index $\tilde{I}$ for all the documents of dataset D is explained as follows:
    (a) *Building Dictionary, W:* Construct the dictionary W by extracting all the unique keywords kw from the input file dataset D.
    (b) *Building Plaintext Index I:* For each keyword kw in W, determine its TF value (1) for each document D_i if it is present in D_i and store it as $I[kw] = [D_i][TF]$. Otherwise, set its TF value to 0, i.e., $I[kw] = [D_i][0]$.
    (c) *Generate Encrypted Index, $\tilde{I}$:* The generated index I is encrypted as follows:

- Each keyword in I is hashed by using a one-way secure hash algorithm SHA-2 with a 256-bit key.
- The TF value of each keyword kw of every document D_i is encrypted using the proposed Enhanced One-to-Many OPE scheme, which is explained in Sect. 4.3.
- The document Ids are not necessary to encrypt as they do not convey any information. The encrypted index generated would then be: $\widetilde{I} = [\widetilde{TF}][D_i]$.

2. *Encrypting Documents:* After generating the encrypted index, the data owner also encrypts his/her documents D using AES algorithm with 128 bit key size.

Retrieval Phase: It includes the following activities to be done by a data user, who will be allowed by the data owner to retrieve documents of his/her interest.

1. *Query_Masking:* The masked query $\widetilde{M_{Q_{kw}}}$ is generated by hashing each keyword of his/her query Q_{kw} using a secure SHA-256 hash function. Then, the data user sends the masked query and the parameter k to the cloud server to retrieve only the relevant top-k documents.
2. *Searching:*
   (a) The cloud server then utilizes the encrypted index $\widetilde{I}$ and adopt the coordinate matching similarity measure [18] to assign the scores to each document D_{id}. Then it sorts the scores of the documents in descending order and sends the top-k of them to the user.
   (b) The data user then uses the corresponding secret key shared by the data owner to decrypt the retrieved documents.

4.3   Enhanced One-to-Many OPE

Algorithm 1: Enhanced One-to-Many OPE

Input: K (Key), D (domain), R (range), *pscore, id(doc), kw*
Output: Cipher text c
1 **while** $|D|! = 1$ **do**
2 $\quad \lfloor \ \{D, R\} = Binary\_Search(K, D, R, pscore)$
3 coin $\leftarrow$ TapeGen(K,$(D,R,1||pscore, id(doc), kw$)
4 $c \xleftarrow{coin} R$
5 **return** c

The proposed Enhanced One-to-Many OPE scheme returns a possible unique ciphertext value c for each keyword kw's TF value, i.e., a plaintext relevance score *pscore* by mapping it to one of the output range of values. The mapping procedure is explained in Algorithm 1. It takes Key (K), input domain (D), output range (R), *pscore*, document identity *id(doc)*, and keyword (kw) and returns a possible unique ciphertext value c. During mapping, the range R

is divided into some non-overlapping interval buckets each with different size. Each bucket contains some range of values. For the given *pscore*, the random-sized bucket is determined using a Binary_Search(.) procedure, which is explained in Algorithm 2. Binary_Search(.) is a recursive procedure, which returns a new domain and a new range of values for a given *pscore* based on an HYGEINV(.) function. HYGEINV(.) is a hypergeometric sampling process that returns an integer value based on the initial domain, range, and middle value. This integer helps Binary_Search(.) in choosing a new domain, and a new range of values for the given *pscore*. In each iteration of binary search, the size of domain D and range R will be reduced by an integer value. Binary_Search(.) stops when the size of the domain becomes 1 where it contains only the given *pscore*. The *pscore* then will be mapped to one of the values in a new range R using a TapeGen(.) function, which is a random coin generator, which generates the seed value. This seed value helps in choosing one of the values in the new range as the ciphertext value c for the given *pscore*.

Algorithm 2: Binary_Search

   Input: K, D, R, *pscore*
   Output: D,R
1 $M \leftarrow length(D)$; $N \leftarrow length(R)$
2 $d \leftarrow min(D) - 1$; $r \leftarrow min(R) - 1$
3 $y \leftarrow r + \text{ceil}(\frac{N}{2})$
4 $coin \xleftarrow{R} \text{TapeGen}(K,(D,R,0\|y))$
5 $x \xleftarrow{R} d + \text{HYGEINV}(coin,M,N,y-r)$
6 **if** $pscore \leq x$ **then**
7 | $D \leftarrow \{d+1,....,x\}$
8 | $R \leftarrow \{r+1,....,y\}$
9 **else**
10 | $D \leftarrow \{x+1,....,d+M\}$
11 | $R \leftarrow \{y+1,....,r+N\}$
12 **return** D,R

---

5   Results and Discussion

The proposed approach has been implemented using Python 3.6 version and tested on Intel i7-4770 CPU system. The experiments are conducted on a Requests for Comments (RFC) [13] dataset. The effectiveness of the proposed approach is compared in terms frequency leakage with the OPE and One-to-Many OPE schemes for different keywords and phrases of the dataset.

5.1   Frequency Analysis Attack

The proposed Enhanced One-to-Many OPE scheme is an extension of One-to-Many OPE [17], which in turn is an extension of OPE [2]. In our experiments, the input domain is the actual plaintext relevance scores, and the output range is set between 0 and $2^{45} - 1$. In OPE, the plaintext score $pscore$, i.e., the TF value of the keyword is mapped to a ciphertext, which is a value within the new range, R (also can be treated as a bucket), which is determined by using the binary search algorithm 2. The selection of ciphertext value within the bucket values for the given $pscore$ is based on the seed value generated by the TapeGen(.). In OPE, this seed value is entirely dependent on the plaintext score $pscore$ due to which the same $pscore$ is mapped to the same ciphertext value. For the given dataset, if the plaintext score is repeated n number of times, then there will be a ciphertext c that will also be repeated n number of times. Thus, frequency information is leaked from OPE encrypted values.

From the RFC dataset, it is found that the keywords "computer", "network", and "communication" are the frequently occurring keywords. The plaintext distribution information of these three keywords is respectively shown in figures Fig. 1(a), 1(b) and 1(c). The values on the x-axis represent the actual plaintext TF values and values on y-axis represent the frequency, i.e., the number of points having the same TF value. As the OPE maps the same plaintext score to the same ciphertext, the distribution of ciphertext values of these keywords would be the same. To infer the plaintext keyword, the cloud server plots the graphs for each encrypted keyword's relevance scores in index. Then, it will note down the encrypted or masked keyword of the index for which the frequency (the repetition of the same encrypted score) is higher than the frequency of scores of other encrypted keywords. This encrypted keyword is more likely to be the frequently occurring keyword, e.g., "computer". Thus, frequency leakage from OPE scores allows the cloud server to deduce plaintext keyword.

In One-to-Many OPE scheme [17], for the given plaintext score $pscore$, it uses both the $pscore$, and document identity $id(doc)$ for generating a different seed for the same $pscore$. Due to this seed, it maps the same plaintext to different ciphertext value within the values of the bucket. There is a scope for frequency leakage with this scheme when two or more keywords have the same score in the same document. This is possible especially when some keywords equally co-occur in the same document, e.g., the keywords of the phrases like { "computer", "network"} or { "communication", "network"}. The cloud server first plots the histogram for each phrase and then identifies the histogram of the phrase in which the frequency of relevance score is higher than the frequency of other possible phrases. The cloud server notes the histogram of this phrase, and the encrypted keywords of this histogram are more likely to be the most co-occurring keywords of the phrases. Thus, the cloud server could infer the phrases' plaintext keywords from the One-to-Many OPE encrypted scores.

The proposed Enhanced One-to-Many OPE scheme, explained in Sect. 4.3, minimizes the frequency leakage of the phrases caused by the One-to-Many-OPE scheme. This approach also uses the same algorithm 2 (Binary search)

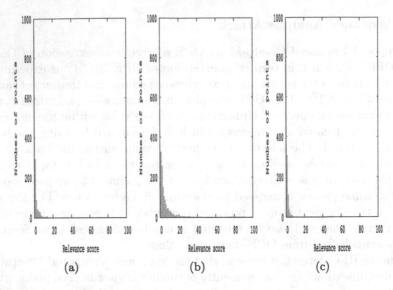

Fig. 1. Plaintext keyword relevance score distribution for keywords: (a) computer (b) network (c) communication

to select the bucket, but the TapeGen(.) here uses *pscore*, document identity, *id*(*doc*) and keyword *kw* for generating a different seed value for the same plaintext score *pscore*. Due to this seed, the same *pscore* of co-occurring keywords will be mapped to different values within the values of the bucket. Thus, it minimizes the frequency leakage of co-occurring keywords of the phrases. This minimization is due to the improvement of the randomness in generating the seed value. Due to this seed, this approach reduces not only the frequency leakage of phrases but also the individual keyword's. To demonstrate the reduction of frequency leakage for an individual keyword, we compare the frequency leakage of One-to-Many OPE encrypted scores, and the proposed Enhanced One-to-Many OPE encrypted scores for the keyword "computer". The distribution information (frequency) of One-to-Many encrypted scores and the proposed Enhanced One-to-many encrypted scores for the keyword "computer" is shown in Fig. 2(a) and 2(b) respectively. The values on x-axis represent the normalized encrypted scores using the min-max normalization approach [10]. The values of the y-axis represent information about the frequency, i.e., the number of points having the same encrypted TF value. Figure 2(b) demonstrates that the frequency leakage for some encrypted scores of the keyword "computer" using the proposed Enhanced One-to-Many OPE scheme is lower than the frequency leakage of One-to-Many OPE scores, shown in Fig. 2(a). As it is already mentioned, the proposed scheme also minimizes the frequency leakage of phrases due to the usage of plaintext score *pscore*, document identity *id*(**doc**) and keyword *kw* for generating a different seed value. Due to this seed, different value within the bucket will be chosen as the ciphertext value for the same plaintext relevance score *pscore*. In RFC dataset, "computer network" and "communication network" are the most

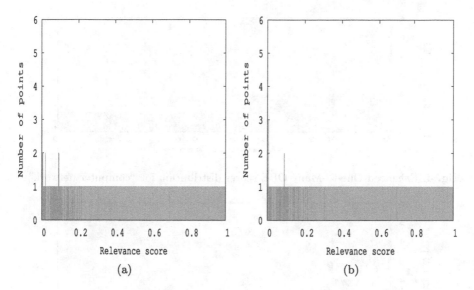

Fig. 2. Encrypted score distribution for "computer": (a) One-to-Many OPE (b) Enhanced One-to-Many OPE

occurring phrases. The proposed Enhanced One-to-Many OPE scheme's frequency leakage is compared with the One-to-Many-OPE scheme to demonstrate the reduction of frequency leakage for the phrases. The distribution information of One-to-Many-OPE encrypted scores and the proposed Enhanced One-to-Many-OPE encrypted scores for the phrase "computer network" is shown in figures Fig. 3 and 4, respectively. It can be observed from Fig. 4 that the frequency leakage of the proposed approach is much lesser than the frequency leakage of the One-to-Many OPE scheme, shown in Fig. 3. Similarly, Fig. 5 and 6 respectively show the distribution information for the phrase "communication network". Figure 6 shows that the frequency leakage of the proposed approach is much lesser than the frequency leakage of the One-to-Many OPE scheme, shown in Fig. 5. Thus, the proposed approach makes it difficult for the cloud server to infer plaintext keywords of the phrases from the Enhanced One-to-Many OPE scores.

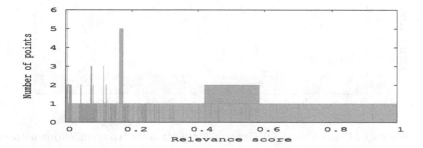

Fig. 3. One-to-Many OPE scores distribution for "computer network"

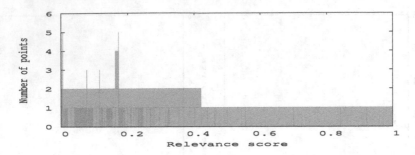

Fig. 4. Enhanced-One-to-Many OPE scores distribution for "computer network"

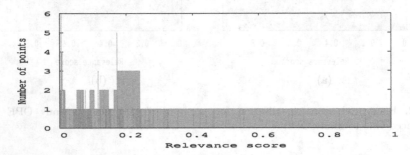

Fig. 5. One-to-Many OPE scores distribution for "communication network"

5.2 Efficiency

The time complexity of mapping a plaintext score *pscore* to a ciphertext *c* is $O(log\ n)$, i.e., $log\ n$ times the Binary_Search(.) process will be called during the mapping. The time cost comparison of mapping a plaintext score, *pscore* to a ciphertext value, *c* using the proposed Enhanced One-to-Many OPE and One-to-Many OPE schemes is shown in Fig. 7. The x-axis values represent domain size and the values on y-axis represent the amount of time taken to map *pscore* to *c*. Y-axis represents the average time of 100 trails for a single mapping operation

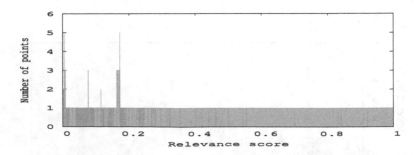

Fig. 6. Enhanced One-to-Many OPE scores distribution for "communication network"

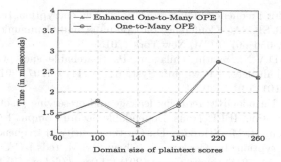

Fig. 7. The time cost comparison of single mapping operation using Enhanced One-to-Many OPE and One-to-Many OPE over a range $|R| = 2^{45}$.

over different domain sizes and a range $|R| = 2^{45}$. From the Fig. 7, it can be observed that efficiency of both the schemes is almost same since both of them call the Binary_Search procedure recursively same number of times.

6 Conclusion

In this paper, an Enhanced One-to-Many OPE scheme is proposed that maps the same plaintext relevance score to a different ciphertext value even when multiple keywords have the same plaintext score within the same document. The proposed scheme may prevent complete frequency leakage if it is used for encrypting information that is moderately distributed. Our future work would be to extend the proposed scheme to minimize the frequency leakage further with the help of constrained random numbers as the seed value in mapping process.

References

1. Bindschaedler, V., Grubbs, P., Cash, D., Ristenpart, T., Shmatikov, V.: The Tao of inference in privacy-protected databases. Proc. VLDB Endow. **11**(11), 1715–1728 (2018)
2. Boldyreva, A., Chenette, N., Lee, Y., O'Neill, A.: Order-preserving symmetric encryption. In: Joux, A. (ed.) EUROCRYPT 2009. LNCS, vol. 5479, pp. 224–241. Springer, Heidelberg (2009). https://doi.org/10.1007/978-3-642-01001-9_13
3. Cash, D., Grubbs, P., Perry, J., Ristenpart, T.: Leakage-abuse attacks against searchable encryption. In: Proceedings of the 22nd ACM SIGSAC Conference on Computer and Communications Security, CCS 2015, pp. 668–679. ACM, New York (2015). https://doi.org/10.1145/2810103.2813700
4. Chenette, N., Lewi, K., Weis, S.A., Wu, D.J.: Practical order-revealing encryption with limited leakage. In: Peyrin, T. (ed.) FSE 2016. LNCS, vol. 9783, pp. 474–493. Springer, Heidelberg (2016). https://doi.org/10.1007/978-3-662-52993-5_24
5. Grubbs, P., Sekniqi, K., Bindschaedler, V., Naveed, M., Ristenpart, T.: Leakage-abuse attacks against order-revealing encryption. In: 2017 IEEE Symposium on Security and Privacy (SP), pp. 655–672, May 2017

6. Kerschbaum, F.: Frequency-hiding order-preserving encryption. In: Proceedings of the 22nd ACM SIGSAC Conference on Computer and Communications Security, CCS 2015, pp. 656–667. ACM, New York (2015)
7. Siva Kumar, D.V.N., Santhi Thilagam, P.: Searchable encryption approaches: attacks and challenges. Knowl. Inf. Syst. **61**(3), 1179–1207 (2018). https://doi.org/10.1007/s10115-018-1309-4
8. Liu, Z., et al.: EncodeORE: reducing leakage and preserving practicality in order-revealing encryption. IEEE Trans. Dependable Secure Comput 1 (2020)
9. Maffei, M., Reinert, M., Schröder, D.: On the security of frequency-hiding order-preserving encryption. In: Capkun, S., Chow, S.S.M. (eds.) CANS 2017. LNCS, vol. 11261, pp. 51–70. Springer, Cham (2018). https://doi.org/10.1007/978-3-030-02641-7_3
10. Margae, S.E., Sanae, B., Mounir, A.K., Youssef, F.: Traffic sign recognition based on multi-block LBP features using SVM with normalization. In: 2014 9th International Conference on Intelligent Systems: Theories and Applications (SITA-2014), pp. 1–7, May 2014. https://doi.org/10.1109/SITA.2014.6847283
11. Naveed, M., Kamara, S., Wright, C.V.: Inference attacks on property-preserving encrypted databases. In: Proceedings of the 22nd ACM SIGSAC Conference on Computer and Communications Security, CCS 2015, pp. 644–655. ACM, New York (2015)
12. Orencik, C., Kantarcioglu, M., Savas, E.: A practical and secure multi-keyword search method over encrypted cloud data. In: 2013 IEEE Sixth International Conference on Cloud Computing, pp. 390–397, June 2013
13. RFC: Request for comments database (2016). https://www.rfc-editor.org/retrieve/bulk/
14. Ristic, M., Noack, B., Hanebeck, U.D.: Secure fast covariance intersection using partially homomorphic and order revealing encryption schemes. IEEE Control Syst. Lett. **5**(1), 217–222 (2021). https://doi.org/10.1109/LCSYS.2020.3000649
15. Roche, D.S., Apon, D., Choi, S.G., Yerukhimovich, A.: POPE: partial order preserving encoding. In: Proceedings of the 2016 ACM SIGSAC Conference on Computer and Communications Security, CCS 2016, pp. 1131–1142. ACM, New York (2016)
16. Swaminathan, A., et al.: Confidentiality-preserving rank-ordered search. In: Proceedings of the 2007 ACM Workshop on Storage Security and Survivability, pp. 7–12. ACM (2007)
17. Wang, C., Cao, N., Ren, K., Lou, W.: Enabling secure and efficient ranked keyword search over outsourced cloud data. IEEE Trans. Parallel Distrib. Syst. **23**(8), 1467–1479 (2012)
18. Witten, I.H., Bell, T.C., Moffat, A.: Managing Gigabytes: Compressing and Indexing Documents and Images, 1st edn. Wiley, New York (1994)

Machine Learning

Access Control Policy Generation from User Stories Using Machine Learning

John Heaps[1]([✉]), Ram Krishnan[2], Yufei Huang[3], Jianwei Niu[1], and Ravi Sandhu[1]

[1] Institute for Cyber Security (ICS), NSF Center for Security and Privacy Enhanced Cloud Computing (C-SPECC), and Department of Computer Science, The University of Texas at San Antonio, San Antonio, USA
{john.heaps,jianwei.niu,ravi.sandhu}@utsa.edu
[2] ICS, C-SPECC, and Department of Electrical and Computer Engineering, The University of Texas at San Antonio, San Antonio, USA
ram.krishnan@utsa.edu
[3] ICS and Department of Electrical and Computer Engineering, The University of Texas at San Antonio, San Antonio, USA
yufei.huang@utsa.edu

Abstract. Agile software development methodology involves developing code incrementally and iteratively from a set of evolving user stories. Since software developers use user stories to write code, these user stories are better representations of the actual code than that of the high-level product documentation. In this paper, we develop an automated approach using machine learning to generate access control information from a set of user stories that describe the behavior of the software product in question. This is an initial step to automatically produce access control specifications and perform automated security review of a system with minimal human involvement. Our approach takes a set of user stories as input to a transformers-based deep learning model, which classifies if each user story contains access control information. It then identifies the actors, data objects, and operations the user story contains in a named entity recognition task. Finally, it determines the type of access between the identified actors, data objects, and operations through a classification prediction. This information can then be used to construct access control documentation and information useful to stakeholders for assistance during access control engineering, development, and review.

Keywords: Access control · Software engineering · Agile development · User stories · Machine learning · Deep learning

1 Introduction

Agile development has gained great popularity in recent years among software development teams. It is able to rapidly produce and update software, and easily

K. Barker and K. Ghazinour (Eds.): DBSec 2021, LNCS 12840, pp. 171–188, 2021.
https://doi.org/10.1007/978-3-030-81242-3_10

react to changes in requirements. However, it has been observed [2, 4, 14, 20] that agile development practices often facilitate the introduction and propagation of access control and other security vulnerabilities. Some such practices include the constant changes in code and requirements which drastically limits the ability to perform security assurance review; frequent code refactoring, changes in functional requirements, and modifications to system design which have a tendency to break security constraints of previously implemented functionality; and the necessity of continuously delivering development iterations on time as well as a push for developing software as quickly as possible often take precedence over time-consuming security assurance activities.

One of the most notable reasons for the proliferation of access control and other security vulnerabilities is that agile development discourages producing comprehensive documentation about the software to be developed [9]. That is, it is normal that no security or access control policy is defined for the software before development begins. A primary reason for this is that agile development allows for changing requirements during development, unlike traditional development models. Since documentation is primarily derived from software requirements, a change in requirements leads to a review and update of all documentation to reflect those changes. Since requirements can change frequently throughout the agile development process, the act of constantly updating development documentation becomes time consuming and burdensome. Agile development elects to forego this work in the interest of time, and instead relies on the expertise of developers to make on-the-fly decisions during development. By pushing this responsibility on the developers and maintaining little to no documentation on decisions made throughout the development process, many opportunities for security vulnerabilities to arise are created.

There have been attempts [3, 5, 15, 17] to address how agile development may be modified to mitigate security issues and bugs by suggesting more documentation be produced by stakeholders. This documentation often focuses on creating and maintaining additional information for user stories, relieving developers from making on-the-fly security decisions. However, the creation and maintenance of the additional user story documentation still requires more time and labor than most development teams are willing to spend.

To help stakeholders mitigate the propagation of access control security vulnerabilities under the agile development model, we propose an automated approach to produce additional documentation so stakeholders have a more holistic view and common understanding of the access control of the software to be developed. Further, it will identify user stories with high ambiguity and allow stakeholders to refine user stories throughout the software development process. In this initial work, we will automatically identify and extract access control information from user stories and then visualize the access control information to stakeholders. This will also give product owners an overview of the access control so they may confirm or indicate changes to it. Ultimately, this approach will relieve developers from making most on-the-fly decisions, help reduce bugs and security vulnerabilities that may be overlooked, save time and money, and better protect the product owner and end users' information.

This initial work will focus specifically on the extraction and presentation of access control information from user stories. This only relates access control information to actors, or users, of the software. Further, due to the ambiguity and limited context of user stories and natural language, in order to determine the exact access control of the system some human involvement is necessary. Ongoing work will incorporate active learning and human interactivity to best refine the access control model.

The contributions we present in the paper are as follows:

- A dataset of over 1600 user stories, labeled for three separate learning tasks related to the extraction of access control information.
- A transformer-based learning model that categorizes user stories into "contains access control information", "does not contain access control information", and "too ambiguous to determine if access control is present".
- A transformer-based learning model that performs a named entity recognition task that predicts if a word in a user story is an "actor" (or end user), a "data object", or an "operation" of the system to be built.
- A transformer-based learning model that predicts the type of access an actor has for a data object present in a user story.

The rest of the paper is organized as: Sect. 2 presents background information on user stories and related work; Sect. 3 describes our data and model on extracting access control information from user stories; Sect. 4 presents the results of our work; and Sect. 5 discusses the conclusion and future work.

2 Background

2.1 User Stories

The agile software development model was conceived in 2001 through a set of tenets and principles [9]. It broke from traditional software models in many ways, but primarily it allowed for changing software requirements throughout the software development process, discouraged spending time on comprehensive documentation, and focused on producing software at regular intervals and as fast as possible. Agile development quickly became popular as it was able to support the ever-faster changing environment of industry.

The only documentation that agile development requires are user stories. User stories define the requirements of the software to be built, and are often written by product owners (rather than software developers) usually making them more ambiguous and abstract than traditional software requirements. They help define software requirements by describing how actors, or end users, will interact with the system. They relate an actor of the system to system data objects and operations. A simple example of a user story would be, "As a system administrator, I want to create user accounts, so that new employees can use the system." The actor of this user story is the "system administrator" and is being related to the "user account" data object. From an access control perspective, this relation would be "create".

```
"As an <ACTOR>, I want to <PERFORM SOME ACTION>, so that <A PURPOSE IS FULFILLED>"

"In order to <FULFILL A PURPOSE> as an <ACTOR>, I want to <PERFORM AN ACTION>."

"As an <ACTOR> <AT SOME TIME> <IN SOME PLACE>, I <PERFORM AN ACTION> for <A PURPOSE>."
```

Fig. 1. Different formats, or templates, for defining user stories.

While not necessary, user stories are often written in some specified format, as shown in Fig. 1. It is expected that the set of user stories changes during development based on the changing needs of the product owner causing different functionality to be integrated into the system with user stories added, deleted, or modified accordingly. Such changes to user stories cause a change to the requirements and access control of the system.

2.2 Related Work

Extracting Information from User Stories. In all previous literature over user stories we reviewed [11,12,19], heuristics and rule-based approaches were used to parse and extract information from user stories as they are written in a similar format. In many datasets that these related works used, it was reported that most of the user stories conformed to defined templates. However, those datasets were not made available as they were proprietary data from industry partners, and in the only public dataset we found almost half the user stories did not conform to defined templates. Heuristics or rule-based approaches would fail for such user stories. We further observed that such approaches were limited in how much information they could extract from user stories. In all cases, such approaches was not able to identify specific data objects or operations.

In the work by Sobieski and Zielinski [19], the authors create a new constrained natural language user story format called "mixfit" that is designed to mark, or qualify, the important words or phrases of a user story (e.g., the actor, operation, data object, etc.). With the important items marked, they can be more easily extracted and utilized when analyzing user stories. However, this requires that stakeholders learn and strictly adhere to the mixfit format, which requires time and training that certain stakeholders (e.g., product owners) cannot invest in.

In the works by Lucassen et al. [12], the authors propose a quality user story framework, which lists criteria on what makes a good user story and what stakeholders should strive for when writing user stories. They also propose a heuristic approach for extracting conceptual models from user stories. However, the model was only tested on user stories that strictly followed user story templates, which is not common in real world datasets. Further, only general requirements information was identified, and was not fine-grained enough to identify unique data objects.

Extracting Access Control Information from Natural Language. These works [1,10,13,18,22] are most closely related to our current research, and focuses on the extraction of access control (or other security information) from a natural language security policy. This work assumes that a natural language policy describing the system's security requirements exists, is complete, and is mostly unambiguous. The general state-of-the-art approach is to perform pre-processing on the natural language text and then use machine learning techniques to extract or analyze security requirements detailed by the policy. However, in the context of agile development, we are not able to make the assumption that such a security policy exists, as the only documentation required by the agile development model is user stories. Further, user stories are often not complete (changing often throughout the development cycle) and are usually purposefully ambiguous as very few design choices are defined before implementation begins. Since agile development has become one of the most popular development models we believe that in many real-world software development scenarios it is not reasonable to assume a security policy exists and instead that only user stories are available as that is what developers more often actually work with.

Because of the ambiguity of user stories, most of the access control information in them is not explicitly stated, having to infer or predict how the actors, data objects, and operations in user stories imply different kinds of access. Further, all user stories are written from the system actors perspective. While actors are related to access control roles or subjects, they do not necessarily maintain a one-to-one translation from the end user of the system to the different types of user roles or subjects within the system. This is different from a natural language policy which is assumed to contain most, if not all, role and subject information.

Further, most of the current work pre-processes the policy text into other formats (such as dependency graphs) and learns patterns or rules based on these alternative formats. However, our approach uses and learns the text directly without the need to convert it. As far as we know, we are also the first to apply transformers to the access control information extraction process.

In the work by Slankas et al. [18], the authors extract access control rules from access control statements. This is achieved through parsing a natural language sentence into a dependency graph and then using machine learning techniques to learn patterns in the graph to identify access control rules. However, this approach has a primary basis in learning grammatical patterns associated with a dependency graph. This causes the approach to identify all instances of the learned patterns as access control rules, but there are many cases where a learned pattern occurs that is not related to access control. The paper attempts to statistically separate which identified patterns are more or less likely to be access control rules using a threshold value. Our approach does not learn such patterns and instead directly predicts on the sentences themselves.

In the work by Alohaly et al. [1], the authors propose to extract subject and object attributes for attribute-base access control from natural language policy. This is achieved by parsing a sentence into a dependency tree based on its grammatical structure and then using convolutional neural networks between grammar relations to predict if a word is an attribute. However, due to a lack of available data, they create a synthetic corpus by artificially injecting attribute information into role-based access control policies; meaning both the corpus injection and machine learning is based on the grammar of sentences.

```
As a camp administrator, I want to be able to add
parents, so that they can enroll their kids at camp.

As an authenticated user, I want to see specific
details on summits, so that I can learn more about
the summit I'm interested in to see if it matches
my interests, register for the event, and get day-
of knowledge to help me get to the location.
```

Fig. 2. A simple and complex example of a user story from the dataset.

3   Approach

We have chosen to use a deep learning approach to be more robust and generalizable in our ability to handle many different formats of user stories. The model identifies general access control information between users (or actors) and the data objects or operations of the system. It is most easily applied to Role-Based or Attribute-Based Access Control. While the access control information can be used in constructing other access control models, it would be difficult to do so without additional context since user stories are written from the actor perspective. The following sections describe the data, labeling process, and model for our approach, which we have made available to the public[1].

3.1   Data

While searching for a dataset of user stories, we found that most papers in the literature use proprietary datasets and were not allowed to publish their dataset along with their work. This unfortunately has revealed a great need for robust datasets of user stories for research involving agile development. We were able to find one dataset published by Dalpiaz et al. [6,7].

The dataset consists of 21 web apps, each with 50–130 user stories for a total of over 1600 user stories. The apps are from various domains, including financial management, medical care, administrative management, and more. The most common types of app were data management platforms, accounting for 12 of the apps in the dataset.

As shown in Fig. 2 user stories can range from the very simple, in the first sentence, to the more complex, in the second sentence. Since many of the user stories were complex statements with more than one actor, data object, or operation, it was difficult to determine a single model that could take a user story as input and produce all necessary access control information as output. For example, the output to the last example would be "(authenticated user, summits, view); (authenticated user, register for the event, access); (authenticated user, event, view); (authenticated user, location, view)", where "authenticated user" is an actor, "view" and "access" are the kind of access to the data object or operation, "register for the event" is an operation, and "summits", "event", and "location" are data objects. The best single model solution we identified was to utilize a neural

---

[1] https://github.com/jheaps/AccessControlPolicyGeneration.

As a camp administrator, I want to be able to remove campers if they don't
attend the camp anymore, so that I can keep the records organized.

As an OlderPerson, I want to use only well-visible buttons.

As an Archivist, I want to see Dates and Extents displayed in both the read
and edit views for Accessions and/or Resources before the list of Subjects.

Fig. 3. User stories containing access control, not containing access control, and ambiguous

translation approach. However, a neural translation model would likely have great difficulty for a number of reasons, such as: listing multiple access control tuples where the order of the tuples is not important and there is constant recurrence of words; analyzing many user stories which do not contain any access control information all of which would be translated to the same word or symbol; and the existence of multiple different user stories that produce the exact same tuples. We therefore constructed a model containing three sub-components to perform the prediction: access control classification - deciding if a user story contains access control information; named entity recognition - identifying the actors and data objects contained in the user story; and access type classification - determining the relation between the actors and the data objects. We used transformers for the model prediction, from the Transformers library [21][2] with PyTorch[3], and also implemented the components using CNN and SVM as baseline comparisons.

Labeling. Each of the three component models to our deep learning approach required a separate labeling of the dataset.

The first component determines if the user story contains access control information with labels: "contains access control" or "does not contain access control". In many cases this was obvious. For example, the first user story in Fig. 3 contains access control information where "camp administrator" is the actor and "campers" is a data object that camp administrator should be able to delete. In contrast, the second user story of Fig. 3 does not contain any access control information. However, there were many user stories where it was difficult to determine if they contained access control information or not. For example, in the third user story in Fig. 3, it is unclear if this is only describing a change to the user interface of the software, in which case it would not contain access control information, or if it is describing a requirement for what data objects an Archivist can view. Further, it is not clear if the data objects are protected or simply available to everyone. During the normal development of a system, a developer would often have to make such a determination themselves. Since we do not know the final decision made about this particular user story, we have decided to include a third label, "unknown", in our labeling. The primary reason for this is to indicate to stakeholders that this is ambiguous or a decision about the access control policy needs to be made here. The earlier the ambiguity about the access control policy can be identified, the

---

[2] https://huggingface.co/transformers/.
[3] https://pytorch.org/.

```
As              Other
a               Other
camp            B-Actor
administrator   I-Actor
,               Other
I               Other
want            Other
to              Other
schedule        Other
events          B-DataObject
.               Other
```

Fig. 4. Named entity recognition example

more decisions about the access control policy can be removed from developers and placed on product owners who should be better able to make such decisions or provide additional context. An "unknown" is treated the same as a "contains access control" from the model's perspective, and is primarily used for presentation and visualization purposes that are described more thoroughly in Sect. 3.3. Further, it will be a great asset during further research when incorporating human aid and interactivity, as described in Sect. 5.

The second component identifies the actors, data objects, and operations in the user story. We have chosen to use a named entity recognition approach to tag words (or sequence of words) that represent actors or data objects in the user story, if they occur. In the example in Fig. 4 the words "camp administrator" are tagged as an actor and "events" as a data object. For labeling, each user story is broken into individual word tokens and labeled with one of seven different labels: "B-Actor" to denote the beginning token of an actor name, "I-Actor" to denote the continuation of actor name, "B-DataObject" to denote the beginning token of a data object name, "I-DataObject" to denote the continuation of a data object name, "B-Operation" to denote the beginning token of an operation, "I-Operation" to denote the continuation of an operation, and "Other" to denote other or a token we do not want to tag.

The third component, determines what type of access exists between the tagged entities in the user story. For the relationship between actors and data objects, the relationship may be: "view", "edit", "create", "delete", or "none". This is a multi-label classification and the user story may imply none or more of the labels between entities. For a simple user story, such as the first user story in Fig. 2, where there is only one actor ("camp administrator") and one data object ("camper") or operation ("remove camper"), it is obvious what actor the access type the label is referring to for the data object and operation. However, there are many user stories in the dataset that contain multiple actors and data objects. In this case, it is difficult to determine which of the actors or data objects would be indicated by a label, or in what order they would be referenced if multiple labels were added to it. To circumvent this problem, we copy the user story for each possible actor and data object present and label then separately with the appropriate label for each couple. Simple user stories with only one actor and data object pair are labeled in the same way. How the presence of the same user story with different labels are input to

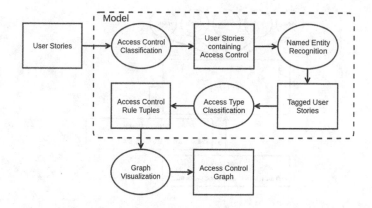

Fig. 5. User story access control extraction and prediction model

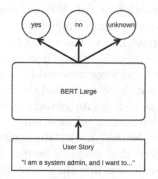

Fig. 6. Access control classification model

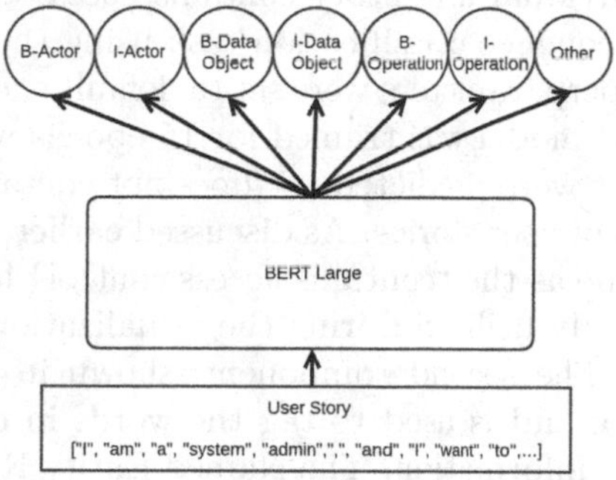

Fig. 7. Named entity recognition model

the model is discussed further in Sect. 3.2. For the operations, we found almost no actor-operation pair where the actor was not allowed access. This seems to be due to the formatting of the user stories in our dataset. User stories are almost always stated in a positive form (i.e., stating that an actor can do something) and are very rarely found in a negative form (i.e., stating that an actor cannot do something). Because of this, there was no way to learn or predict if a actor had "access" or "no access" to an operation as almost all user stories were formatted in a way that implied the actor had "access".

3.2 Model

We utilized the Transformers library to implement our models. The input, output, and flow of the component models are shown in Fig. 5, with rectangles representing input/output and ovals representing operations or learning tasks.

The first component, shown in Fig. 6, is the Access Control Classification task which takes the initial set of all user stories and predicts whether each

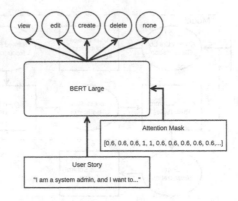

Fig. 8. Access type classification model

individual user story contains access control information or not. We performed a sequence classification task using the BERT Large [8] transformer model. All hyperparameters were set to default except the dropout rate which was set to 0.4. The model was trained for 15 epochs with a batch size of 30. Those user stories that were predicted as "does not contain access control" were removed from the set of user stories. As discussed earlier, the "unknown" label is treated the exact same as the "contains access control" label during modeling and prediction, and is only utilized during the visualization of the access control information.

The second component, shown in Fig. 7, is the Named Entity Recognition task and is used to tag the words in each user story that contains access control information. The Named Entity Recognition task identifies the actors, data objects, and operations in each user story. Each user story is tokenized into individual words and then a classification task is performed on each word to determine how it should be tagged. We use the BERT Large transformer model with all hyperparameters set to default, except the dropout rate which was set to 0.5. The model was trained for 15 epochs with a batch size of 30.

The final component, shown in Fig. 8, is the Access Type Classification and is used to identify the type of access ("view", "edit", "create", "delete", or "none") a given actor has for a given data object within each user story. We performed a multi-label sequence classification task using the BERT Large transformer model with all hyperparameters set to default, with the exception of the dropout rate which was set to 0.5. The model was trained for 15 epochs with a batch size of 30. This gives the final information to produce a list of tuples, "(actor, data object, access type)", that define the access control of the system. As mentioned in Sect. 3.1, if a user story has multiple actor and data object couples, the same user story is repeated and uniquely labeled for each unique couple. This poses a problem in that the same input implies multiple different outputs, but given the same input a model will always predict the same output (after training is completed). While the most important tokens in each user story is the unique actor and data object couple, we do not want to lose the context of the rest of the tokens in the user story that are vital to determining the type of access between the actor and data object. That is, we wish to emphasize the actor and data object tokens and de-emphasize (but not ignore) the other tokens in the user story. To achieve this we utilize the attention, or padding, mask of the transformer model. We see this as a logical and natural extension to the functionality and purpose of the attention mask. Normally utilized to negate the values of padding tokens so they do not affect the computations and results of a transformer's prediction, the attention mask can use values between $[0, 1]$. The different values in the mask will reduce the values of certain tokens toward the model's prediction while placing greater value on others. As can be seen in Fig. 8, the partial user story has the actor "system admin". If we assume the user story has multiple data objects (user accounts, permission files, etc.), then we will need to predict the type of access between each system admin and data object couple. The normal attention mask for an input would be a 1 for each token and 0 for any padding, if present. So if no padding existed it would be a vector of all ones the same size as the number of tokens in the user story. To emphasize the related actor and data object, the corresponding indices of "system admin" and a data object will be left as a 1 in the attention mask, but the rest of the tokens will have their value changed to less than 1 but greater than 0. In Fig. 8 this is shown in the attention mask where "system admin" is represented by ones and the rest of the tokens (that are shown) are 0.6. In this way, the same input can be used to predict multiple different outputs as those tokens deemed more important to the current prediction will be given greater value. As a final note, the value chosen for this example to represent de-emphasized tokens was 0.6. This is not necessarily the optimal value to achieve the best results with the model and likely multiple different values will need to be trained to determine the optimum performance.

3.3   Access Control Presentation

The final step in our approach is to present the access control information to stakeholders. We first perform some logical reduction of the set of access control tuples. In some cases, the same access control information is predicted from different user stories and so duplicates are removed. Also different user stories may imply different access control relationships between the same actor and data object and are combined into a single tuple.

We believe that one common and useful presentation of the access control information in software documentation is in graphical form. We have written a script to transfer the set of tuples to graphviz[4] format where actors, data objects, and operations are vectors and the types of access between them are edges.

Any access control tuples that were predicted as "unknown" by the first model component are colored in red, and those predicted as "contains access control" are colored in black. This is to stress the ambiguity of the access control information in the related user stories. This should prompt discussion and refinement of those user stories with stakeholders early on in production so that

---

[4] https://graphviz.org/.

Table 1. Precision, Recall, and F1 Score for the Access Control Classification and Named Entity Recognition tasks across three testing apps.

App name	Metric	ACC score	NER score
Frictionless	Precision	92.3% ± 1.8	88.2% ± 2.9
	Recall	89.7% ± 2.1	86.4% ± 4.4
	F1 Score	91.0% ± 2.0	87.3% ± 4.7
Alfred	Precision	79.1% ± 3.4	80.8% ± 4.7
	Recall	86.6% ± 2.7	80.1% ± 6.1
	F1 Score	82.7% ± 3.0	83.8% ± 5.3
CamperPlus	Precision	80.2% ± 2.5	84.4% ± 5.3
	Recall	88.3% ± 3.2	76.0% ± 4.1
	F1 Score	84.1% ± 2.8	80.0% ± 4.6

decisions about access control can be made in advance. This will save stakeholders time and money and relieve developers from the responsibility of making on-the-fly decisions about access control policy.

The presentation of access control information is not limited to graphical form: question and answer systems, rule lists, and (with limited human involvement) generating an access control natural language policy and policy specification can be achieved. This is further discussed as ongoing and future work in Sect. 5.

4   Evaluation

For our evaluation, we performed training, validation, and testing using, roughly, a 80%, 10%, 10% split, respectively. The testing set was created by removing all the user stories of one app from the rest of the dataset so that we can check the performance of the model on a set of user stories from an app that the model has never seen before. For validation, 10% of the remaining user stories were taken after being shuffled. The training set consisted of all other apps' user stories not in the testing or validation sets. We performed three separate training, validation, and testing runs for each component. Each run used a different app for the testing phase: Frictionless for data management platform and data archiving, Alfred for medical and elderly care, and CamperPlus for administrative management. In general, we observed that Frictionless performed better than the other apps during testing. This is likely due to the fact that the data management platform category represents over half of the apps in the training set. In contrast, there are very few apps in the training set that belong to the same category as Alfred and CamperPlus.

4.1   Access Control Classification

As seen in Table 1, Frictionless outperformed Alfred and CamperPlus by a margin of 7 and 9 percentage points, respectively. We analyzed those user stories that

Table 2. Precision, Recall, and F1 Score for the access type classification task across three testing apps.

App name	Metric	F1 Score
Frictionless	View	86.4% ± 4.2
	Edit	84.6% ± 5.5
	Create	81.1% ± 4.6
	Delete	81.7% ± 3.7
	None	82.2% ± 4.2
Alfred	View	80.6% ± 3.8
	Edit	79.8% ± 4.3
	Create	75.6% ± 5.7
	Delete	75.3% ± 6.0
	None	80.5% ± 5.3
CamperPlus	View	83.2% ± 5.1
	Edit	79.3% ± 3.6
	Create	76.5% ± 4.9
	Delete	75.6% ± 3.9
	None	79.9% ± 4.6

were predicted incorrectly and found some commonalities. In general, it seemed that non-functional user stories (i.e., user stories that describe attributes of the system, but not a functional usage of the system, for example "As an Older Person, I want to use only well-visible buttons") were more difficult for the model to categorize. This is likely due to much fewer non-functional user stories being present in the dataset. Other user stories that were predicted poorly contained acronyms or uncommon names of file types or 3rd party software. Again, since the model has not seen such words before, it was likely difficult to predict over.

4.2 Named Entity Recognition

As seen in Table 1, Frictionless again outperformed Alfred and CamperPlus of about 3 and 7 percentage points, respectively. The named entity recognition task suffered from the same main setback as the access control classification task, with most unique or uncommon words leading to a decrease in performance and were the most common missed words. Interestingly, there seemed to be some words common to multiple apps, but with different labels. For example, "team" was an actor name in some apps but a data object in others. It seems it was difficult for the model to reconcile this.

4.3 Access Type Classification

As seen in Table 2, Frictionless also outperformed Alfred and CamperPlus in the access type classification. The problem of a lack of data can definitely be

Table 3. Transformers, CNN, and SVM models and their average F1 scores for the three model components

Model	Component	F1 Score
Transformers	Access Control Classification	91.9% ± 2.0
	Named Entity Recognition	87.3% ± 3.4
	Access Type Classification	83.2% ± 4.4
CNN	Access Control Classification	84.3% ± 4.1
	Named Entity Recognition	86.7% ± 3.6
	Access Type Classification	79.1% ± 5.4
SVM	Access Control Classification	84.4% ± 1.3
	Named Entity Recognition	69.8% ± 3.9
	Access Type Classification	73.2% ± 4.3

seen here, as "create" and "delete" had much fewer occurrences than "view" and "edit", and did much poorer amongst all three apps.

We tested how well the average F1 score of all access categories performed at different attention mask values of de-emphasis (as described in Sect. 3.2). We tested mask values of 0.25, 0.5, 0.6, 0.75, and 0.9. For Fictionless, the best mask value was 0.6, Alfred performed best at 0.75, while CamperPlus did best at 0.6 as well.

4.4 SVM and CNN Comparison

In Table 3, the average F1 score for all components for each the Transformers, CNN, and SVM models are shown. Transformers performed the best in all categories, with CNN as a close second. Interestingly, CNN performed almost just as well as Transformers in the named entity recognition task, where SVM performed the worst in that category. We see that the Transformers model performed better than CNN and SVM as baseline measures, but not as significantly as we would have hoped.

We believe the primary approach to increasing the Transformer model's (or any model's) accuracy is a much larger and more robust dataset. It is clear that with 1600 example user stories the models were able to learn in all three components but the learning was quite limited, both in the size of the dataset as well as the variety of types of apps.

4.5 Graph Visualization

Finally, we have performed an end-to-end run of one of the testing apps, CamperPlus, through the full model and have visualized a sample of the graph in Fig. 9. We have chosen only a sample as the full graph was too large to insert into the paper, and these showed the more interesting results. In the graph, the different

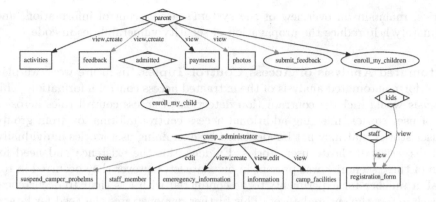

Fig. 9. Graph visualization sample of CamperPlus app

types of shapes denote the type of entity it is: diamonds are actors, rectangles are data objects, and ovals are operations. Only data objects have a type of access associated with them, as the operations were all permitted access, as described in Sect. 3.1. The red lines are those user stories that were predicted as "unknown", and would prompt stakeholders to take a closer look at those user stories related to them. In future work, human interactivity would allow a stakeholder to confirm or invalidate the "unknown" predictions, as described in Sect. 5. In some cases the model worked well. We can see that the "parent", "camp administrator", "staff", and "kids" actors were all predicted correctly, and many of the data objects and operations were correctly predicted as such. However, "admitted" was not an actor (or a data object or operation). We believe it was predicted as an actor because it may be close enough to "administrator" which was labeled as an actor. The "suspend camper problems" should just be "suspend camper" and be an operation, not a data object. The "information" data object was likely supposed to be "medical information" as that was a data object missing from the full graph. While we tried to remove most plural words after the final output of the model, we can see that some were missed, such as "enroll my child" and "enroll my children".

We can see that the graph is not perfect, which was expected given the F1 scores from testing. However, there is a significant portion of the access control graph that is correct, and with the help of human aid shows great potential for assisting stakeholders during the agile development software process.

5   Conclusion

This paper describes an automated approach to extracting access control information from user stories through the labeling of a dataset of 1600 user stories and a learning model based on transformers with three components: access control classification, named entity recognition, and access type classification. The resulting list of tuples was visualized to help stakeholders better refine user stories, maintain an overview of the system's access control information, and ultimately help reduce the propagation of security vulnerabilities in code.

Automated Analysis of Access Control Tuples. In future work we plan to perform automated analysis of the extracted access control information. This analysis would include: contradiction detection of access control rules across a set of user stories, inferring additional access control information from groups of user stories that may not be identified by examining user stories individually, and detecting duplicate user stories. Evidence for the existence and need for each of these scenarios was present in our dataset, however they occurred in too small a number for experiments to be conducted with less than 10 occurrences of each across the entire dataset. This further demonstrates the need for larger, more robust user story datasets.

Integration of Human Interactivity. We recognize that there is little additional information about access control that can be extracted or determined directly from user stories in a fully automated approach. That is, it is difficult, or impossible, to determine a software's exact access control from user stories without human involvement. We plan to extend our approach to make it interactive with stakeholders so that they can help refine the access control information by providing additional context, such as specific roles and attributes of the system. Humans will be able to validate (or invalidate) ambiguous relations that were marked as "unknown" or were incorrectly predicted. This will also allow for active learning [16] with the models to better refine prediction results. While this interactivity will require some continual review from humans, the vast majority of the maintenance of access control information and documentation will still be performed automatically by our approach. This should keep human involvement at a minimum, increasing development time as little as possible.

Tracking the Agile Development Cycle. A primary aspect of agile development is changing requirements through the modification of the set user stories, which occurs throughout the agile development cycle. We plan to show how our approach can handle changing user stories throughout the development process.

Additional Security Policy Generation. Finally, while this work describes an automated approach to the construction of access control documentation, other security and system documentation could be inferred or extracted from user stories. For example, privacy requirements, interaction and use of external technologies and third party applications, and generally ensuring security best practices based on security decisions and implementations can all be handled in a mostly automated approach to ensure that stakeholders have necessary documentation when developing the system.

Acknowledgments. We would like to thank the CREST Center For Security And Privacy Enhanced Cloud Computing (C-SPECC) through the National Science Foundation (NSF) (Grant Award #1736209), the NSF Division of Computer and Network Systems (CNS) (Grant Award #1553696), the and NSF Division of Computing and Communication Foundations (Grant Award #2007718) for their support and contributions to this research.

References

1. Alohaly, M., Takabi, H., Blanco, E.: A deep learning approach for extracting attributes of ABAC policies. In: Proceedings of the 23nd ACM on Symposium on Access Control Models and Technologies, pp. 137–148 (2018)
2. Bartsch, S.: Practitioners' perspectives on security in agile development. In: 2011 Sixth International Conference on Availability, Reliability and Security, pp. 479–484. IEEE (2011)
3. Ben Othmane, L., Angin, P., Weffers, H., Bhargava, B.: Extending the agile development process to develop acceptably secure software. IEEE Trans. Dependable Secure Comput. **11**(6), 497–509 (2014)
4. Beznosov, K., Kruchten, P.: Towards agile security assurance. In: Proceedings of the 2004 Workshop on New Security Paradigms, pp. 47–54 (2004)
5. Boström, G., Wäyrynen, J., Bodén, M., Beznosov, K., Kruchten, P.: Extending XP practices to support security requirements engineering. In: Proceedings of the 2006 International Workshop on Software Engineering for Secure Systems, pp. 11–18 (2006)
6. Dalpiaz, F.: Requirements data sets (user stories). Mendeley Data (2018). https://doi.org/10.17632/7zbk8zsd8y.1
7. Dalpiaz, F., van der Schalk, I., Lucassen, G.: Pinpointing ambiguity and incompleteness in requirements engineering via information visualization and NLP. In: Kamsties, E., Horkoff, J., Dalpiaz, F. (eds.) REFSQ 2018. LNCS, vol. 10753, pp. 119–135. Springer, Cham (2018). https://doi.org/10.1007/978-3-319-77243-1_8
8. Devlin, J., Chang, M.W., Lee, K., Toutanova, K.: BERT: pre-training of deep bidirectional transformers for language understanding. arXiv preprint arXiv:1810.04805 (2018)
9. Fowler, M., Highsmith, J., et al.: The agile manifesto. Softw. Dev. **9**(8), 28–35 (2001)
10. Karimi, L., Aldairi, M., Joshi, J., Abdelhakim, M.: An automatic attribute based access control policy extraction from access logs. arXiv preprint arXiv:2003.07270 (2020)
11. Lucassen, G., Dalpiaz, F., van der Werf, J.M.E., Brinkkemper, S.: Improving agile requirements: the quality user story framework and tool. Requirements Eng. **21**(3), 383–403 (2016). https://doi.org/10.1007/s00766-016-0250-x
12. Lucassen, G., Robeer, M., Dalpiaz, F., Van Der Werf, J.M.E., Brinkkemper, S.: Extracting conceptual models from user stories with visual narrator. Requirements Eng. **22**(3), 339–358 (2017). https://doi.org/10.1007/s00766-017-0270-1
13. Narouei, M., Takabi, H., Nielsen, R.D.: Automatic extraction of access control policies from natural language documents. IEEE Trans. Dependable Secure Comput. **17**, 506–517 (2020)
14. Oueslati, H., Rahman, M.M., ben Othmane, L.: Literature review of the challenges of developing secure software using the agile approach. In: 2015 10th International Conference on Availability, Reliability and Security, pp. 540–547. IEEE (2015)

15. Pohl, C., Hof, H.J.: Secure scrum: development of secure software with scrum. arXiv preprint arXiv:1507.02992 (2015)
16. Settles, B.: Active learning. In: Synthesis Lectures on Artificial Intelligence and Machine Learning, vol. 6, no. 1, pp. 1–114 (2012)
17. Siponen, M., Baskerville, R., Kuivalainen, T.: Integrating security into agile development methods. In: Proceedings of the 38th Annual Hawaii International Conference on System Sciences, p. 185a. IEEE (2005)
18. Slankas, J., Xiao, X., Williams, L., Xie, T.: Relation extraction for inferring access control rules from natural language artifacts. In: Proceedings of the 30th Annual Computer Security Applications Conference, pp. 366–375 (2014)
19. Sobieski, Ś., Zieliński, B.: User stories and parameterized role based access control. In: Bellatreche, L., Manolopoulos, Y. (eds.) MEDI 2015. LNCS, vol. 9344, pp. 311–319. Springer, Cham (2015). https://doi.org/10.1007/978-3-319-23781-7_25
20. Wäyrynen, J., Bodén, M., Boström, G.: Security engineering and eXtreme programming: an impossible marriage? In: Zannier, C., Erdogmus, H., Lindstrom, L. (eds.) XP/Agile Universe 2004. LNCS, vol. 3134, pp. 117–128. Springer, Heidelberg (2004). https://doi.org/10.1007/978-3-540-27777-4_12
21. Wolf, T., et al.: Transformers: state-of-the-art natural language processing. In: Proceedings of the 2020 Conference on Empirical Methods in Natural Language Processing: System Demonstrations, pp. 38–45 (2020)
22. Xiao, X., Paradkar, A., Thummalapenta, S., Xie, T.: Automated extraction of security policies from natural-language software documents. In: Proceedings of the ACM SIGSOFT 20th International Symposium on the Foundations of Software Engineering, pp. 1–11 (2012)

PERUN: Confidential Multi-stakeholder Machine Learning Framework with Hardware Acceleration Support

Wojciech Ozga[1,2]($\boxtimes$), Do Le Quoc[3], and Christof Fetzer[1,4]

[1] TU Dresden, Dresden, Germany
[2] IBM Research Europe—Zurich, Rüschlikon, Switzerland
woz@zurich.ibm.com
[3] Huawei Munich Research Center, München, Germany
[4] Scontain UG, Dresden, Germany

Abstract. Confidential multi-stakeholder machine learning (ML) allows multiple parties to perform collaborative data analytics while not revealing their intellectual property, such as ML source code, model, or datasets. State-of-the-art solutions based on homomorphic encryption incur a large performance overhead. Hardware-based solutions, such as trusted execution environments (TEEs), significantly improve the performance in inference computations but still suffer from low performance in training computations, *e.g.*, deep neural networks model training, because of limited availability of protected memory and lack of GPU support.

To address this problem, we designed and implemented PERUN, a framework for confidential multi-stakeholder machine learning that allows users to make a trade-off between security and performance. PERUN executes ML training on hardware accelerators (*e.g.*, GPU) while providing security guarantees using trusted computing technologies, such as trusted platform module and integrity measurement architecture. Less compute-intensive workloads, such as inference, execute only inside TEE, thus at a lower trusted computing base. The evaluation shows that during the ML training on CIFAR-10 and real-world medical datasets, PERUN achieved a $161\times$ to $1560\times$ speedup compared to a pure TEE-based approach.

Keywords: Multi-stakeholder computation · Machine learning · Confidential computing · Trusted computing · Trust management

1 Introduction

Machine learning (ML) techniques are widely adopted to build functional artificial intelligence (AI) systems. For example, face recognition systems allow paying at supermarkets without typing passwords; natural language processing systems allow translating information boards in foreign countries using smartphones;

Do Le Quoc performed this work at TU Dresden.

© IFIP International Federation for Information Processing 2021
Published by Springer Nature Switzerland AG 2021
K. Barker and K. Ghazinour (Eds.): DBSec 2021, LNCS 12840, pp. 189–208, 2021.
https://doi.org/10.1007/978-3-030-81242-3_11

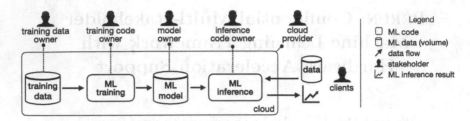

Fig. 1. Stakeholders share source code, data, and computing power to build an ML application. They need a framework to establish mutual trust and share data securely.

medical expert systems help to detect diseases at an early stage; image recognition systems help autonomous cars to identify road trajectory and traffic hazards. To build such systems, multiple parties or stakeholders with domain knowledge from various science and technology fields must cooperate since machine learning is fundamentally a multi-stakeholder computation, as shown in Fig. 1. They would benefit from sharing their intellectual property (IP) – private training data, source code, and models – to jointly perform machine learning computations only if they can ensure their IP remains confidential.

Training Data Owner. ML systems rely on *training data* to build *inference models*. However, the data is frequently sensitive and cannot be easily shared between disjoint entities, like in the case of the healthcare data used for training diagnostic models contain privacy-sensitive patient information. The strict data regulations, such as general data protection regulation (GDPR) [1], impose an obligation on secure data processing. Specifically, the training data must be under the training data owner's control and must be protected while at rest, during transmission, and training computation.

Training Code Owner. The training code owner implements a *training algorithm* that trains an *inference model* over the training data. The *training code* (*e.g.*, Python code) typically contains an optimized training model architecture and tuned parameters that build the business value and the inference model quality. Thus, the training code is considered as confidential as the training data. The training requires high computing power, and, as such, it is economically justifiable to delegate its execution to the cloud. However, in the cloud, users with administrative access can easily read the training service source code implemented in popular programming languages, such as Python.

Model Owner. The inference model is the heart of any inference service. It is created by training the model with a large amount of training data. This process requires extensive computing power and is time-consuming and expensive. Thus, the model owner, a training code owner, or a third party that buys the model, must protect the model's confidentiality. The trained models may reveal the privacy of the training data [3]. Several works [3,13] demonstrated that extracted images from a face recognition system look suspiciously similar to images from the underlying training data.

Inference Code Owner. The inference code is an AI service allowing clients to use the inference model on a business basis. The inference code is frequently developed using Python or JavaScript and hosted in the cloud. Thus, the confidentiality of the code and the integrity of the computation must be protected against an adversary controlling computing systems executing the AI service.

Inference Data Owner. The inference data owner is a client of an AI service. He wants to protect his input data. Imagine a person sending an X-ray scan of her brain to a diagnostic service to check for a brain tumor. The inference data, e.g., a brain's scan, is privacy-sensitive and must not be accessible by the AI service provider.

To build an AI service, stakeholders must trust that others follow the rules protecting each other's IP. However, it is difficult to establish trust among them. First, some stakeholders might collude to gain advantages over others [41]. Second, even a trustworthy stakeholder might lack expertise in protecting their IP from a skilled attacker gaining access to its computing resources [12,43]. We tackle the following problem: How to allow stakeholders to jointly perform machine learning to unlock all AI benefits without revealing their IP?

Recent works [33,37] demonstrated that cryptographic techniques, such as secure multi-party computation [56] and fully homomorphic encryption [15], incur a large performance overhead, which currently prevents their adoption for computing-intensive ML. Alternative approaches [34,42] adopted trusted execution environments (TEEs) [36] to build ML systems showing that TEEs offer orders of magnitude faster ML computation, at the cost of weaker security guarantees compared to pure cryptographic solutions. Specifically, the pure cryptographic solutions compute on encrypted data, while in TEE-based approaches, a trusted ML software processes the plaintext data in a CPU-established execution environment (called *enclave*), which is isolated from the untrusted operating system and administrator. Although promising for the ML inference, TEEs still incur considerable performance overhead for memory-intensive computations, like deep training, because of the limited memory accessible to the enclave and lack of support for hardware accelerators, like graphical processing units (GPUs). Thus, since TEEs alone are not enough for the ML training processes, we raise the question: What trade-off between security and performance has to be made to allow the ML training to access hardware accelerators?

We propose PERUN, a framework allowing stakeholders to share their code and data only with certain ML applications running inside an enclave and on a trusted OS. PERUN relies on encryption to protect the IP and on a trusted key management service to generate and distribute the corresponding cryptographic keys. TEE provides confidentiality and integrity guarantees to ML applications and to the key management service. Trusted computing technologies [14] provide integrity guarantees to the OS, allowing ML computations to access hardware accelerators. Our evaluation shows that PERUN achieves 0.96× of native performance execution on the GPU and a speedup of up to 1560× in training a real-world medical dataset compared to a pure TEE-based approach [34].

Altogether, we make the following contributions:

- We designed a secure multi-stakeholder ML framework that: *(i)* allows stakeholders to cooperate while protecting their IP (Sect. 3.1, Sect. 3.2), *(ii)* allows stakeholders to select trade-off between the security and performance, allowing for hardware accelerators usage (Sect. 3.3, Sect. 3.4).
- We implemented the PERUN prototype (Sect. 4) and evaluated it using real-world datasets (Sect. 5).

2 Threat Model

Stakeholders are financially motivated businesses that cooperate to perform ML computation. Each stakeholder delivers an input (*e.g.*, input training data, code, and ML models) as its IP for ML computations. The IP must remain confidential during ML computations. The stakeholders have limited trust. They do not share their IP directly, but they encrypt them so that only other stakeholders' applications, which source code they can inspect under a non-disclosure agreement or execute in a sandbox, can access the encryption key to decrypt it.

An adversary wants to steal a stakeholder's IP when it resides on a computer executing ML computation. Such a computer might be provisioned in the cloud or a stakeholder's data center, *e.g.*, a hybrid cloud model. In both cases, an adversary has no physical access to the computer. For this, we rely on state-of-the-art practices controlling and restricting access to the data center to trusted entities.

However, an adversary might exploit an OS misconfiguration or use social engineering to connect to the OS remotely. We assume she can execute privileged software to read an ML process's memory after getting administrative access to the OS executing ML computation. One of the mitigation techniques used in PERUN, integrity measurement architecture (IMA) [45], effectively limits software that can execute on the computer under the assumption that this software, which is considered trusted, behaves legitimately also after it has been loaded to the memory, *i.e.*, an adversary cannot tamper with the process' code after it has been loaded to the memory. This might be achieved using existing techniques, like enforcing control flow integrity [27], fuzzing [57], formally proving the software implementation [58], using memory-safe languages [35], using memory corruption mitigation techniques, like position-independent executables, stack-smashing protection, relocation read-only techniques, or others.

The CPU with its hardware features, hardware accelerators, and secure elements (*e.g.*, TPM) are trusted. We exclude micro-architectural and side-channel attacks, like Foreshadow [51] or Spectre [30]. We rely on the soundness of the cryptographic primitives used within software and hardware components.

3 Design

Our objective is to provide an architecture that: (i) supports multi-stakeholder ML computation, (ii) requires zero code changes to the existing ML code, (iii) allows for a trade-off between security and performance, (iv) uses hardware accelerators for computationally-intensive tasks.

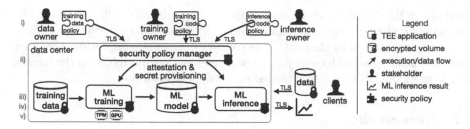

Fig. 2. PERUN framework supports multi-stakeholder ML computation. Stakeholders trust the security policy manager. Inside security policies, they define which stakeholder's application can access a cryptographic key allowing decryption of confidential code or data. TEE protects code, data, and cryptographic keys.

3.1 High-Level Overview

Figure 2 shows the PERUN framework architecture that supports multi-stakeholder computation and the use of dedicated hardware accelerators. The framework consists of five components: (i) *stakeholders*, the parties who want to perform ML jointly while keeping their IP protected; (ii) *security policy manager*, a key management and configuration service that allows stakeholders to share IPs for ML computations without revealing them; (iii) *ML computation* including training and inference; (iv) *GPU*, hardware accelerators enabling high-performance ML computation; and (v) *TEE* and *TPM*, *secure elements* enabling confidentiality and integrity of ML computations on untrusted computing resources.

To allow multiple stakeholders to perform ML and keep their IP confidential, we propose that the IP remains under the stakeholder's control. To realize that idea, we design the security policy manager that plays the role of the *root of trust*. Stakeholders establish trust in this component using the *remote attestation* mechanism, like [25], offered by a TEE. The TEE, *e.g.*, Intel software guard extensions (SGX) [11], guarantees the confidentiality and integrity of processed code and data. After stakeholders ensure the security policy manager executes in the TEE, they submit security policies defining access control to their encryption keys. Each stakeholder's IP is encrypted with a different key, and the security policy manager uses security policies to decide who can access which keys. From a technical perspective, the security policy manager generates the keys inside the TEE and sends them only to authenticated ML computations executing inside the TEE. Thus, these keys cannot be seen by any human.

Depending on individual stakeholders' security requirements, PERUN offers different throughput/latency performances for ML computations. For stakeholders willing strong integrity and confidentiality guarantees, PERUN executes ML computations only inside TEEs enclaves, *i.e.*, input and output data, code, and models never leave the enclave. For stakeholders accepting a larger trusted computing base in exchange for better performance, PERUN enables trusted computing technologies [14] to protect ML computations while executing them on hardware accelerators, *e.g.*, GPU. Specifically, it uses IMA, which is an integrity enforcement mechanism that prevents adversaries from running arbitrary software on the OS, *i.e.*, software that allows reading data residing in the main memory or being transferred to or processed by the GPU. The security policy manager verifies that such a mechanism is enabled by querying a secure element compatible with the trusted platform module (TPM) [19] attached to the remote computer.

3.2 Keys Sharing

Stakeholders use security policies to share encryption keys protecting their IP. For example, the training data owner specifies in his security policy that he allows the ML computation of the training code owner to access his encryption key to decrypt the training data. The security policy manager plays a key role in the key sharing process. It generates an encryption key inside the TEE and securely distributes it to ML computations accordingly to the security policy. The training code owner cannot see the shared secret in the example above because it is transferred only to his application executing inside the TEE.

To provision ML computations with encryption keys, the security policy manager authenticates them using a remote attestation protocol offered by a TEE engine, *e.g.*, the SGX remote attestation protocol [25]. During the remote attestation, the TEE engine provides the security policy manager with a cryptographic measurement of the code executing on the remote platform. The cryptographic measurement – the output of the cryptographic hash function over the code loaded by the TEE engine to the memory – uniquely identifies the ML computation, allowing the security policy manager to authorize access to the encryption key based on the ML computation identity and stakeholder's security policies.

3.3 Security Policy and Trade-Offs

PERUN relies on security policies as a means to define dependencies among stakeholders' computation and shared data.

Listing 1.1 shows an example of a policy. The policy has a unique name (line 1), typically combining a stakeholder's name and its IP name. The name is used among stakeholders to reference *volumes* containing code, input, or output data. A volume is a collection of files encrypted with an encryption key managed by the security policy manager. Only the authorized ML computations have access to the key required to decrypt the volume and access the IP. To prevent an adversary from changing the policy, the stakeholder embeds his public key inside the policy (line 21). The security policy manager accepts only policies containing a valid signature issued with a corresponding stakeholder's private key.

The ML computation definition consists of a command required to execute the computation inside a container (line 2) and a cryptographic hash over the source code content implementing the ML computation (line 8). The security policy manager uses the hash to authenticate the ML computation before providing it with the encryption key.

Listing 1.1. Security policy example

```
 1 name: training_owner/training_code
 2 command: python /app/training.py
 3 volumes:
 4   - path: /training_data
 5     import: training_data_owner/training_data_service
 6   - path: /inference_model
 7     export: inference_owner/inference_service
 8 integrity_hash: {"0a11...bb3f"}
 9 operating_system:
10   certificate_chain: |-
11     -----BEGIN CERTIFICATE-----
12     # certificate chain allowing
13     # verification of the secure
14     # element manufacturer
15     -----END CERTIFICATE-----
16   integrity:
17     measure0: e0f1...4be6
18     measure1: ae44...3a6e
19     measure2: 3d45...796d
20 stakeholder: |-
21   -----BEGIN PUBLIC KEY-----
22   # the policy owner's public key
23   -----END PUBLIC KEY-----
```

The policy allows selecting trade-offs between security and performance. For example, a training code owner who wants to use the GPU to speed up the ML training computation might define conditions under which he trusts the OS. In such a case, a stakeholder defines a certificate chain permitting to verify the authenticity of a secure element attached to the computer (line 10) and expected integrity measurements of the OS (lines 16–19). The security policy manager only provisions the ML computations with the encryption key if the OS integrity (kernel sources and configuration) are trusted by the stakeholder. Specifically, the OS integrity measurements reflect what kernel code is running and whether it has enabled the required security mechanisms. Only then, the ML computations can access the confidential data and send it to the outside of the TEE, *e.g.*, GPU.

We discuss now and evaluate later (Sect. 5.2) two security levels that are particularly important for the ML computation. The first one, the *high-assurance security level*, fits well the inference because it offers strong security guarantees provided by the TEE, allowing the inference model to execute in an untrusted data center controlled by an untrusted operator. It comes at performance limitation, which is acceptable for inference because, typically, inference operates on much smaller data than ML training and does not need access to hardware accelerators. The high-assurance security level offers confidentiality and integrity of code and data at rest and in runtime. The trusted computing base (TCB) is low; It includes only the inference model executing inside the TEE, the hardware providing the TEE functionality, and the key distribution process. The second security level, the *high integrity level*, fits well the ML training because it enables access to hardware accelerators required for intensive computation. It comes at the cost of a larger TCB compared to the high-assurance security level because the code providing access to the hardware accelerators, *i.e.*, an operating system, must be trusted. PERUN relies on the TPM to establish trust with load-time kernel integrity and on IMA to extend this trust to the OS runtime integrity.

3.4   Hardware ML Accelerators Support

Typically, ML computations (*e.g.*, deep neural networks training) are extremely intensive because they must process a large amount of input data. To decrease the computation time, popular ML frameworks, such as TensorFlow [2], support hardware accelerators, such as GPUs or Google tensor processing units (TPUs). Unfortunately, existing hardware accelerators do not support confidential computing, thus not offering enough security guarantees for the multi-stakeholder ML computation. For example, an adversary who exploits an OS misconfiguration [55] can launch arbitrary software to read data transferred to the GPU from any process executing in the OS. Even if ML computations execute inside the TEE enclaves, an adversary controlling the OS can read the data when it leaves the TEE, *i.e.*, it is transferred to the GPU or is processed by the GPU. Because of this, we design PERUN to support additional security mechanisms protecting access to the data (also code and ML models) while being processed out of the TEE. This also allows stakeholders to trade-off between security level and performance they want to achieve when performing ML computations.

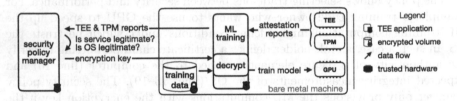

Fig. 3. The high-level overview of PERUN supporting secure computation using hardware accelerator, *e.g.*, the GPU. PERUN performs both the SGX and TPM attestation before provisioning the ML code with cryptographic keys. The successful TPM attestation informs that the legitimate OS with enabled integrity-enforcement mechanisms controls access to the GPU.

Figure 3 shows how PERUN enables hardware accelerator support. ML computations transfer to the security policy manager a report describing the OS's integrity state. The report is generated and cryptographically signed by a secure element, *e.g.*, a TPM chip, physically attached to the computer. The security policy manager authorizes the ML computation to use the encryption key only if the report states that the OS is configured with the required security mechanisms. Precisely, the integrity enforcement mechanism, such as integrity measurement architecture (IMA) [45], controls that the OS executes only software digitally signed by a stakeholder. Even if an adversary gains root access to the system, she cannot launch arbitrary software that allows her to sniff on the communication between the ML computation and the GPU, read the data from the main memory, or reconfigure the system to disable security mechanisms or load a malicious driver. This also allows PERUN to mitigate software-based microarchitectural and side-channel attacks [9,16,51,54], which are vulnerabilities of TEEs.

To enable hardware accelerator support, a stakeholder specifies expected OS integrity measurements inside the security policy (Listing 1.1, lines 9–19) and certificates allowing verification of the secure element identity. The OS integrity measurements are cryptographic hashes over the OS's kernel loaded to the memory during the boot process. A secure element collects such measurements during the boot process and certifies them using a private key linked to a certificate issued by its manufacturer. The certificate and integrity measurements are enough for the security policy manager to verify that the IMA enforces the OS integrity.

Although the hardware accelerator support comes at the cost of weaker security guarantees (additional hardware and software must be trusted when compared to a pure TEE-based approach), it greatly improves the ML training computation's performance (see Sect. 5.2).

3.5 Zero Code Changes

PERUN framework requires zero code changes to run existing ML computations, thus providing a practical solution for legacy ML systems. To achieve it, PERUN adapts platforms supporting running legacy applications inside the TEE, such as SCONE [5] or GrapheneSGX [49]. These platforms allow executing unmodified code inside the TEE by recompiling the code using dedicated cross-compilers or running them with a modified interpreter executing in the TEE.

3.6 Policy Deployment and Updates

A stakeholder establishes a transport layer security (TLS) connection to the security policy manager to deploy a policy. During the TLS handshake, the stakeholder verifies the identity of the security policy manager. The security policy manager owns a private key and corresponding certificate signed by an entity trusted by a stakeholder. For example, such a certificate can be issued by a TEE provider who certifies that given software running inside a TEE and identified by a cryptographic hash is the security policy manager. Some TEE engines, such as SGX, offer such functionality preventing even a service administrator from seeing the private key [29]. For other TEEs, a certificate might be issued by a cloud provider operating the security policy manager as part of cloud offerings.

PERUN requires that the security policy manager authorizes changes to the deployed policy. Otherwise, an adversary might modify the stakeholder's policy allowing malicious code to access the encryption key. In the PERUN design, the stakeholder includes his public key inside the digitally signed security policy. Since then, the security policy manager accepts changes to the policy only if a new policy has a signature issued with the stakeholders' private key corresponding to the public key present in the existing policy. By having a public key embedded in the security policy, other stakeholders can verify that the policy is owned by the stakeholder they cooperate with. The details of the policy security manager regarding key management, high availability, tolerance, and protection against rollback attacks are provided in [18].

4 Implementation

We implemented the PERUN prototype based on TensorFlow version 2.2.0 and the SCONE platform [5] because SCONE provides an ecosystem to run unmodified applications inside a TEE. We also rely on the existing key management system provided by the SCONE [50] and its predecessor [18] to distribute the configuration to applications. We rely on Intel SGX [11] as a TEE engine because it is widely used in practice.

Our prototype uses a TPM chip [19] to collect and report integrity measurements of the Linux kernel loaded to the memory during a trusted boot [46] provided by tboot [10] with Intel trusted execution technology (TXT) [17]. The Linux kernel is configured to enforce the integrity of software, dynamic libraries, and configuration files using Linux IMA [45], a Linux kernel's security subsystem. Using the TPM chip, verifies that the kernel is correctly configured and interrupts its execution when requirements are not met.

We use an nvidia GPU as an accelerator for ML computation. The ML services are implemented in Python using TensorFlow framework, which supports delegating ML computation to the GPU.

4.1 Running ML Computations Inside Intel SGX

To run unmodified ML computations inside the SGX enclaves, we use the SCONE cross-compiler and SCONE-enabled Python interpreters provided by SCONE as Docker images. They allow us to build binaries that execute inside the SGX enclave or run Python code inside SGX without any source code changes.

The SCONE wraps an application in a dynamically linked loader program (*SCONE loader*) and links it with a modified C-library (*SCONE runtime*) based on the musl libc [39]. On the ML computation startup, the SCONE loader requests SGX to create an isolated execution environment (enclave), moves the ML computation code inside the enclave, and starts. The SCONE runtime, which executes inside the enclave along with the ML computations, provides a sanitized interface to the OS for transparent encryption and decryption of data entering and leaving the enclave. Also, the SCONE runtime provides the ML computations with its configuration using configuration and attestation service (CAS) [50].

4.2 Sharing the Encryption Key

We implement the security policy manager in the PERUN architecture using the CAS, to generate, distribute, and share encryption keys between security policies. We decided to use the CAS because it integrates well with SCONE-enabled applications and implements the SGX attestation protocol [25]. Other key management systems supporting the SGX attestation protocol might be used [8, 32] but require additional work to integrate them into SCONE.

We create a separate CAS policy for each stakeholder. The policy contains an identity of the stakeholder's IP (data, code, and models) and its access control and configuration. It is uploaded to CAS via mutual TLS authentication using a stakeholder-specific private key corresponding to the public key defined inside the policy. This fulfills the Perun requirement of protecting unauthorized stakeholders from modifying policies. The IP identity is defined using a unique per application cryptographic hash calculated by the SGX engine over the application's pages and their access rights. The SCONE provides this value during the application build process. The CAS allows for the specification of the encryption key as a program argument, environmental variable, or indirectly as a key related to an encrypted volume. Importantly, the CAS allows defining which policies have access to the key. Thus, with the proper policy configuration, stakeholders share keys among enclaves as required in the Perun architecture.

Our prototype uses the CAS encrypted volume functionality, for which the SCONE runtime fetches from the CAS the ML computation configuration containing the encryption key. Specifically, following the SGX attestation protocol, the SCONE runtime sends to CAS the SGX attestation report in which the SGX hardware certifies the ML computation identity. The CAS then verifies that the report was issued by genuine SGX hardware and the ML computation is legitimate. Only afterward, it sends to the SCONE runtime the encryption key. The SCONE runtime transparently encrypts and decrypts data written and read by the ML computation from and to the volume. The ML computations, e.g., training and inference authorized by stakeholders via policies, can access the same encryption key, thus gaining access to a shared volume.

4.3 Enabling GPU Support with Integrity Enforcement

Our prototype implementation supports delegating ML computations to the GPU under the condition that the integrity-enforced OS handles the communication between the enclave and the GPU. The integrity enforcement mechanism prevents intercepting confidential data that leaves the enclave because it limits the OS functionality to a subset of programs essential to load the ML computation and the GPU driver. Thus, a malicious program cannot run alongside the ML computations on the same computing resources. We use trusted boot and TPM to verify it, i.e., that the remote computer runs a legitimate Linux kernel with enabled integrity enforcement that limits software running on the computer to the required OS services, the GPU driver, and ML computations.

Trusted Boot. Trusted computing technologies define a set of technologies that measure, report, and enforce kernel integrity. Specifically, during the computer boot, we rely on a trusted bootloader [10], which uses a hardware CPU extension [17] to measure and securely load the Linux kernel to an isolated execution environment [46]. The trusted bootloader measures the kernel integrity (a cryptographic hash over the kernel sources) and sends the TPM chip measurements.

The TPM stores integrity measurements in a dedicated tamper-resistant memory called platform configuration registers (PCRs). A PCR value cannot be set to an arbitrary value. It can only be extended with a new value using a cryptographic hash function: $PCR\_extend = hash(PCR\_old\_value \parallel data\_to\_extend)$. This prevents tampering with the measurements after they are extended to a PCR. The TPM implements a TPM attestation protocol [20] in which it uses a private key known only to the TPM to sign a report containing PCRs. Our prototype uses the TPM attestation protocol to read the TPM report certifying that an ML computation executes on an integrity-enforced OS, *i.e.*, a Linux kernel with enabled IMA.

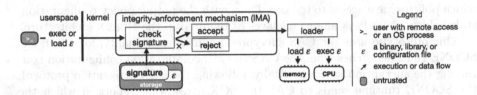

Fig. 4. The kernel integrity-enforcement system authenticates a file by checking its digital signature before loading it to the memory.

Integrity Enforcement. IMA is a kernel mechanism that authenticates files before allowing them to be loaded to the memory. Figure 4 shows how the IMA works. A process executing in userspace requests the kernel to execute a new application, load a dynamic library, or read a configuration file. IMA calculates the cryptographic hash over the file's content, reads the file's signature from the file's extended attribute, and verifies the signature using a public key stored in the kernel's *ima keyring*. If the signature is correct, IMA extends the hash to a dedicated PCR and allows the kernel to continue loading the file.

Trusted Boot Service. Because SCONE is proprietary software, we could not modify the SCONE runtime to provide the CAS with the TPM report. Instead, we implemented this functionality in a *trusted boot service* that uses the TPM to verify that the ML computations execute in the integrity-enforced OS.

The CAS performs the SGX attestation of the trusted boot service and provisions it with the TPM certificate as well as a list of the kernel integrity measurements. The trusted boot service reads the integrity measurements stored in PCRs using the TPM attestation protocol. The TPM genuineness is ensured by verifying the TPM certificate using a certificate chain provided by the CAS. The Linux kernel integrity is verified by comparing the integrity measurements certified by the TPM with the measurements read from the CAS.

We implemented the trusted boot service as an additional stage in the ML data processing. It enables other ML computations to access the confidential data only if the OS state conforms to the stakeholder's security policy. It copies the confidential data from an encrypted volume of one ML computation to a volume accessible to another ML computation after verifying the kernel integrity using the TPM. Our implementation is complementary with Linux unified key setup

(LUKS) [7]. LUKS allows the kernel to decrypt the file system only if the kernel integrity has not changed. This prevents accessing the trusted boot service's volume after modifying the kernel configuration, *i.e.*, disabling the integrity-enforcement mechanism.

5   Evaluation

Testbed. Experiments were executed on a ASUS Z170-A mainboard equipped with an Intel Core i7-6700K CPU supporting SGXv1, Nvidia GeForce RTX 2080 Super, 64 GiB of RAM, Samsung SSD 860 EVO 2 TB hard drive, Infineon SLB 9665 TPM 2.0, a 10 Gb Ethernet network interface card connected to a 20 Gb/s switched network. Hyper-threading is enabled. The enclave page cache (EPC) is configured to reserve 128 MB of RAM. CPUs are on the microcode patch level 0xe2. We run Ubuntu 20.04 with Linux kernel 5.4.0-65-generic. Linux IMA is enabled. The hashes of all OS files are digitally signed using a 1024-bit RSA asymmetric key. The signatures are stored inside files' extended attributes, and the certificate signed by the kernel's build signing key is loaded to the kernel's keyring during initrd execution.

Datasets. We use two datasets: (i) the classical CIFAR-10 image dataset [31], and (ii) the real-world medical dataset [47].

5.1   Attestation Latency

We run an experiment to measure the overhead of verifying the OS integrity using the TPM. Precisely, we measure how much time it takes an application implementing the trusted boot service to receive configuration from the security policy manager, read the TPM, and verify the OS integrity measurements.

The security policy manager executes on a different machine located in the same data center. It performs the SGX attestation before delivering a configuration containing two encryption keys – a typical setup for ML computations – and measurements required to verify the OS integrity. The security policy manager and the trusted boot service execute inside SCONE-protected Docker containers.

Table 1 shows that launching the application inside a SCONE-protected container takes 1573 ms. Running the same application that additionally receives the configuration from the security policy manager incurs 118 ms overhead. Additional 719 ms are required to read the TPM quote, verify the TPM integrity and authenticity, and compare the read integrity measurements with expected values provided by the security policy manager. As we show next, 2.5 s overhead required to perform SGX and TPM attestation is negligible considering the ML training execution time.

5.2   Security and Performance Trade-Off

To demonstrate the advantage of PERUN in allowing users to select the trade-off between security and performance, we compare the performance of different security levels provided by PERUN and the pure SGX based system

called SecureTF [34]. We run the model training using the following setups:
(i) only CPU (*Native*); (ii) GPU (*Native GPU*); (iii) PERUN, IMA enabled
(PERUN+*IMA*); (iv) PERUN, IMA and SGX enabled (PERUN+*IMA*+*SGX*); (v)
PERUN with GPU, IMA enabled (PERUN+*IMA*+*GPU*).

Table 1. End-to-end latency of verifying software authenticity and integrity using SGX
and TPM attestation. Mean latencies are calculated as 10% trimmed mean from ten
independent runs. *sd* stands for standard deviation.

	Execution time
Application in a container	1573 ms (sd=16 ms)
+ SGX attestation	1691 ms (sd=37 ms)
+ SGX and TPM attestation	2410 ms (sd=33 ms)

The *Native* and *Native GPU* levels represent scenarios where no security
guarantees are provided. PERUN+*IMA* and PERUN+*IMA*+*GPU* represent the
high integrity level (Sect. 3.3) in which ML training can execute directly on the
CPU or GPU (high performance) while require to extend trust to the operating
system (large TCB). Finally, PERUN+*IMA*+*SGX* represents the high-assurance
security level where all computations are performed inside the TEE (limited
performance) but requires a minimal amount of trust in the remote execution
environment (low TCB). In all setups, the trusted boot service executes inside
the enclave.

CIFAR-10 Dataset. We perform training using the CIFAR-10 dataset, a convolutional neural network containing four *conv* layers followed by two fully connected layers. We use *BatchNorm* after each conv layer. We apply the *ADAM*
optimization algorithm [28] with the learning rate set to 0.001.

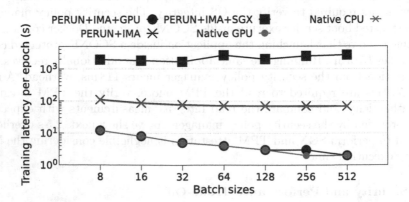

Fig. 5. The CIFAR-10 training latency comparison among different security levels
offered by PERUN. Mean latencies are calculated from five independent runs.

Figure 5 shows the training latency, and Fig. 6 shows the PERUN speedup depending on setups and batch sizes. At the high-assurance security level (PERUN +*IMA*+*SGX*), PERUN achieves almost the same performance as the pure SGX-based system, secureTF. This is because the training data is processed only inside the enclave, and SGX performs compute-intensive paging caused by the limited EPC size (128 MB) that cannot accommodate the training computation data (8 GB). PERUN +*IMA*+*GPU* and PERUN +*IMA* achieve 1321× and 40× speedup when relying just on the high integrity level compared to secureTF (batch size of 512). With these setups, the PERUN performance is similar to native systems (~0.96× of native latency) because the integrity protection mechanism performs integrity checks only when it loads files to the memory for the first time, leading to almost native execution afterward.

Real-World Medical Dataset. Next, we evaluate PERUN using a large-scale real-world medical dataset [47]. The dataset contains a wide range of medical images, including images of cancer and tumor treatment regimens for various parts of the human body, *e.g.*, brain, colon, prostate, liver, and lung. It was created via CT or MRI scans by universities and research centers from all around the world. We perform training over the brain tumor images dataset (6.1 GB) using the 2-D U-Net [44] TensorFlow architecture from Intel AI [4]. It makes use of the *ADAM* optimizer that includes 7 760 385 parameters with 32 feature maps. We set the learning rate to 0.001 and the batch size to 32.

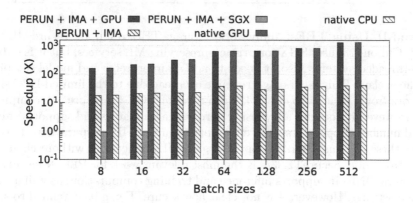

Fig. 6. The CIFAR-10 training speedup of evaluated systems in comparison to PERUN with the highest security level (PERUN + IMA + SGX).

Table 2 shows that at the high-assurance security level (the data is processed entirely inside the enclave), PERUN +*IMA*+*SGX* achieves the same performance as the referenced SGX-based system. However, when relying just on the high integrity level to protect the data, PERUN +*IMA*+*GPU* and PERUN +*IMA* achieve a speedup of 1559× and 47× compared to secureTF, respectively. We maintain the accuracy of 0.9875 in all experiments (dice coef: 0.5503, soft dice coef: 0.5503).

6   Related Work

Secure Multi-party Computation. Although cryptographic schemes, such as secure multi-party computation (MPC) and fully homomorphic encryption, are promising to secure multi-stakeholder ML computation, they have limited application in practice [42,52]. They introduce high-performance overhead [26,33,37,38,42], which is a limiting factor for computing-intensive ML, and require to heavily modify existing ML code. Furthermore, they do not support all ML algorithms, such as, deep neural networks. Some of them also require additional assumptions, like MPC protocol requiring a subset of honest stakeholders. Unlike PERUN, most of them lack support for training computation.

Table 2. The training latency comparison among different security levels of PERUN, secureTF, and native. The results were obtained from a single run.

System	Latency per epoch	Speedup
Native CPU	5 h 26 min 14 sec	47×
Native GPU	9 min 54 sec	1561×
PERUN+IMA	5 h 26 min 17 sec	47×
PERUN+IMA+SGX	257 h 27 min 49 sec	~ 1×
PERUN+IMA+GPU	9 min 55 sec	1560×
secureTF	257 h 43 min 53 sec	(baseline)

Secure ML Using TEEs. Many works leverage TEE to support secure ML [21, 23,42]. Chiron [23] uses SGX for privacy-preserving ML services, but it is only a single-threaded system. Also, it needs to add an interpreter and model compiler into the enclave. This incurs high runtime overhead due to the limited EPC size. The work from Ohrimenko et al. [42] also relies on SGX for secure ML computations. However, it does not allow using hardware accelerators and supports only a limited number of operators—not enough for complex ML computations. In contrast to these systems, PERUN supports legacy ML applications without changing their source code. SecureTF [34] is the most relevant work for PERUN because it also uses SCONE. It supports inference and training computation, as well as distributed settings. However, it is not clear how secureTF can be extended to support secure multi-stakeholders ML computation. Also, secureTF does not support hardware accelerators, making it less practical for training computation. Other works [6,40,48] use SGX and untrusted GPUs for secure ML computations. They split ML computations into trusted parts running in the enclave and untrusted parts running in the GPU. However, they require changing the existing code and do not support multi-stakeholder settings.

Trusted GPUs. Although trusted computation on GPUs is not commercially available, there is ongoing research. HIX [24] enables memory-mapped I/O access from applications running in SGX by extending an SGX-like design with duplicate versions of the enclave memory protection hardware. Graviton [53] proposes hardware extensions to provide TEE inside the GPU directly. Graviton requires modifying the GPU hardware to disable direct access to the critical GPU interfaces, e.g., page table and communication channels from the GPU driver. Telekine [22] restricts access to GPU page tables without trusting the kernel driver, and it secures communication with the GPU using cryptographic schemes. The main limitation of these solutions is that they require hardware modification of the GPU design, so they cannot protect existing ML computations, and they also do not support multi-stakeholder ML computations.

7   Conclusion

PERUN allows multiple stakeholders to perform ML without revealing their intellectual property. It provides strong confidentiality and integrity guarantees at the performance of existing TEE-based systems. With the help of trusted computing, PERUN permits utilizing hardware accelerators, reaching native hardware-accelerated systems' performance at the cost of a larger trusted computing base. When training an ML model using real-world datasets, PERUN achieves $0.96\times$ of native performance execution on the GPU and a speedup of up to $1560\times$ compared to the state-of-the-art SGX-based system.

Acknowledgment. We thank the anonymous reviewers for their insightful comments and suggestions as well as Maksym Planeta and Hieu Le for their feedback and help. This work has received funding from the European Union's Horizon 2020 research and innovation program under the AI-Sprint project (ai-sprint-project.eu), grant agreement No 101016577.

References

1. Regulation (EU) 2016/679 of the European parliament and of the council of 27 April 2016 on the protection of natural persons with regard to the processing of personal data and on the free movement of such data, and repealing directive 95/46/EC (2016)
2. Abadi, M., et al.: TensorFlow: a system for large-scale machine learning. In: 12th USENIX Symposium on Operating Systems Design and Implementation (OSDI 2016), pp. 265–283 (2016)
3. Abadi, M., et al.: Deep learning with differential privacy. In: Proceedings of the 2016 ACM SIGSAC Conference on Computer and Communications Security (CCS 2016), pp. 308–318 (2016)
4. Intel AI: Deep Learning Medical Decathlon Demos for Python. https://github.com/IntelAI/unet/. Accessed Feb 2021
5. Arnautov, S., et al.: SCONE: secure linux containers with Intel SGX. In: 12th USENIX Symposium on Operating Systems Design and Implementation (OSDI 2016), pp. 689–703 (2016)
6. Asvadishirehjini, A., Kantarcioglu, M., Malin, B.: GOAT: GPU Outsourcing of Deep Learning Training With Asynchronous Probabilistic Integrity Verification Inside Trusted Execution Environment. arXiv preprint arXiv:2010.08855 (2020)
7. Broz, M.: LUKS2 on-disk format specification, version 1.0.0. In: LUKS Documentation (2018)

8. Chakrabarti, S., Baker, B., Vij, M.: Intel SGX Enabled Key Manager Service with OpenStack Barbican. arXiv preprint arXiv:1712.07694 (2017)
9. Chen, G., Chen, S., Xiao, Y., Zhang, Y., Lin, Z., Lai, T.H.: SgxPectre: stealing intel secrets from SGX enclaves via speculative execution. In: 2019 IEEE European Symposium on Security and Privacy (EuroS&P), pp. 142–157 (2019)
10. Intel Corporation: Trusted Boot (tboot). https://sourceforge.net/projects/tboot/. Accessed May 2021
11. Costan, V., Devadas, S.: Intel SGX explained. IACR Cryptol. ePrint Arch. **2016**(86), 1–118 (2016)
12. Emont, J., Stevens, L., McMillan, R.: Amazon Investigates Employees Leaking Data for Bribes. https://www.wsj.com/articles/amazon-investigates-employees-leaking-data-for-bribes-1537106401. Accessed Feb 2021
13. Fredrikson, M., Jha, S., Ristenpart, T.: Model inversion attacks that exploit confidence information and basic countermeasures. In: Proceedings of the 22nd ACM SIGSAC Conference on Computer and Communications Security (CCS 2015), pp. 1322–1333 (2015)
14. Gallery, E., Mitchell, C.J.: Trusted computing: security and applications. Cryptologia **33**(3), 217–245 (2009)
15. Gentry, C.: Fully homomorphic encryption using ideal lattices. In: Proceedings of the 41st Annual ACM Symposium on Theory of Computing, pp. 169–178 (2009)
16. Goetzfried, J., Eckert, M., Schinzel, S., Mueller, T.: Cache attacks on Intel SGX. In: Proceedings of the 10th European Workshop on Systems Security (EuroSec 2017) (2017)
17. Greene, J.: Intel trusted execution technology: hardware-based technology for enhancing server platform security. Intel Corporation (2010)
18. Gregor, F., et al.: Trust management as a service: enabling trusted execution in the face of Byzantine stakeholders. In: 2020 50th Annual IEEE/IFIP International Conference on Dependable Systems and Networks (DSN 2020), pp. 502–514 (2020)
19. Trusted Computing Group: TPM Library Specification, family "2.0", level 00, revision 01.38. In: TCG Resources, TPM 2.0 Library (2016)
20. Trusted Computing Group: TCG Trusted Attestation Protocol (TAP) Information Model for TPM Families 1.2 and 2.0 and DICE Family 1.0. Version 1.0, Revision 0.36 (2019)
21. Grover, K., Tople, S., Shinde, S., Bhagwan, R., Ramjee, R.: Privado: Practical and Secure DNN Inference with Enclaves. arXiv preprint arXiv:1810.00602 (2018)
22. Hunt, T., Jia, Z., Miller, V., Szekely, A., Hu, Y., Rossbach, C.J., Witchel, E.: Telekine: secure computing with cloud GPUs. In: 17th USENIX Symposium on Networked Systems Design and Implementation (NSDI 2020) (2020)
23. Hunt, T., Song, C., Shokri, R., Shmatikov, V., Witchel, E.: Chiron: Privacy-preserving Machine Learning as a Service. arXiv preprint arXiv:1803.05961 (2018)
24. Jang, I., Tang, A., Kim, T., Sethumadhavan, S., Huh, J.: Heterogeneous isolated execution for commodity GPUs. In: Proceedings of the Twenty-Fourth International Conference on Architectural Support for Programming Languages and Operating Systems (ASPLOS 2019), pp. 455–468 (2019)
25. Johnson, S., Scarlata, V., Rozas, C., Brickell, E., Mckeen, F.: Intel software guard extensions: EPID provisioning and attestation services. White Paper **1**(1–10), 119 (2016)
26. Juvekar, C., Vaikuntanathan, V., Chandrakasan, A.: GAZELLE: a low latency framework for secure neural network inference. In: Proceedings of the 27th USENIX Conference on Security Symposium (USENIX Security), pp. 1651–1668 (2018)

27. Khandaker, M.R., Liu, W., Naser, A., Wang, Z., Yang, J.: Origin-sensitive control flow integrity. In: 28th USENIX Security Symposium (USENIX Security 2019), pp. 195–211 (2019)
28. Kingma, D.P., Ba, J.: Adam: a method for stochastic optimization. arXiv preprint arXiv:1412.6980 (2014)
29. Knauth, T., Steiner, M., Chakrabarti, S., Lei, L., Xing, C., Vij, M.: Integrating remote attestation with transport layer security. arXiv preprint arXiv:1801.05863 (2018)
30. Kocher, P., et al.: Spectre attacks: exploiting speculative execution. In: 2019 IEEE Symposium on Security and Privacy (SP), pp. 1–19. IEEE (2019)
31. Krizhevsky, A., Hinton, G.: Learning multiple layers of features from tiny images. Technical report, Citeseer (2009)
32. Kumar, A., Kashyap, A., Phegade, V., Schrater, J.: Self-Defending Key Management Service (SDKMS) with Intel Software Guard Extensions (SGX). White Paper (2018)
33. Kumar, N., Rathee, M., Chandran, N., Gupta, D., Rastogi, A., Sharma, R.: CrypTFlow: secure TensorFlow inference. In: IEEE Symposium on Security and Privacy (S&P 2020), pp. 336–353 (2020)
34. Le Quoc, D., Gregor, F., Arnautov, S., Kunkeland, R., Bhatotia, P., Fetzer, C.: secureTF: a secure TensorFlow framework. In: Proceedings of the 21th International Middleware Conference (Middleware), pp. 44–59 (2020)
35. Matsakis, N.D., Klock, F.S.: The rust language. ACM SIGAda Ada Lett. **34**, 103–104 (2014)
36. McKeen, F.: Innovative instructions and software model for isolated execution. In: Proceedings of the 2nd International Workshop on Hardware and Architectural Support for Security and Privacy (HASP 2013) (2013)
37. Mishra, P., Lehmkuhl, R., Srinivasan, A., Zheng, W., Ada Popa, R.: Delphi: a cryptographic inference service for neural networks. In: 29th USENIX Security Symposium (USENIX Security 2020), pp. 2505–2522 (2020)
38. Mohassel, P., Zhang, Y.: SecureML: a system for scalable privacy-preserving machine learning. In: 2017 IEEE Symposium on Security and Privacy (S&P 2017), pp. 19–38 (2017)
39. muslc: musl libc. https://musl.libc.org. Accessed Feb 2021
40. Ng, L.K., Chow, S.S., Woo, A.P., Wong, D.P., Zhao, Y.: Goten: GPU-outsourcing trusted execution of neural network training and prediction. In: 35th AAAI Conference on Artificial Intelligence (2019)
41. Noor, T.H., Sheng, Q.Z., Zeadally, S., Yu, J.: Trust management of services in cloud environments: obstacles and solutions. ACM Comput. Surv. **46**, 1–30 (2013)
42. Ohrimenko, O., et al.: Oblivious multi-party machine learning on trusted processors. In: Proceedings of the 25th USENIX Conference on Security Symposium, pp. 619–636 (2016)
43. Reuters: Ex-Microsoft employee charged with leaking trade secrets to blogger. https://www.reuters.com/article/us-microsoft-tradesecret-idUSBREA2J07K20140320. Accessed Feb 2021
44. Ronneberger, O., Fischer, P., Brox, T.: U-Net: convolutional networks for biomedical image segmentation. In: Navab, N., Hornegger, J., Wells, W.M., Frangi, A.F. (eds.) MICCAI 2015. LNCS, vol. 9351, pp. 234–241. Springer, Cham (2015). https://doi.org/10.1007/978-3-319-24574-4_28
45. Sailer, R., Zhang, X., Jaeger, T., Van Doorn, L.: Design and implementation of a TCG-based integrity measurement architecture. In: USENIX Security Symposium, pp. 223–238 (2004)

46. Shin, J., et al.: TCG D-RTM Architecture, Document Version 1.0.0. Trusted Computing Group (2013)
47. Simpson, A.L., et al.: A large annotated medical image dataset for the development and evaluation of segmentation algorithms. arXiv preprint arXiv:1902.09063 (2019)
48. Tramèr, F., Boneh, D.: Slalom: fast, verifiable and private execution of neural networks in trusted hardware. In: 7th International Conference on Learning Representations (ICLR) (2019)
49. Tsai, C.C., Porter, D.E., Vij, M.: Graphene-SGX: a practical library OS for unmodified applications on SGX. In: Proceedings of the 2017 USENIX Conference on USENIX Annual Technical Conference (USENIX ATC 2017), pp. 645–658 (2017)
50. Scontain UG: SCONE Configuration and Attestation Service (CAS). https://sconedocs.github.io/CASOverview/. Accessed Feb 2021
51. Van Bulck, J., et al.: Foreshadow: extracting the keys to the intel SGX kingdom with transient out-of-order execution. In: 27th USENIX Security Symposium (USENIX Security 2018), pp. 991–1008 (2018)
52. Volgushev, N., Schwarzkopf, M., Getchell, B., Varia, M., Lapets, A., Bestavros, A.: Conclave: secure multi-party computation on big data. In: Proceedings of the 14th EuroSys Conference (EuroSys 2019) (2019)
53. Volos, S., Vaswani, K., Bruno, R.: Graviton: trusted execution environments on GPUs. In: Proceedings of the 13th USENIX Conference on Operating Systems Design and Implementation (OSDI 2018), pp. 681–696 (2018)
54. Weisse, O., et al.: Foreshadow-NG: breaking the virtual memory abstraction with transient out-of-order execution. Technical report (2018)
55. Xu, T., et al.: Do not blame users for misconfigurations. In: Proceedings of the Twenty-Fourth ACM Symposium on Operating Systems Principles (SOSP 2013) (2013)
56. Yao, A.C.: Protocols for secure computations. In: 23rd IEEE Annual Symposium on Foundations of Computer Science (SFCS 1982), pp. 160–164 (1982)
57. Zeller, A., Gopinath, R., Böhme, M., Fraser, G., Holler, C.: The fuzzing book (2019)
58. Zinzindohoué, J.K., Bhargavan, K., Protzenko, J., Beurdouche, B.: HACL*: a verified modern cryptographic library. In: Proceedings of the 2017 ACM SIGSAC Conference on Computer and Communications Security, pp. 1789–1806 (2017)

PDF Malware Detection Using Visualization and Machine Learning

Ching-Yuan Liu[1]([envelope]), Min-Yi Chiu[2], Qi-Xian Huang[2], and Hung-Min Sun[1]

[1] Department of Computer Science, National Tsing Hua University, Hsinchu, Taiwan
joeyliu7878@gapp.nthu.edu.tw, hmsun@cs.nthu.edu.tw
[2] Institute of Information Systems and Applications, National Tsing Hua University, Hsinchu, Taiwan
s106065803@m106.nthu.edu.tw, xiangg800906three@gapp.nthu.edu.tw

Abstract. Recently, as more and more disasters caused by malware have been reported worldwide, people started to pay more attention to malware detection to prevent malicious attacks in advance. According to the diversity of the software platforms that people use, the malware also varies pretty much, for example: Xcode Ghost on iOS apps, FakePlayer on Android apps, and WannaCrypt on PC. Moreover, most of the time people ignore the potential security threats around us while surfing the internet, processing files or even reading email. The Portable Document Format (PDF) file, one of the most commonly used file types in the world, can be used to store texts, images, multimedia contents, and even scripts. However, with the increasing popularity and demands of PDF files, only a small fraction of people know how easy it could be to conceal malware in normal PDF files. In this paper, we propose a novel technique combining Malware Visualization and Image Classification to detect PDF files and identify which ones might be malicious. By extracting data from PDF files and traversing each object within, we can obtain the holistic tree-like structure of PDF files. Furthermore, according to the signature of the objects in the files, we assign different colors obtained from SimHash to generate RGB images. Lastly, our proposed model trained by the VGG19 with CNN architecture achieved up to 0.973 accuracy and 0.975 F1-score to distinguish malicious PDF files, which is viable for personal, or enterprise-wide use and easy to implement.

Keywords: Malware detection · PDF malware · Malware visualization · Machine learning

1 Introduction

The goal of malware detection is to identify whether a file is malicious or belongs to a certain malware family. Though various malware detection techniques have been proposed, they can be divided into two categories: static analysis and dynamic analysis. Since static analysis focuses on the content or the signatures of

© IFIP International Federation for Information Processing 2021
Published by Springer Nature Switzerland AG 2021
K. Barker and K. Ghazinour (Eds.): DBSec 2021, LNCS 12840, pp. 209–220, 2021.
https://doi.org/10.1007/978-3-030-81242-3_12

a certain malware, it has the advantage of being more time-efficient while at the cost of inaccuracy. Dynamic analysis observes and records a malware's behavior, which makes it more accurate but usually is time-consuming. Researchers have applied static and dynamic analysis techniques to examine common file formats in several platforms, such as executable files on ×64 machines or .apk files on Android phones. In this paper, we only focus on Portable Document Format (PDF), which is supported by most platforms.

PDF, released in 1987 by Adobe, is currently one of the most used file formats worldwide. The main reason is that PDF files can be recognized and human-readable in most settings. Moreover, they can contain texts as well as various types of information, such as images, audios, scripts, or other files. Some PDF files even have forms or buttons inside, making them more interactive and richful. Due to their interactiveness and richness, some PDF files have been exploited by hackers to conduct malicious behaviors such as executing embedded javascript while opening the document, moving the mouse cursor to download and launch the malware to steal sensitive information of users, or encrypt certain types of files in the machine [1,2]. Evidence shows that emails with malicious PDF attachments have caused severe damages to businesses and governments [3]. Therefore, the techniques of how to detect malicious PDF files in advance before opening them have been proposed and gained more and more attention nowadays.

Our proposed method is a novel technique that employs the image classification for PDF malware detection. A PDF file usually consists of multiple objects, which can be analogous to nodes in a huge tree structure that each node has a parent node and multiple children nodes. Taking advantage of this characteristic of PDF, we trace the tree-like structure to extract all objects in PDF files by a specific order. After applying the malware visualization technique, we build a dataset containing images of PDF files. With those images obtained from PDF files at hand, we can then train our malware detection model to distinguish the malicious PDF files from the benign ones.

The rest of the paper is organized as follows. In Sect. 2 we will briefly describe the structure and the characteristics of PDF files. Section 3 will be discussing the related works of malware detection on PDF files. More detail of our experiment and proposed technique will be discussed in Sect. 4. We will present our result in Sect. 5. And finally, our conclusions will be narrated in Sect. 6.

2   Background

PDF files are normally divided into four sections: Header, Body, 'xref' Table, and Trailer as shown in Fig. 1.

Fig. 1. The main structure of PDF.

The *header* specifies the PDF version of the file and some hidden characters to inform software to recognize the file as the format of PDF. The *body*, as the biggest part of a PDF file, contains all the main contents of the file, including all the objects. Each object may contain multiple tags, and indirect objects to more detailedly describe the corresponding object. For several files with a special purpose, objects may also attach streams with information of text, images, or even scripts. The main purpose of the *'xref' table* is to store the relative address of objects in each section in order to optimize the efficiency of object finding. Lastly, the *trailer*, which is the starting point to read a PDF file, instructs the object number of the root object, the size of the PDF file, and some additional information, including the most important part about the address of 'xref' table.

Because of the flexibility PDF provides, attackers utilize a variety of methods to sneak any kind of malicious scripts or make the target files obfuscate to evade malware detection tools. Such as dividing a file into multiple sections to create multiple 'xref' tables, deleting size or root object's information in trailer, or utilize one or more encoding methods to encrypt scripts inside an object. These alterations will not damage the readability of a PDF file, but make it vulnerable for attackers to steal sensitive information.

To reduce the detection error caused by an attacker's obfuscation, our method focuses on visualizing PDF files to seek the relative pattern of malicious files despite them having been obfuscated.

3   Related Work

More and more researchers utilize machine learning in the field of malware detection. However, only a few specialists take advantage of image and object detection, which results in better efficiency and accuracy.

O'Shaughnessy et al. [4] utilized byte plot technique to fill the bytes of executable files into RGB images with three kinds of Space-Filling Curve patterns, including Z-order, Gray-code, and Hilbert curves. They extracted image features with Local Binary Patterns (LBP), Gabor filters, and Histogram of Gradients (HOG) and trained them with K-Nearest Neighbor (KNN), Random Forest (RF), and Decision Trees (DT) models. After all, the results of the proposed method were compared with GIST Byte Plot Method [5].

Fu et al. [6] focused on analyzing files of PE format. After section division, which divided a file into code section, data section, and other natural sections, they calculated the entropy value, byte value, and relative section size of the sections independently. Besides image malware visualization, they also recorded the ASCII strings that appeared in the file. With the image features and the string features, they trained their method with multiple classifiers in order to achieve high accuracy and performance.

Bhodia et al. [7] transferred binary files into grayscale images directly. After feature extraction they trained the dataset with a simple KNN classifier. Darus et al. [8] also transformed files into grayscale images, but they focused on .apk files for Android applications.

Although Kapoor et al. [9] did not do malware visualization, they extracted the control flow of malwares while dividing codes into multiple basic blocks. By traversing the basic code blocks, extracting features and opcode sequence, they achieve high accuracy with multiple machine learning classifiers. Han et al. [10] also focused on analyzing the control flow of malwares. But after parsing, extracting features, and hashing with specific hash functions, they filled in colors for each pixel accordingly and created a unique RGB image for every malware sample. After the process, they successfully differentiated different malware families in a vast malware dataset.

While researchers mentioned above focused on a wide variety of file formats, such as PE files, normal executable files, mobile applications, etc., few of them noticed the dangers of PDF files [11–13].

Filiol et al. [14] and Maiorca et al. [15] did detailed investigation of multiple malicious PDF files detection techniques proposed in recent years. According to the investigation, we can acknowledge that even fewer researchers combine malware visualization with PDF malware detection.

The most inspiring research was done by Corum et al. [16]. Through byte plot and Markov plot they directly transferred raw PDF files into grayscale images and utilized SIFT and ORB to extract keypoints of the images as the main features of the files. Besides keypoint features, they also extract texture features of images, including LBP, local entropy, and Gabor filter. By comparing multiple combinations of keypoint features and texture features, their method results in an accurate model for object detection. While directly extracting features from grayscale images without examining the structure or even contents in PDF files, their method lacks a reasonable explanation.

Through the proposed methods mentioned above, we can easily notice that there are multiple malware visualization techniques done to all kinds of file formats, but fewer of them focused on PDF files. With all the inspiration from those researches, this paper contributed a novel malware visualization technique combined with machine learning which is tailored for PDF files.

4   Methodology

4.1   Overview

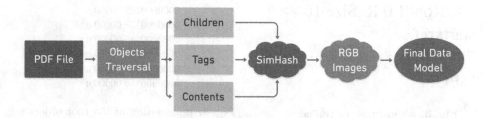

Fig. 2. An overview of the proposed method.

In our observation, an increasing amount of PDFs containing various mutations of malicious scripts has been exploited by hackers nowadays to lure the victims to open or browse the file, which usually comes as an attachment of email. Those malicious PDFs contain similar patterns of scripts, structure of tags and contents in objects, and tree-like composition order.

Our proposed method consisting of several processing steps is shown in Fig. 2. First, we traverse the whole PDF file by its structure defined in the PDF specification. Next, we extract three main features, including children, tags, and contents. In step three, we apply the SimHash algorithm in order to compare the similarity between image results. After hashing, we assign different colors according to the hash value of every object and generate an RGB image for each PDF file. Finally, we train our malware detection model with the images obtained from the previous step. Details will be described in the following sections.

4.2   PDF Objects Traversal

According to the PDF structure mentioned in section II, a standard PDF structure can be divided into four parts: header, body, 'xref' table and trailer [17]. The correct way to parse a PDF file should start from the trailer to the header.

From the trailer, we can obtain the detailed information of the file, including which object is the root, how many objects are there in the file, and the relationships among the objects as shown in Fig. 3. Therefore, attackers will not let such information be easily discovered, they will try to obfuscate or even remove the information which in fact will not affect the readability of the file. Apart from the trailer, we can also get important information by reading the two numbers in the first line of the 'xref' table as shown in Fig. 4, which indicate the first object and the size of the section, respectively. Due to the flexibility characteristic of a PDF file, there might be multiple 'xref' tables and trailers, which in turn results in multiple root objects for different sections.

trailer

<< /Root 1 0 R /Size 10 >>

startxref

7989

%EOF

Fig. 3. An example of trailer.

```
xref
0 10
0000000000 65535 f
0000000017 00000 n
0000000109 00000 n
0000000169 00000 n
0000000267 00000 n
0000000328 00000 n
0000000686 00000 n
0000000789 00000 n
0000006784 00000 n
0000006886 00000 n
```

Fig. 4. Information of the root object and total objects in the section can be observed from the first line of 'xref' Table.

After identifying the root object, we start the process of object traversal. During the traversal, we extract the three main features from every object as shown in Fig. 5.

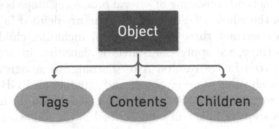

Fig. 5. The three main features we extract from each object.

According to the PDF structure, under an object there might be multiple data types that start with "/" to describe it. We treat them as the tags of an object. Next, some tags can be followed by other objects called indirect objects that contain more information about the tag. Those indirect objects are treated as the object's children as shown in Fig. 6.

Finally, many malicious PDF files can contain scripts which will be triggered when opening the files, moving the mouse cursor on the document, or clicking any (in)visible button in the files. In order to take the plain instead of encrypted contents as features, streams with suspicious properties such as "JS",

4 0 obj <<[/Type]/Action /S /JavaScript /JS [5 0 R] >> endobj

Fig. 6. Sample of tags (red rectangle) and children (yellow rectangle) relatively. (Color figure online)

"JavaScript", "MacroForm" and "XFA" are decompressed and saved as objects' contents.

By doing depth first search (DFS) on all children objects, we can simulate the tree-like structure of the PDF file. Afterwards, tags and contents of every object in a PDF file will be extracted by the order of the tree-like structure of the PDF file. The complete procedure is shown in Algorithm 1.

Algorithm 1. PDF objects traversal

1: Obtain root objects through trailer and 'xref' table
2: **for** all root objects **do**
3:     **if** has contents with malicious tags **then**
4:         do flatedecode to decompress contents
5:     Put children in traversing queue
6:     Do DFS to traverse all the objects

4.3   Malware Visualization

After the preprocessing described above, a dataset containing features such as tags and contents of every object in a single PDF file is created. To find out the similarity of features among PDF files, tags and contents are combined into one string, which is computed by SimHash and stored as strings in binary format.

SimHash is a local sensitive function and is adopted by Google to distinguish duplicate web pages. The main characteristic of SimHash is that though the original texts are similar, the output after hashing will not be completely different like other hash functions.

Through SimHash, we can obtain a number which can be translated into binary representation. To prove that SimHash suits our experiment, we can know that the hamming distance is related to the similarity between two texts by observing the binary values. Afterwards, the binary values will be divided into three sections with equivalent length in order to retrieve the relative values of RGB pixels to form a color image as shown in Fig. 7.

To create an image formed by multiple RGB pixel blocks, two parameters are employed: color, which is given by the output value of SimHash; size, the height of an image block. To obtain the size, besides tags' length, we also have to retrieve the correct length of contents. To avoid some extremely lengthy scripts

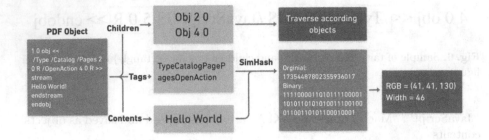

Fig. 7. Detailed example of feature extraction.

with unused padding words that attackers intend to evade malware detection, we reduce the contents' length by a specific ratio which we obtained by averaging all contents' length in our PDF dataset. Then the tags' and contents' length of a single object are accumulated and are made to be related to the height of a color block.

By iterating the steps shown in Algorithm 2, an RGB image can be formed with multiple color blocks with specific order representing different objects and the relative order in a PDF file. Lastly, for better classification, images are scaled to "200 × 150" as shown in Fig. 8.

Algorithm 2. Malware Visualization

1: **for** every traversed object **do**
2:     Do SimHash with Tags and Contents
3:     Acquire output from SimHash and transform into binary
4:     Divide binary string into 3 equal parts to get relative RGB color
5:     Reduce content length with specific ratio
6:     Accumulate Tags Length and Content Length to get the height of color block
7: Draw images with all the RGB color and Height
8: Scale images into 200 × 150

---

5   Experimental Result

To prove whether our result matches the assumption that currently most malicious PDFs are generated by certain manner. We utilize Structural Similarity Index Measure (SSIM), which is good at comparing the similarity by recognizing the whole structure of images.

First, we randomly chose a malicious image and a benign image. Then, we compare the chosen images with both malicious and benign datasets independently. Finally, the average SSIM value between each dataset is acquired as shown in Fig. 9. From the result, the similarity between malicious PDFs can

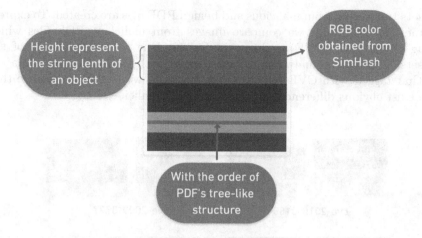

Fig. 8. Sample of the resulting image.

be acknowledged, while benign images seem to have more diversity related to malicious images. We can tell that malicious PDFs are much more similar than benign images, which implies our assumption and method are reasonable. The failure of the malware detection may be caused by the certain amount of similarities among the malicious and benign images.

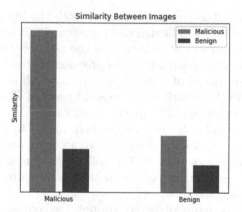

Fig. 9. SSIM comparison between malicious and benign images.

Our datasets consisting of 9000 malicious and benign PDF files each are from Contagio, which is a collection of malware samples and benign samples for comparison. After the malware visualization process described above, two image

218   C. Liu et al.

datasets transferred from malicious and benign PDF files are created. To express
our method concretely, we compare images from malicious PDF files which
belong to the same CVE as shown in Fig. 10. Due to the limitation of the
datasets, most of the malwares are not tagged with CVE numbers while the
rest that tagged with CVE are outdated. However, we can still recognize that
there exist obvious differences in patterns among malicious images.

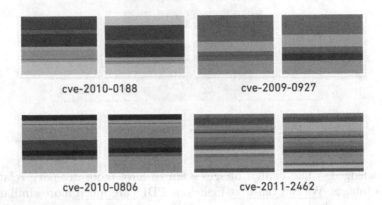

cve-2010-0188   cve-2009-0927

cve-2010-0806   cve-2011-2462

Fig. 10. Patterns between malicious PDF files from the same CVE can be easily
observed.

To further examine the proposed method, both the image datasets, which
contain more than 9000 files of benign and malicious PDF files each, are further
divided into training and validation sets with the ratio of 60% and 40% respectively. In order to prove the feasibility of our proposed method, the images will be
trained with a built-in model of Tensorflow library, which is the VGG19 model.

The VGG19 model [18], which was proposed by Visual Geometry Group from
Oxford University, is specialized in image classification. With multiple smaller
convolution layers instead of larger ones, VGG19 results in higher non-linearity
and fewer parameters required, making it more accurate and more efficient.

With the characteristics that VGG19 offered, we utilize it to prove our
method further. The performance of our method is measured by using two primary metrics: accuracy and F1-score.

The accuracy can be easily observed through the confusion matrix shown as
Fig. 11. To sum up, our proposed method results in high accuracy of 0.973 and
relative F1-score of 0.975.

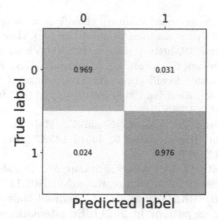

Fig. 11. Confusion matrix of the experimental result.

6 Conclusion

With portability, flexibility, and stability, Portable Document Format (PDF) becomes one of the most used file formats worldwide. People utilize PDF files to store documents, submit resumes or even transport sensitive information. The advantages and popularity also turn PDF files into huge targets for malicious attackers. Our proposed method is based on the idea of maximizing the utilization of PDF format's characteristics. First of all, after objects traversing and features extraction, we applied SimHash to retrieve relative RGB color values. When finishing the preprocessing to every PDF file, image datasets of benign and malicious PDF files are created. At last, the datasets are trained by the VGG19 model to achieve 0.973 accuracy and 0.975 F1-score. Our results show that the proposed method received relatively excellent performance. It indicates that there are some patterns for sure to recognize benign or malicious PDF files by turning them into images with reasonable methods.

For further research, besides optimizing our proposed method, distinguishing different malware families between a huge dataset of malwares will also be our main focus.

References

1. SentinelOne: Malicious PDFs - Revealing the Techniques Behind the Attacks. https://www.sentinelone.com/blog/malicious-pdfs-revealing-techniques-behind-attacks/. Accessed 27 Mar 2019
2. Cybersecurity Insiders: Cyber Attack with Ransomware hidden inside PDF Documents. https://www.cybersecurity-insiders.com/cyber-attack-with-ransomware-hidden-inside-pdf-documents/
3. Kaspersky: Top 4 dangerous file attachments. https://www.kaspersky.com/blog/top4-dangerous-attachments-2019/27147/. Accessed 31 May 2019

4. O'Shaughnessy, S.: Image-based malware classification: a space filling curve approach. In: 2019 IEEE Symposium on Visualization for Cyber Security (VizSec), pp. 1–10. IEEE, October 2019. https://doi.org/10.1109/VizSec48167.2019.9161583

5. Nataraj, L., Karthikeyan, S., Jacob, G., Manjunath, B.S.: Malware images: visualization and automatic classification. In: Proceedings of the 8th International Symposium on Visualization for Cyber Security, pp. 1–7, July 2011. https://doi.org/10.1145/2016904.2016908

6. Fu, J., Xue, J., Wang, Y., Liu, Z., Shan, C.: Malware visualization for fine-grained classification. IEEE Access **6**, 14510–14523 (2018). https://doi.org/10.1109/ACCESS.2018.2805301

7. Bhodia, N., Prajapati, P., Di Troia, F., Stamp, M.: Transfer learning for image-based malware classification. arXiv preprint arXiv:1903.11551 (2019)

8. Darus, F.M., Ahmad, N.A., Ariffin, A.F.M.: Android Malware classification using XGBoost on data image pattern. In: 2019 IEEE International Conference on Internet of Things and Intelligence System (IoTaIS), pp. 118–122. IEEE, November 2019. https://doi.org/10.1109/IoTaIS47347.2019.8980412

9. Kapoor, A., Dhavale, S.: Control flow graph based multiclass malware detection using bi-normal separation. Defence Sci. J. **66**(2), 138–145 (2016). https://doi.org/10.14429/dsj.66.9701

10. Han, K., Kang, B., Im, E.G.: Malware analysis using visualized image matrices. Sci. World J. **2014**, 106–120 (2014). https://doi.org/10.1155/2014/132713

11. Laskov, P., Šrndić, N.: Static detection of malicious JavaScript-bearing PDF documents. In: Proceedings of the 27th Annual Computer Security Applications Conference, pp. 373–382, December 2011. https://doi.org/10.1145/2076732.2076785

12. Maiorca, D., Ariu, D., Corona, I., Giacinto, G.: A structural and content-based approach for a precise and robust detection of malicious PDF files. In: 2015 International Conference on Information Systems Security and Privacy (ICISSP), pp. 27–36. IEEE, February 2015

13. Smutz, C., Stavrou, A.: Malicious PDF detection using metadata and structural features. In: Proceedings of the 28th Annual Computer Security Applications Conference, pp. 239–248, December 2012. https://doi.org/10.1145/2420950.2420987

14. Blonce, A., Filiol, E., Frayssignes, L.: Portable document format (pdf) security analysis and malware threats. In: Presentations of Europe BlackHat 2008 Conference, March 2008

15. Maiorca, D., Biggio, B.: Digital investigation of PDF files: unveiling traces of embedded malware. IEEE Secur. Privacy **17** (2017). https://doi.org/10.1109/MSEC.2018.2875879

16. Corum, A., Jenkins, D., Zheng, J.: Robust PDF malware detection with image visualization and processing techniques. In: 2019 2nd International Conference on Data Intelligence and Security (ICDIS), pp. 108–114. IEEE, June 2019. https://doi.org/10.1109/ICDIS.2019.00024

17. Whitington, J.: PDF Explained: The ISO Standard for Document Exchange, 1st edn. O'Reilly Media, Newton (2011)

18. Simonyan, K., Zisserman, A.: Very deep convolutional networks for large-scale image recognition CORR.abs/1409.1556. arXiv preprint arXiv:1409.1556

Deep Learning for Detecting Network Attacks: An End-to-End Approach

Qingtian Zou[1](✉), Anoop Singhal[2], Xiaoyan Sun[3], and Peng Liu[1]

[1] The Pennsylvania State University, State College, USA
{qzz32,pxl20}@psu.edu
[2] National Institute of Standards and Technology, Gaithersburg, USA
anoop.singhal@nist.gov
[3] California State University, Sacramento, USA
xiaoyan.sun@csus.edu

Abstract. Network attack is still a major security concern for organizations worldwide. Recently, researchers have started to apply neural networks to detect network attacks by leveraging network traffic data. However, public network data sets have major drawbacks such as limited data sample variations and unbalanced data with respect to malicious and benign samples. In this paper, we present a new end-to-end approach to automatically generate high-quality network data using protocol fuzzing, and train the deep learning models using the fuzzed data to detect the network attacks that exploit the logic flaws within the network protocols. Our findings show that fuzzing generates data samples that cover real-world data and deep learning models trained with fuzzed data can successfully detect real network attacks.

Keywords: Network attack · Protocol fuzzing · Deep learning

1 Introduction

Cyberattacks happen constantly with growing complexity and volume. As one of the most prevalent ways to compromise enterprise networks, network attack remains a prominent security concern. It can lead to serious consequences such as large-scale data breaches, system infection, and integrity degradation, particularly when network attacks are employed in attack strategies such as advanced persistent threats (APT) [11,20]. Among the different types of network attacks, the *logic-flaw-exploiting network attacks*, which exploit the logic flaws within the protocol specifications or implementations, are very commonly seen. Detecting logic-flaw-exploiting network attacks is very important considering their common presence in APT campaigns. However, it is still a very challenging problem.

Network attack detection methods can mainly be classified into two categories: *host-independent* methods and *host-dependent* methods. The former solely relies on the network traffic, while the latter [8] depends on additional data

© IFIP International Federation for Information Processing 2021
Published by Springer Nature Switzerland AG 2021
K. Barker and K. Ghazinour (Eds.): DBSec 2021, LNCS 12840, pp. 221–234, 2021.
https://doi.org/10.1007/978-3-030-81242-3_13

collected on the victim hosts. The host-dependent methods have some evident drawbacks: they have fairly high deployment costs and operation costs; they are error-prone due to necessary manual configuration by human administrators. Therefore, host-independent detection methods are highly desired as they can decrease deployment and operation costs while reducing the attack surface of detection system. Unfortunately, we found that the existing host-independent methods, including the classical intrusion detection approaches, often fall short in detecting some well-known and commonly used network attacks.

Recently there is a trend for using machine learning (ML) and deep learning (DL) techniques to detect network attacks. Nevertheless, the DL approaches could also achieve mixed results [6], if they do not address the following two challenges. The first challenge is *useful data sets*. Neural networks require high-quality data and correct labels, which are hard to obtain in real world. Real-world network traffic is often flooded with benign packets, which makes labeling very difficult. Although public data sets [1,5,12,15,16] for network attacks are available, they are barely useful in detecting logic-flaw-exploiting network attacks due to unbalancing and different focuses. The second challenge is to *identify appropriate neural networks and train the models*. There are a variety of neural network architectures, including multi-layer perceptron (MLP), convolutional neural network (CNN), recurrent neural network (RNN), etc., which have different characteristics and capabilities. Questions such as which architecture works best for network attack detection, and how to tune the hyper-parameters within models for optimization, are not yet answered.

In this paper, we propose an end-to-end approach to detect the logic-flaw-exploiting network attacks. The end-to-end approach means it starts with acquiring data and ends with detecting attacks using the trained neural networks. To address the data generation challenge, we propose a new protocol fuzzing-based approach to generate the network traffic data. With protocol fuzzing, a large variety of *malicious* network packets for a chosen network attack can be generated at a fast speed. Since the network packets are all generated from the chosen network attacks, they can be labeled as malicious packets automatically without much human efforts. Protocol fuzzing can also generate data with more variations than real world data, or even data that are not yet observed in real world. Moreover, these merits remain when protocol fuzzing is leveraged to generate the needed *benign* network packets. It should be noted that our method is different from data synthesis. Data synthesis is to enhance existing data [9], while our method is to generate new data.

To address the neural network model training challenge, we propose the following procedures: 1) For network attacks (PtH) where we can identify fields of interest, we directly examine the data, and then propose the suitable data representation and neural network architecture. 2) For other network attacks that the field of interests are not obvious, such as DNS cache poisoning and ARP poisoning attacks, we apply different neural network architectures to find out the ones with best performance. We propose to use accuracy, F1 score, detection rate, and false positive rate as the metrics to evaluate the neural networks. All models are trained on the data set with fuzzing involved. We then select the models that work best and evaluate them further on both the fuzzing data set and real attack data set with no fuzzing involved.

The main contributions of this work include: 1) Proposing a DL based end-to-end approach to detect the logic-flaw-exploiting network attacks; 2) Proposing protocol fuzzing to automatically generate high-quality network traffic data for applying DL techniques; 3) Proposing and evaluating neural network models for logic-flaw-exploiting network attack detection; 4) Demonstrating the effectiveness of our approach with three classical logic-flaw-exploiting network attacks, including PtH attack, DNS cache poisoning attack, and ARP poisoning attack.

The remaining of the paper is structured as follows. Section 2 discusses related works. Section 3 presents our experiments setup. Section 4 presents evaluations. Section 5 presents some discussions. Section 6 presents conclusions.

2   Related Work

The research community has been tackling the network attack detection problem from different perspectives with both classical and novel approaches.

Traditional Network Attack Detection Approaches. Traditionally, people usually detect network attacks with approaches such as signature-based, rule-based, and anomaly detection-based methods. In the past, signature-based intrusion detection system (IDS) usually manually crafted signatures [17], which heavily depends on manual efforts. The current techniques focus more on automatic generation of signatures [10]. However, signatures need to be constantly updated to align with new attacks and signature-based detection can be easily evaded by slightly changing the attack payload. Similar problems also exist for rule-based methods [4], which constantly need updates to the rules. As for anomaly detection-based methods, although they require much less manual efforts for updating, they tend to raise too many false positives [2].

Traditional ML and DL for Network Attack Detection. Network attacks are essential for APTs. Some common network attack types include probing, DoS, Remote-to-local, etc. Both traditional ML and DL methods have been adopted for network attack detection. Some focus on one type of network attack and perform binary classifications. For example, MADE [14] employs ML to detect malware C&C network traffic, Ongun et al. [13] employs ML to detect botnet traffic, and DeepDefense [18] employs DL to detect distributed DoS (DDoS) attacks. Others [6,19] try multi-class classifications, which include one benign class and multiple malicious classes for different kinds of network attacks. The above-mentioned research works all use public data sets.

Network Data Sets for Training and Testing Detection Models. To apply DL for network attack detection, data sets are required. Commonly used public data sets include KDD99 [15], NSL-KDD [5], UNSW-NB15 [12], CICIDS2017 [16], and CSE-CIC-IDS2018 [1]. The public data sets are all generated in test-bed environments, with simulated benign and malicious activities.

However, we found that a main "missing piece" in these public data sets is that they do not focus on logical-protocol-flaw-exploiting network attacks such as PtH. (These data sets focus on worms, Botnets, backdoors, and DoS/DDoS.) Moreover, we observe that the data generation methodology used in these data sets, if being employed, would very likely result in data sets that fail to meet the unique training data requirements (e.g., the diversity in training data) of deep learning.

Protocol Fuzzing. Fuzzing is originally a black-box software testing technique, which reveals implementation bugs by feeding mutated data. A key function of fuzzers is to generate randomized data which still follows the original semantics. There are tools for building flexible and security-oriented network protocol fuzzers, such as SNOOZE [3]. Network protocol fuzzing frameworks such as AutoFuzz [7] were also presented. They either act as clients, constructing packets from the beginning, or act as proxies, modifying packets on the fly. We use protocol fuzzing for a different purpose to directly generate high-quality data sets for training neural networks. Instead of using the tools/frameworks mentioned earlier, we prepare our own fuzzing scripts for this specific purpose.

3   Data Generation and Detection Model Training

Since the available public data sets are barely useful for detecting the logic-flaw-exploiting network attacks, we generate comprehensive benign and malicious data sets from scratch. We have performed data generation for all three demonstration attacks including PtH, DNS cache poisoning, and ARP poisoning. ARP poisoning attack only requires one malicious packet for a successful attack, so we call it the single-packet attack. PtH and DNS cache poisoning attacks, however, need multiple malicious packets for one successful attack, so we call them multi-packet attacks. Due to page limits, we only discuss data generation details about multiple-packet attacks in this section because they are more complicated than single-packet attacks. All attacks are carried out thousands of times so that a fair amount of malicious data can be collected. Benign data generation also lasts long enough to gather the commensurate amount of data compared to malicious data. The network packet capturing is performed at the victim's side.

3.1   Protocol Fuzzing and the Implementation

In client-server enterprise computing, the server-side protocol implementations are often complex and error-prone. Hence, there is a need to thoroughly test the server-side implementation. Protocol fuzzing tools [3,7] are usually functioning at the client side to trigger unexpected errors on the server programs. A main difference between protocol fuzzing and software fuzzing is that the protocol specification, especially its state transition diagram, will be used to guide the fuzzing process. In this way, stateful fuzzing tests could be performed.

This paper leverages protocol fuzzing to change the contents of network packets, specifically, the values of some fields in the packets. If a field is to be fuzzed, it will be assigned with pre-determined values, rather than the values chosen by the network client program. The fuzzed fields are chosen based on the following steps: 1) All fields in the packet of the attack-specific protocol are considered. 2) One field on the list will be picked and fuzzed by assigning pre-determined values, rather than values that are normally provided by the network programs. 3) The success rate of the attack after fuzzing the field will be monitored. If the attack success rate is above 50%, it confirms that this field can be fuzzed. 4) After one field is fuzzed, the above steps will be repeated for the next field on the list, while keeping the already fuzzed field(s) still fuzzed.

An additional benefit of protocol fuzzing is that it can generate and cover malicious data samples which may otherwise be overlooked. In deep learning, the changed values for the fuzzing fields may make the malicious data samples misclassified as benign. With protocol fuzzing, if the malicious data are generated in attacks, they'll be labeled as malicious automatically. Thus, these malicious data samples won't be omitted in the malicious data set.

3.2   PtH

PtH Attack. PtH is a well-known technique for lateral movement. In remote login, plain text passwords are usually converted to hashes for authentication. Some authentication mechanisms only check whether hashes or the calculation results of them matches or not. PtH relies on these vulnerable mechanisms to impersonate normal users with dumped hashes. We assume that: (a) normal users use benign client programs that are usually authenticated through more reliable mechanisms other than just using hashes, and that (b) attackers cannot get the plain text passwords and have to rely on hashes to impersonate a normal user. We can capture the network packets at the server side and find out which kind of authentication mechanism is used by a user: the more reliable mechanism, or the vulnerable mechanism using only hashes. The login sessions using those vulnerable authentication mechanisms can then be identified as PtH attack.

Windows remote login processes, if not properly configured, can use such vulnerable authentication mechanisms. Windows remote login can be divided into three stages, protocol and mechanism negotiation (initial communication), authentication, and task-specific communication (afterwards communication). Each stage contains multiple network packets, and hashes are used in the authentication stage for impersonation. The authentication stage can be viewed as a sequence made up of client's authentication request, server's challenge, client's challenge response and server's authentication response. The client first sends a session setup request to the server; then the server responds to the client with a challenge; on receiving the challenge, the client uses the challenge and hashes to do calculations and sends back the result in challenge response packet; finally, the server verifies the result and sends back authentication response indicating whether authentication succeeds or not.

Data Generation. We set up a Windows 2012 Server R2 as the victim server machine, a Windows 7 as the user client machine, and another Kali Linux as the attacker machine. The data sets are automatically generated by protocol fuzzing, and the protocol of interest here is Server Message Block (SMB), or a newer version of it, denoted as SMB2. SMB/SMB2 provides functions including file sharing, network browsing, printing, and inter-process communication over a network. In our data generation, more than 15 fields are fuzzed in each SMB/SMB2 packets, including SMB flags, SMB capabilities, and fields in SMB header, etc. We leverage the PtH script in Metasploit Framework to launch the attack. The process is to start the Metasploit Framework, set exploit parameters, start the exploitation, and then wait 25 s while monitoring the attack status. If the waiting time is too short, the attack may be stopped before completion. While the console is waiting at the foreground, the exploitation is ongoing at the background. Network packets in all the three stages, initial communication, authentication, and afterwards communication, are fuzzed. After the exploitation, based on whether the attack succeeds or not, we may continue to establish C&C, like what a real attacker will do. (The C&C network traffic are mainly TCP packets, which are not used for attack detection. Details are discussed later.) Finally, we quit all possibly established sessions and the Metasploit Framework, and then either freshly start another fuzzing iteration to generate more data or stop. The sign of a successful PtH attack is an established reverse shell, which can be observed at the attacker's side.

The same fuzzing method has also been applied in the generation of benign data. We first prepare a list of normal commands, including files reading, writing, network interactions, etc. For each benign fuzzing iteration, we randomly choose a command from the list, and then use valid username, plain-text password, and tool to log in to the server and execute the command.

Due to fuzzing, not all PtH attempts or benign access attempts can be guaranteed to succeed. For failed PtH attempts, we remove them from malicious data because they do not generate real malicious impact, and they cannot be categorized as benign either because they are generated with attacker tools for malicious purpose. For failed benign accesses, we keep them in benign data, because normal users can also have failed logins due to mistakes like typos.

In one PtH attack, there are packets for initial communications, authentication and afterwards communications. One data sample consists of multiple packets, and those packets may come from one, two, or all of the three stages above. Besides, one complete PtH attack or benign activity most certainly contains more packets than one data sample can represent. When labeling, if the session is malicious, then all data samples generated from this session is labeled malicious, and the same is also true for the benign cases.

Detections. To detect PtH attack with neural networks, we have two key insights that help determine the representation of data samples: 1) Network communication for authentication is actually a sequence of network packets in certain order. An earlier packet can affect the packet afterwards. For example, the first several packets between a server and a client may be used to communicate and determine which protocol to use (e.g. SMB or SMB2), and packets afterwards will use the decided protocol. The attack is to get authenticated by the server, which requires a sequence of packets to accomplish. Therefore, each data sample should be a sequence of packets, rather than an individual packet. 2) PtH relies on authentication mechanisms that legitimate users usually don't use. The network packets for the benign and malicious authentication are different. Since both authentication methods use SMB/SMB2 packets, the differences between them thus exist in the fields of the SMB/SMB2 layer. Therefore, data in SMB/SMB2 layer is used for PtH detection. In addition, the differences of field values between benign and malicious authentication will be helpful to distinguish them. For this attack, we choose Long-short term memory (LSTM) as the architecture for the neural network.

Fields of interest in PtH reside in the SMB/SMB2 layer. They are cmd (2), flags (4), and NT_status (4). Numbers indicate the field lengths in bytes. For example, the cmd field in IP layer has 2 bytes.

3.3 DNS Cache Poisoning

DNS Cache Poisoning. A major functionality of DNS is to provide the mapping between the domain names and IP addresses. When a client program refers to a domain name, the domain name needs to be translated to an IP address. The DNS servers are responsible to perform such translation.

DNS cache poisoning attack can target local DNS servers. When the local DNS server receives a query which it does not have the corresponding records (first stage), it will inquire the global DNS server (second stage). On receiving the response (third stage), the local DNS server saves this record in its cache to avoid inquiring the global DNS again when receiving the same query. It then forwards the response to the user machine (fourth stage). However, the DNS server cannot verify the response at the third stage, and this is where the attacker can fool the local DNS server. Pretending as the global DNS server, the attacker can send a spoofed DNS response to the local DNS server with falsified DNS records. If the fake response arrives earlier than the real one, the local DNS server will save the falsified record to its cache and forward it to the user machine. When new queries about the same domain name comes in, the local DNS server will not query the global DNS server again because the corresponding record has been cached. Consequently, it will answer the user machine with the falsified record, until the record expires or the cache is flushed.

Data Generation. For this attack, ten fields, such as time to live values in different layers, are fuzzed. The test bed contains three machines: a local DNS server whose DNS cache is flushed periodically, a user machine which sends out DNS queries to the local DNS server periodically, and an attacker machine which sniffs for DNS requests sent by the local DNS server and answers them with spoofed responses as in the attack scenario, or does nothing otherwise.

In the malicious scenario, we make the user machine ask for the IP address of one specific domain name from the local DNS server using command *dig*. The domain name is one that does not have a corresponding record on the local DNS server, thus enabling the DNS cache poisoning attack towards it. The attacker machine sniffs for DNS queries with that specific domain name sent out from the local DNS server, and responds them with fuzzed DNS responses with falsified IP addresses. Then the DNS cache gets poisoned and the user machine gets the falsified DNS record. We keep the user machine sending out DNS queries periodically, so that the above process repeats many times and a large amount of data can be generated. However, as discussed earlier, if the local DNS server has the record for the domain name in its cache, it will not send out DNS queries for it. This is why we flush the DNS cache of the local DNS server, so that it remains vulnerable in different iterations. If the attack is successful, the falsified IP addresses can be seen on the results of *dig*.

In the benign scenario, we prepare a list containing 4098 domain names. In each iteration, the user machine randomly chooses one domain name from the list, and sends a request to the local DNS server. To resemble the malicious scenario, the cache of local DNS server is also flushed periodically so that the local DNS server always needs to communicate with the global DNS server.

The domain name used in the malicious scenario and the domain names used in the benign scenario do not overlap. Both the domain names and the IP addresses (falsified or genuine) are excluded during training, which can be treated as signatures. Because DNS cache poisoning is a multi-packet attack, the labeling to data samples is also based on sessions, similar to PtH attack.

Detections. Network packets from DNS cache poisoning attack form sessions which consist of queries and answers. Therefore, each data sample should include data from multiple network packets. In addition, it is not clear which fields may be of importance, so we need to investigate the packet content, rather than simply generalizing the packets with packet types as we did in PtH detection. The data samples are processed to be image-like. That is, each row represent one packet, and each element in the row represent one byte in that packet. We use a convolutional neural network (CNN) to do the classifications, which has been proven to work well in image classification problems. The labeling is done towards each data sample, which is the entire matrix, rather than an individual packet. Matrices generated from malicious data are labeled as malicious, and matrices from benign data are labeled as benign. Similar to PtH detection, we have trained a series of neural networks with different settings for comparisons.

Fields of interest in DNS cache poisoning attack reside in the IP, UDP and DNS layers. In the IP layer, fields of interest are Version (4/8), IHL (4/8), DSF (1), TLen (2), ID (2), Flags (3/8), FragOff (13/8), TTL (1), port (1), and chksum (2). In the UDP layer, fields of interest are src_port (2), dst_port (2), hd_len (2), and chksum (2). In the DNS layer, fields of interest are TID (2), flags (2), q (2), AnRR (2), AuRR (2), and AdRR (2). Numbers indicate the field lengths in bytes. For example, the FragOff field in IP layer has 13/8 bytes, meaning that this field consists of 13 bits (one byte equals to eight bits).

4   Evaluations

This section provides the evaluation results of the three demonstration attacks on the selected best-performing and best-detecting models. For comparison with DL models, we have also trained traditional ML models, including k-nearest neighbor (kNN) models, support vector machine (SVM) models with various kernels, decision tree (DT) models, and random forest (RF) models. They are trained, selected, and evaluated on the same data sets. For PtH and ARP poisoning, the traditional ML models' data samples and features are the same as those for DL models. However, for DNS cache poisoning, the same data sample and feature cannot be used because the input space is too large for traditional ML models to handle. Therefore, we employed principal component analysis (PCA) for dimension reduction, and only select the top-rated one-fifth PCA features. On average, they can explain about 97.09% of the original data.

4.1   Model Selection

For model selection, we consider not only the perspective of neural network performance, but also the perspective of security. We use accuracy Acc and F1 score ($F1$), two commonly used metrics, to measure the classification, and use detection rate (DR) and false positive rate (FPR) for attack detection effectiveness. DR shows the detector's ability of detecting attacks. FPR shows how likely the detector raises false alarms. We call the best-performing model as the one that gets the highest average of Acc and $F1$, denoted as $P = \frac{Acc+F1}{2}$, and the best-detecting model as the one that gets the highest average of DR and $1 - FPR$, denoted as $D = \frac{DR+1-FPR}{2}$. If FPR cannot be calculated (no benign data sample), we let $D = DR$. We simply take the average because all the chosen metrics are equally important for evaluations.

The generated fuzzing data set is randomly split into two parts: 80% as the training set, and 20% as the test set. The training set is then further randomly split into four parts of about the same size, upon which 4-fold cross-validation is employed to avoid over-fitting. All the reported results are the average results among four folds. The best-performing and best-detecting models are selected based on the average P and D results on the validation set across all four folds.

4.2   Data Sets

Table 1 shows the data set statistics. The data set contains fuzzed set (split into training set and test set) and non-fuzzed set (real attack set). A data set with sufficient and balanced data samples is essential for training the models effectively. Lack of training data can result in poor results, while biased data sets may result in biased models. If the fuzzing data set is already balanced, we directly use all the data samples without balancing. Otherwise, we perform data

Table 1. Data set statistics.

Attacks	Set	Size	Benign to malicious ratio
ARP poisoning	Training	9584	1.005:1
	Test	2400	0.982:1
	Real attack	17471	0:1
PtH* (best-performing)	Training	3932	1.364:1
	Test	983	1.329:1
	Real attack	214	0:1
PtH* (best-detecting)	Training	2556	0.974:1
	Test	640	0.839:1
	Real attack	192	0:1
DNS cache poisoning*	Training	30928	1.003:1
	Test	7732	0.988:1
	Real attack	263	0:1

*For multi-packet attacks, we only list the data set statistics corresponding to the best-performing or best-detecting models.

set balancing first. Specifically, if the benign data sets have significantly more data samples than the malicious data sets, we down-sample the benign data sets to match the size of malicious data sets, and vice versa.

4.3   Best-Performing Models

Table 2 presents the evaluation results on the best-performing models for each network attack. All models get acceptable to good results on training set and test set. **For multi-packet attacks, DL models are substantially better than traditional ML models, especially on real attack set.** In PtH detection, the LSTM model achieves near 99% accuracy on the real attack set, while ML models cannot reach 1/4 accuracy. In DNS cache poisoning detection, the CNN model's accuracy on the real attack set is 100%, while ML model can reach about 47% accuracy at most. Selected DL models' F1 scores are also far better than those of traditional ML models. For ARP poisoning detection, DL models do not have many advantages over traditional ML models, and all models' performances downgrade on real attack set comparing to those of training set and test set. The reason is that the real attack set for ARP poisoning is generated on a different LAN, with different valid MAC and IP addresses.

Table 2. Evaluation results on best-performing models.

Attacks	DL or ML	Model type[1]	Training set		Test set		Real attack set	
ARP	DL	MLP	99.91%	0.9991	99.75%	0.9975	72.84%	0.8429
		CNN	99.94%	0.9994	99.79%	0.9979	73.02%	0.8441
		RNN	99.91%	0.9991	99.75%	0.9975	72.83%	0.8428
		LSTM	99.91%	0.9991	99.75%	0.9975	72.83%	0.8428
	ML	kNN	99.90%	0.9990	99.93%	0.9993	81.99%	0.9010
		SVM-Linear	99.87%	0.9987	99.90%	0.9990	72.83%	0.8428
		SVM-Poly	99.96%	0.9996	99.93%	0.9993	72.83%	0.8428
		SVM-Radial	99.97%	0.9997	99.93%	0.9993	72.83%	0.8428
		DT	99.84%	0.9984	99.90%	0.9990	82.35%	0.9032
		RF	99.97%	0.9997	99.93%	0.9993	72.83%	0.8428
PtH	DL	LSTM-P	98.45%	0.9865	98.07%	0.9831	98.96%	0.9948
	ML	kNN	96.77%	0.9682	96.53%	0.9658	23.44%	0.3797
		SVM-Linear	96.89%	0.9694	96.72%	0.9674	13.02%	0.2304
		SVM-Poly	97.75%	0.9779	94.69%	0.9479	23.44%	0.3797
		SVM-Radial	98.07%	0.9810	93.72%	0.9378	18.23%	0.3084
		DT	94.70%	0.9467	95.44%	0.9533	18.23%	0.3084
		RF	100.00%	1.0000	97.99%	0.9798	14.06%	0.2466
DNS	DL	CNN	99.87%	0.9987	99.73%	0.9973	100.00%	1.0000
	ML	kNN	98.67%	0.9867	98.35%	0.9834	0.00%	0.0000
		SVM-Linear	96.01%	0.9608	95.17%	0.9527	0.00%	0.0000
		SVM-Poly	99.63%	0.9963	98.70%	0.9870	0.00%	0.0000
		SVM-Radial	100.00%	1.0000	98.66%	0.9867	0.00%	0.0000
		DT	87.01%	0.8771	86.88%	0.8754	47.01%	0.6395
		RF	100.00%	1.0000	97.50%	0.9752	34.19%	0.5096

[1] For multi-packet attacks, only proposed DL models are presented.

4.4 Best-Detecting Models

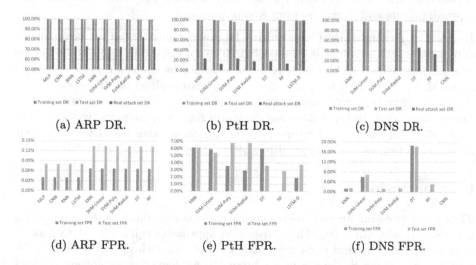

(a) ARP DR.   (b) PtH DR.   (c) DNS DR.

(d) ARP FPR.   (e) PtH FPR.   (f) DNS FPR.

Fig. 1. Evaluation results on the best-detecting models.

Figure 1 presents the evaluation results of best-detecting models. FPRs on real attack sets are not presented because there is no negative data sample, so FPR cannot be calculated. Similar to the best-performing case, all models get acceptable to good results on training and test set. For single-packet attack detection, DL models do not have many advantages over ML models. **For multi-packet attacks, DL models are better than ML models, especially on real attack set.** Because there is no negative data sample in the real attack set, $DR = Acc$. As for FPR, although it cannot be calculated in the real attack set, results show that DL models achieve generally lower FPRs comparing to ML models on the training and test sets.

5   Discussions and Limitations

Lack of Efficiency: Training a neural network requires a large amount of data samples. However, the number of data samples can be affected in many ways. On one hand, protocol fuzzing in nature cannot guarantee that all malicious/benign activities are successful. On the other hand, the time consumed by each benign/-malicious activity cannot be overlooked. Take PtH as an example, we spent about 4 days running 5,000 attack iterations, of which 611 failed. The total amount of network packets captured is 497,956, of which 103,718 are related packets. However, the final number of data samples is only in the thousands.

Neural Networks for Various Network Attacks: Though we have verified our idea on three chosen network attacks, we trained separate neural networks for different attacks. We can not train a generic neural network to detect various network attacks. It is difficult to train such a neural network because different network attacks have different characteristics, which may need different data representations and neural network architectures.

Impact of Probability Threshold: The raw outputs for output layers of the detection neural networks are the probabilities for the data sample to be benign or malicious, which add up to 1. If the probability for malicious class is beyond a threshold (e.g., 0.5), then the data sample is classified as malicious. When the probability threshold increases, the model is more likely to classify a data sample as benign, and thus decrease detection rates and false positive rates. The probability threshold can be tuned depending on whether the defender prefers higher detection rates or lower false positive rates.

6   Conclusion

This paper presents an end-to-end approach to detect the logic-flaw-exploiting network attacks using DL. The end-to-end approach begins with data generation and collection, and ends with attack detection with neural networks. We address two major challenges in applying DL for logic-flaw-exploiting network attack detection: the generation of useful data sets and the training of appropriate neural network models. We show the effectiveness of our approach with three specific demonstration attacks, including PtH, DNS cache poisoning, and ARP poisoning. We have generated high quality network traffic data using protocol fuzzing, trained neural networks with generated data, and evaluated the trained models from the perspective of both neural network performance and attack detection. We have also discussed the limitations of our experiments and approach.

Disclaimer

This paper is not subject to copyright in the United States. Commercial products are identified in order to adequately specify certain procedures. In no case does such identification imply recommendation or endorsement by the National Institute of Standards and Technology, nor does it imply that the identified products are necessarily the best available for the purpose.

Acknowledgement. This work was supported by NIST 60NANB20D180.

References

1. IDS 2018: Datasets: Research: Canadian Institute for Cybersecurity: UNB, January 2020. https://www.unb.ca/cic/datasets/ids-2018.html. Accessed 4 July 2020
2. Amini, M.: RT-UNNID: a practical solution to real-time network-based intrusion detection using unsupervised neural networks. Comput. Secur. **25**(6), 459–468 (2006)
3. Banks, G., Cova, M., Felmetsger, V., Almeroth, K., Kemmerer, R., Vigna, G.: SNOOZE: toward a stateful NetwOrk prOtocol fuzZEr. In: Katsikas, S.K., López, J., Backes, M., Gritzalis, S., Preneel, B. (eds.) ISC 2006. LNCS, vol. 4176, pp. 343–358. Springer, Heidelberg (2006). https://doi.org/10.1007/11836810_25
4. Choi, J.: A method of DDoS attack detection using HTTP packet pattern and rule engine in cloud computing environment. Soft. Comput. **18**(9), 1697–1703 (2014). https://doi.org/10.1007/s00500-014-1250-8
5. Dhanabal, L., Shantharajah, S.: A study on NSL-KDD dataset for intrusion detection system based on classification algorithms. Int. J. Adv. Res. Comput. Commun. Eng. **4**(6), 446–452 (2015)
6. Faker, O., Dogdu, E.: Intrusion detection using big data and deep learning techniques. In: ACMSE 2019 - Proceedings of the 2019 ACM Southeast Conference (2019)
7. Gorbunov, S., Rosenbloom, A.: AutoFuzz: automated network protocol fuzzing framework. Int. J. Comput. Sci. Netw. Secur. **10**(8), 239–245 (2010)
8. Goswami, S.: An unsupervised method for detection of XSS attack. IJ Netw. Secur. **19**(5), 761–775 (2017)
9. Jan, S.T.: Throwing darts in the dark? Detecting bots with limited data using neural data augmentation. In: The 41st IEEE S&P (2020)
10. Kaur, S., Singh, M.: Automatic attack signature generation systems: a review. IEEE Secur. Priv. **11**(6), 54–61 (2013)
11. Milajerdi, S.M.: HOLMES: real-time APT detection through correlation of suspicious information flows. In: 2019 IEEE Symposium on Security and Privacy (SP). IEEE (2019)

12. Moustafa, N., Slay, J.: UNSW-NB15: a comprehensive data set for network intrusion detection systems (UNSW-NB15 network data set). In: 2015 Military Communications and Information Systems Conference, MilCIS 2015 - Proceedings. Institute of Electrical and Electronics Engineers Inc., December 2015
13. Ongun, T.: On designing machine learning models for malicious network traffic classification. arXiv:1907.04846 [cs, stat], July 2019
14. Oprea, A.: MADE: security analytics for enterprise threat detection. In: Proceedings of the 34th Annual Computer Security Applications Conference, ACSAC 2018. Association for Computing Machinery, December 2018. https://doi.org/10.1145/3274694.3274710
15. Pfahringer, B.: Winning the KDD99 classification cup: bagged boosting. ACM SIGKDD Explor. Newsl. 1(2), 65–66 (2000)
16. Sharafaldin, I.: Toward generating a new intrusion detection dataset and intrusion traffic characterization. In: ICISSP 2018 - Proceedings of the 4th International Conference on Information Systems Security and Privacy, January 2018
17. Taylor, C.: Low-level network attack recognition: a signature-based approach. In: IEEE Proceedings of the PDCS 2001 (2001)
18. Yuan, X., Li, C., Li, X.: DeepDefense: identifying DDoS attack via deep learning. In: 2017 IEEE International Conference on Smart Computing, SMARTCOMP 2017 (2017)
19. Zhang, Y.: PCCN: parallel cross convolutional neural network for abnormal network traffic flows detection in multi-class imbalanced network traffic flows. IEEE Access 7, 119904–119916 (2019)
20. Zou, Q.: An approach for detection of advanced persistent threat attacks. IEEE Ann. Hist. Comput. 53(12), 92–96 (2020)

Potpourri I

Potpourri

Not a Free Lunch, But a Cheap One: On Classifiers Performance on Anonymized Datasets

Mina Alishahi[✉] and Nicola Zannone

Eindhoven University of Technology, Eindhoven, The Netherlands
{m.sheikhalishahi,n.zannone}@tue.nl

Abstract. The problem of protecting datasets from the disclosure of confidential information, while published data remains useful for analysis, has recently gained momentum. To solve this problem, anonymization techniques such as k-anonymity, ℓ-diversity, and t-closeness have been used to generate anonymized datasets for training classifiers. While these techniques provide an effective means to generate anonymized datasets, an understanding of how their application affects the performance of classifiers is currently missing. This knowledge enables the data owner and analyst to select the most appropriate classification algorithm and training parameters in order to guarantee high privacy requirements while minimizing the loss of accuracy. In this study, we perform extensive experiments to verify how the classifiers performance changes when trained on an anonymized dataset compared to the original one, and evaluate the impact of classification algorithms, datasets properties, and anonymization parameters on classifiers' performance.

Keywords: Privacy-preserving · k-anonymity · ℓ-diversity · t-closeness · Classifiers comparison

1 Introduction

Classification is the task of identifying to which category (class) a new observation belongs based on a training set of observations whose category membership is known beforehand. Nowadays, data classification algorithms (classifiers) are widely used in many real-world applications, including but not limited to, face and speech recognition, text analysis, fraud and anomaly detection, recommendation system, weather forecasting, and medical image analysis [1,27].

Classifiers are typically trained over a corpus of training data that is directly accessible by the data analyzer. However, in many real-world scenarios the training data is generated and governed by different entities who are unwilling to share their data with the analyzer. This is because the data might contain privacy-sensitive information, and its disclosure might raises privacy concerns [23].

© IFIP International Federation for Information Processing 2021
Published by Springer Nature Switzerland AG 2021
K. Barker and K. Ghazinour (Eds.): DBSec 2021, LNCS 12840, pp. 237–258, 2021.
https://doi.org/10.1007/978-3-030-81242-3_14

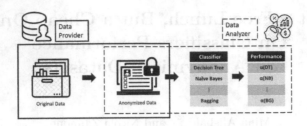

Fig. 1. Reference architecture.

To solve this issue, a large body of research has investigated how to train a practically useful classifier, while preserving individuals' privacy. Existing solutions can be categorized into two main groups. One category comprises cryptographic-based approaches in which the classifier model is securely computed. The main drawback of these approaches is that they are not scalable, and they are designed only for a specific classification algorithm [10]. The other category of solutions comprises data anonymization techniques, in which the values in the data are replaced with a more general representation before the dataset is published. This study focuses on the second category, *i.e.*, the application of data anonymization techniques in data classification, where one entity (data provider) owns the data and the other entity (data analyzer) is interested in training a classifier on this data. The data analyzer does not know which classifier outperforms the other classifiers on the shared (anonymized) data. This knowledge would enable the analyzer to decide which classifier to be trained based on the anonymization technique employed, the dataset properties, and the desired performance metric. The problem addressed in this work can be defined as follows.

Problem Statement: A data provider wants to release a dataset T to a data analyzer for modeling a classifier on this data. Each record $\vec{x}_i$ in T is an $(n+1)$-dimensional vector $\vec{x}_i = (v_1, v_2, \ldots, v_n, C_i)$, where the first n elements are the attribute-values and the last element is the class label of that record. The data provider wants to protect the dataset against linking an individual to sensitive information using an anonymization approach. Consider, for instance, a health center wants to share the patients' records with a medical research center for identifying the causes and symptoms of a new disease. However, the shared information might potentially raise the patients' privacy concerns. Thus, the health center only is ready to share the data if no individual record can be linked to the corresponding patient, *i.e.*, dataset can only be revealed in anonymized version.

The data analyzer who has access to the anonymized version of data is interested in training a classifier. In our example, the medical research center wants to model a classifier over the symptoms of a disease to predict whether a new patient suffers from this disease or not. However, the data analyzer has no knowledge which classifier should be selected on the published anonymized data. An overview of this communication model with the following two entities is presented in Fig. 1:

- *Data provider* who shares the anonymized table of data respecting his/her privacy requirements, *e.g.*, the published table satisfies 3-anonymity.
- *Data analyzer* who uses the anonymized table of data as training dataset to train a classifier.

We assume that the data analyzer has time limitation and thus is not able to evaluate the performance of different classifiers on the published anonymized data.

Our Contribution: To solve the aforementioned problem, we investigate the classifiers performance on anonymized datasets via answering the following research questions:

RQ1: How the performance of different classification algorithms changes when trained on anonymized datasets?

RQ2: Which classifiers are more affected by the employment of anonymization techniques?

RQ3: Which dataset properties affect the performance of classifiers trained on anonymized datasets?

RQ4: How the classifiers performance is affected by changing the anonymization parameters?

To answer these questions, we compute the performance of eight well-known classification algorithms, namely Decision Tree, Naïve Bayes, k Nearest Neighbors, Support Vector Machine, Random Forest, Logistic Regression, AdaBoost, and Bagging, over 10 benchmark datasets (original and anonymized versions). The performance of classifiers is measured using accuracy, precision, recall, and F1-score metrics. The selected anonymization approaches are k-anonymity [21,25], ℓ-diversity [15], and t-closeness [12]. The contribution of this work can be summarized as follows:

- We provide insight on the difference between classifiers performance trained over original and anonymized datasets (RQ1). We show that some classifiers significantly outperform the others in this regard.
- We compare the classifiers performance on anonymized datasets and highlight which classifiers outperform the others in terms of the associated performance metric and anonymization approach (RQ2). We show that this outperformance is statistically significant and provide insight on the origin of this difference.
- We investigate the impact of dataset properties, *i.e.*, dataset size, the number of attributes, and the number of class labels on classifiers performance (RQ3). We show which dataset property and to what extent has impact on classifiers performance on anonymized dataset.
- We evaluate the effect of anonymization parameters on classifiers performance through the enforcement of different values of k, ℓ, and t (RQ4). We show that the variation of anonymization parameter has negligible impact on the trend of classifiers performance.

- Based on our experimental results, we draw recommendations to guide data providers and analyzers in the selection of the classification algorithm to be used (cheap lunch).

Outline: The remainder of this paper is organized as follows. The next section presents the background. Section 3 explains our methodology and the setup of the experiments. Section 4 presents the experimental results, whereas Sect. 5 discusses our findings. Section 6 discusses related work. Finally, Sect. 7 concludes the paper and provides directions for future work.

2   Preliminaries

This section introduces the anonymization techniques and classification algorithms considered in this work.

2.1   Anonymization Techniques

For our study we consider three well-known anonymization techniques: k-anonymity [21,25], ℓ-diversity [15], and t-closeness [12]. We assume a dataset comprising a set of records, where each record corresponds to one individual. Each record is described by a number of attributes, which can be divided into three categories: 1) *identifiers* that univocally identify the individuals, *e.g.*, social security number, 2) *quasi-identifier* attributes whose values taken together can be used to potentially identify an individual, *e.g.*, zip-code, birth-date, and gender, 3) *sensitive* attributes that an adversary is not allowed to discover the values of that attribute for any individual, *e.g.*, a patient's disease or an employer's salary.

k-**anonymity:** A release of data is said to satisfy k-anonymity if each record in the release cannot be distinguished from at least $k - 1$ other records in the release with respect to quasi-identifiers [25]. k-anonymity is susceptible to some attacks, *e.g. homogeneity* and *background knowledge* attacks.

ℓ-**diversity:** The ℓ-diversity model addresses some of the weaknesses of k-anonymity. In particular, k-anonymity does not protect the values of sensitive attributes, specifically when the values in a group are identical. To address this drawback, the ℓ-diversity model introduces constraints on intra-group diversity for sensitive attributes [15]. The ℓ-diversity does not consider the semantic closeness of the distinct values in a sensitive attribute. This problem is addressed by t-closeness.

t-**closeness:** An equivalence group is said to satisfy t-closeness if the distance between the distribution of a sensitive attribute in this group and the distribution of the attribute in the whole table is not greater than a given threshold t. A dataset is said to satisfy t-closeness if all equivalence classes satisfy t-closeness [12].

2.2   Classification Algorithms

We investigate the performance of eight well-known classification algorithms, namely Decision Tree (DT), Naïve Bayes (NB), k-Nearest Neighbors (kNN), Support Vector Machine (SVM), Random Forest (RF), Logistic Regression (LR), AdaBoost (AB), and Bagging (BG). Next, we briefly introduce these classifiers (for more detail refer to [1]).

Decision Trees (DT) are classification algorithms with a tree-based structure drawn upside down with its root at the top. Each internal node represents a test/condition based on which the tree splits into branches/edges. The end of the branch that does not split anymore (respecting some stopping criteria) is the decision/leaf. The paths from root to leaf represent classification rules. One advantage of DTs is the comprehensibility of the classification structures. This enables the analyzer to verify which attributes determined the final classification. The drawback is that DTs might be non-robust for datasets with a large number of attributes.

Naïve Bayes (NB) algorithms are statistical classifiers based on the Bayes Theorem for calculating probabilities and conditional probabilities. It makes use of all attributes contained in the data, and analyses them individually as though they are equally important and independent (naïve assumption) from each other. Naïve Bayes model is easy to build and particularly useful for very large data sets.

k-Nearest Neighbors (kNN) algorithm is a non-parametric instance-based model that classifies a new instance based on the class of the majority of its k nearest neighbors w.r.t. a given training dataset. To obtain the nearest neighbors for each data point, kNN uses a measure to compute the distance between pairs of data points (*e.g.*, Euclidean distance). The advantage of kNN lies in its simplicity, while computation time is usually high since all training data has to be revisited for classifying a new instance.

Support Vector Machine (SVM) is a linear modelling with instance-based learning. The algorithm selects a small number of critical boundary instances from each category (class labels) and builds a linear discriminate function that separates them as widely as possible. In the case that no linear separation is possible, the technique of kernel is used to automatically project the training instances into a higherdimensional space and to learn a separator in that space. The SVMs have the advantage of generalization, and also standout for their robustness to high dimensional data. The drawback of the SVMs is the difficulty of model interpretation and the sensibility to parameter tuning.

Random Forest (RF) is an ensemble learning method for classification which operates by constructing a multitude of decision trees at training time and outputting the class that is the mode of the classes of the individual trees. RF corrects for decision trees' habit of overfitting to their training set. The RF has also been recognised to be among the most accurate classifiers.

Logistic Regression (LR) is a statistical model that is largely employed in statistical data analysis to classify binary dependent variables. In regression analysis, logistic regression (or logit regression) is used to estimate the parameters of a logistic model returning the probability of occurrence of a class. To this end, the LR classifier builds a logit variable that contains the natural log of the odds of the class occurring or not. Then, a maximum likelihood estimation algorithm is applied to estimate the probabilities.

AdaBoost (AB) is a technique that builds an ensemble of classifiers (generally Decision Trees) sequentially, one classifier at a time, until the predefined number of classifiers is reached. Each subsequent classifier is trained on a set of samples with weights to emphasize the instances misclassified by the previous classifiers. The ensemble decision is made by weighted voting. The weights are determined by the individual accuracy. AB has been found to be very useful but too sensitive to noise in the data.

Bagging (Bootstrap Aggregating) (BG) is an ensemble learning model designed to improve the stability and accuracy of classification algorithms. It considers several homogeneous weak learners, where each weak learner has been trained independently from the others. Then, the bagging model combines them with an averaging process.

3   Experimental Methodology

The experimental analysis aims to compare the classifiers performance when they are trained over anonymized datasets. This comparison evaluates how the employment of anonymization techniques affect classifiers performance (**RQ1**), and investigate how the choice of classification algorithm (**RQ2**), dataset properties (**RQ3**), and anonymization parameters (**RQ4**) influence the classifiers performance. Next, we present the experimental setup, the datasets used for the experiments and the evaluation approach.

Experimental Setup: We use the implementation of classification algorithms provided by Scikit-learn library[1] with their default parameters. Turning a dataset into an anonymized (k-anonymous, ℓ-diverse, and t-close) dataset is a complex problem in which finding the optimal partition is an NP-hard problem. In this study, we employ the *Mondrian* algorithm, which uses a greedy search algorithm to recursively partition the domain space into regions [11].[2]

Datasets: The classifiers were trained over ten datasets selected from the UCI Repository.[3] Table 1 summarizes the statistics of the selected datasets.

Adult: The dataset contains 48842 instances described by 14 attributes (both numerical and categorical) such as *age*, *occupation*, *education*, and *working class*.

---
[1] https://scikit-learn.org.
[2] The code used for our experiments is available at https://github.com/minaalishahi/classifiersperformance.
[3] https://archive.ics.uci.edu/ml/datasets/.

Table 1. Datasets information

Dataset	# Attributes	# Labels	# Instances
Adult	8	2	48000
Credit	15	239	690
Absent	21	17	740
Derma	33	6	366
Wine	12	2	4898
Network	22	4	1075
Bank	64	2	10,503
Optical	64	10	5600
Diabet	20	2	1151
Heart	13	2	299

The class attribute represents their income, which has two possible values: '>50K' and '<50K'.

Credit: The dataset contains 690 instances of clients' information at a bank described by 15 attributes (both numerical and categorical) such as *age* and *background behavior*, which are used to predict the score of a requester with 239 possibilities (based on this score it is decided whether the credit card application should be accepted, revised, or denied).

Absenteeism at Work (Absent): The dataset contains 740 instances described by 21 attributes (both categorical and numerical), *e.g.*, *age*, *education*, *average workload per day*, and *social smoker*. The class label denotes the hours that a new employee might be absent in a month in the future. While the absent hours can vary from one hour to 160 h in a month, the current dataset only shows 17 distinct values for absent hours (from 20 to 36 h).

Dermatology (Derma): This dataset contains 366 instances described by 33 numerical attributes such as *age*, *family history*, *knee and elbow involvement*, which are used to predict the type of Eryhemato-Squamous disease as a real problem in dermatology (skin disorders). The majority of attributes take their values from the set $\{0, 1, 2, 3\}$.

Wine Quality (Wine): The dataset contains 4898 instances of wine samples described by 12 numerical attributes such as *pH value*, *citric acid*, *total sulfur dioxide*, which are used to predict the wine quality (good or bad).

Optical Burst Switching (OBS) Network (Network): The dataset contains 1075 instances of Burst Header Packet (BHP) flood attacks in Optical Burst Switching networks (OBS), described by 22 numerical attributes such as *Average Delay Time per Second*, *Percentage of Lost Packet rate*, to classify the strategy against an attack according to network nodes behavior into four classes as NB-No Block, Block, No Block, NB-Wait (NB = Not Behaving correctly).

Polish Companies Bankruptcy (Bank): The dataset contains 10503 instances of emerging markets around the world described by 64 numerical attributes such as *current assets/short-term liabilities*, *profit on sales/total sales*, to predict whether a Polish company will face bankruptcy or not.

Optical Digits (Optic): The dataset contains 5620 handwritten digits written by 43 persons. Each record is a matrix of 8×8 where each element is an integer in the range $\{0, \ldots, 16\}$ and the class label is one of the integer number in the set $\{0, \ldots, 9\}$.

Diabetic Retinopathy Debrecen (Diabet): The dataset contains 1151 instances of Messidor image set described by 20 attributes such as *the diameter of the optic disc* and *quality assessment*. All features represent a detected lesion, or a descriptive feature of an anatomical part or an image-level descriptor. The class label represents whether an image shows signs of diabetic retinopathy or not.

Heart Failure Clinical Record (Heart): The dataset collected the medical records of 299 patients who had heart failure during a pre-determined number of days (say follow-up period). The dataset is described by 13 attributes such as *age*, to predict whether a patient dies (Boolean) during the follow-up period.

Note that for all datasets, it is assumed that the explicit identifier attributes have been removed from the data, the attribute representing the class label is considered as the sensitive attribute, and the remaining attributes are considered quasi-identifier attributes. In all datasets, categorical attributes either were removed or the categorical values were replaced with numerical values (when the conversion is a valid assumption). This is because in the application of anonymization approaches over categorical attributes, in general, the presence of an expert in the field is required to provide the taxonomy trees for generalization.

Evaluation Approach: We assess the classifiers performance in terms of *Accuracy, Precision, Recall,* and *F1-score*, which are defined based on True Positives (TP), True Negatives (TN), False Positives (FP), and False Negatives (FN) values. True Positives (TP) are the correctly predicted positive values, *i.e.,* the value of actual class is positive and the value of predicted class is also positive. True Negatives (TN) are the correctly predicted negative values, *i.e.,* the value of actual class is negative and value of predicted class is also negative. False Positives (FP) are the non-correctly predicted positive values, *i.e.,* the value of actual class is negative and the predicted class is positive. False Negatives (FN) are the non-correctly predicted negative values, *i.e.,* the value of actual class is positive but the predicted class is negative. To compute these values for datasets with multiple class labels, we first compute the TP, TN, FP, and FN values for each individual class label against the remaining class labels. Then, the average of these values over all class labels is returned as final TP, TN, FP, and FN. Accuracy is the ratio of correctly predicted observations over the total number of observations. Precision is the ratio of correctly predicted positive observations to the total number of predicted positive observations. Recall is the ratio of correctly predicted positive observations to all observations in actual class positive. F1-Score is the weighted average of Precision and Recall. Therefore, this score takes both false positives and false negatives into account. Formally:

$$\text{Accuracy} = \frac{TP + TN}{TP + FP + FN + TN} \qquad \text{Precision} = \frac{TP}{TP + FP}$$

$$\text{Recall} \quad = \frac{TP}{TP + FN} \qquad \qquad \text{F1-Score} = 2 \cdot \frac{\text{Recall} \cdot \text{Precision}}{\text{Recall} + \text{Precision}}$$

To answer research questions RQ1, RQ2, and RQ3, we compute classifier performance in terms of accuracy, precision, recall, and F1-scores on the original, 3-anonymity, 2-diversity, and 0.2-closeness datasets. To answer RQ4, we compute classifier accuracy when anonymization parameters, *i.e.*, k, ℓ, and t, vary.

Criteria for RQ1: To investigate to what extent the use of anonymization techniques affect a classifier performance, we compare the performance of classifiers trained on anonymized datasets with the one of classifiers trained on the corresponding original datasets by computing the *performance ratio*. For each considered performance metric, the performance ratio is computed by dividing the performance of a classifier trained over an anonymized dataset by the performance of the classifier trained over the corresponding original datasets. As we are interested on the average performance of classification algorithms, we aggregate performance metrics over the 10 datasets.

To verify whether the differences between classification algorithms are statistically significant, we use a non-parametric statistical test, named *Wilcoxon* test [26]. The Wilcoxon test can be adapted to our problem as follows.

Definition 1 (Wilcoxon test). *Given two classification algorithms, let d_i be the signed difference between the performance scores of the classifiers obtained by applying each algorithm on a given dataset for a given privacy level. The differences d_i ($1 \le i \le N$ where N is the number of anonymized datasets to which the classification algorithms are applied) are ranked based on the absolute values (average rank is assigned for equal performances). Let R^+ denote the sum of the ranks for datasets and privacy level on which $d_i > 0$, and let R^- be the sum of the ranks for datasets and privacy level on which $d_i < 0$ (dividing the sum of the ranks for which $d_i = 0$ evenly), i.e.,*

$$R^+ = \sum_{d_i > 0} rank(d_i) + \frac{1}{2} \sum_{d_i = 0} rank(d_i) \,, \quad R^- = \sum_{d_i < 0} rank(d_i) + \frac{1}{2} \sum_{d_i = 0} rank(d_i)$$

Let $T = \min(R^+, R^-)$, then

$$z = \frac{T - \frac{1}{4}N(N+1)}{\sqrt{\frac{1}{4}N(N+1)(2N+1)}}$$

is approximately distributed normally. Under this condition, the difference between the accuracy distribution of the two classification algorithms is statistically significant (i.e., the null hypothesis is rejected) if the p-value is less than or equal to a given significance level σ. In our experiments, we require a 95% confidence interval, which corresponds to $\sigma = 0.05$ (i.e., the null-hypothesis can be rejected if z is smaller than -1.96).

Criteria for RQ2: To investigate the impact of the adoption of anonymization techniques on classifiers performance, we first compute the average of classifiers' performance for each performance metric over the 10 datasets. Then, we employ a non-parametric statistical test, named the *Friedman* test [5], on these results. This test was designed to compare classification algorithms over multiple datasets, and the outcome determines whether the algorithms are equal in terms of performance or not. If the classification algorithms exhibit different performance, *p*-values (in *Holm* methodology) are used to order them based on their performance. The Friedman test can be adapted to our problem as follows.

Definition 2 (Friedman Test). *Given n classification algorithms and m datasets, let r_{ij} denote the rank of j-th algorithm on the i-th dataset. The Friedman test compares the average ranks of algorithms, i.e., $R_j = \frac{1}{m}\sum_i r_{ij}$. Under the null-hypothesis stating that all algorithms are equivalent (their average ranks R_j are close), the Friedman statistic*

$$\chi_F^2 = \frac{12m}{n(n+1)}\left(\sum_j R_j^2 - \frac{n(n+1)^2}{4}\right)$$

is distributed according to the well-known χ_F^2 distribution with $n-1$ degrees of freedom, for n and m big enough ($n \geq 5$, $m \geq 10$). To decide if the classifiers' performance is significantly different, the Friedman test is used as:

$$F_F = \frac{(m-1)\chi_F^2}{m(n-1) - \chi_F^2}$$

where the probability distribution can be approximated by a F-distribution with $n-1$ and $(n-1)(m-1)$ degrees of freedom. The table of critical values can be found in statistical books [9]. The difference between the performance distribution of all classification algorithms is statistically significant (i.e., the null hypothesis is rejected) if the p-value is less than or equal to a given significance level σ. In our experiments, we require a 95% confidence interval, which corresponds to $\sigma = 0.05$.

If the null-hypothesis is rejected, we need to determine where the differences truly came from. To answer this question, generally, a post-hoc statistical test named the *Nemenyi* test is used. However, in some cases, the Nemeneyi test is not able to detect why the Friedman test has rejected the null-hypothesis. In this regard, the tests like *Bonferroni* or *Holm* are more powerful in detecting where the difference comes from using a control variable [5,8]. In this study, we use the average performance of classifiers as the required control variable. This test assigns a score to classifiers by comparing the respective *p*-values of a classifier compared to the others such that the classifier with higher score has the better performance for that specific assessment.

Table 2. Classifier performance ratio.

Anonymity	Metric	Classifier							
		DT	NB	kNN	SVM	RF	LR	AB	BG
3-anonymity	Accuracy	0.93	0.90	1.00	1.03	0.85	0.91	**1.10**	0.84
	Precision	0.88	0.89	1.04	1.09	0.90	0.90	**1.20**	0.86
	Recall	0.87	0.80	1.00	1.03	0.85	0.91	**1.10**	0.84
	F1 score	0.87	0.81	1.03	1.11	0.87	0.90	**1.22**	0.85
2-diversity	Accuracy	0.83	0.86	0.93	0.93	0.75	0.86	**1.05**	0.75
	Precision	0.77	0.86	0.95	0.96	0.77	0.85	**1.08**	0.75
	Recall	0.78	0.77	0.93	0.93	0.75	0.86	**1.05**	0.75
	F1 score	0.77	0.79	0.95	0.99	0.76	0.85	**1.15**	0.75
0.2-closeness	Accuracy	0.71	0.61	0.70	0.71	0.62	0.68	**0.77**	0.62
	Precision	0.55	0.54	0.66	0.68	0.54	0.61	**0.71**	0.54
	Recall	0.65	0.57	0.70	0.71	0.62	0.68	**0.77**	0.62
	F1 score	0.59	0.57	0.69	0.72	0.58	0.63	**0.76**	0.58

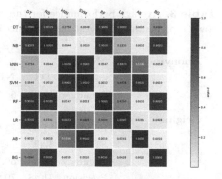

Fig. 2. Heatmap of Wilcoxon test.

Criteria for RQ3: Research question RQ3 aims to understand the effect of dataset properties on the performance of classifiers when trained on anonymized datasets. To this end, we investigate how the classifier accuracy, precision, recall, and F1-scores vary over anonymized datasets with different sizes, the number of attributes, and the number of class labels. To assess this variation, for each dataset and each performance metric we compute the average performance over all classification algorithms. To determine whether the impact of a specific dataset property is significant, we compare the distribution of classifiers' performance based on aggregated results for that specific property.

Criteria for RQ4: To investigate the effect of anonymization parameters, *i.e.*, $k, \ell,$ and t in k-anonymity, ℓ-diversity, and t-closeness on classifiers performance, we compute classifier accuracy, precision, recall, and F1-scores on anonymized datasets for $k \in \{3, 4, 5, 6\}$, $\ell \in \{2, 3, 4, 5\}$, and $t \in \{0.2, 0.3, 0.4, 0.5\}$.

It is worth noting that the selection of value ℓ from the set $\{2, 3, 4, 5\}$ requires that the dataset under analysis contains at least 5 distinct class labels. Out of the 10 selected datasets, the datasets satisfying this requirement (*i.e.*, with more than five class labels) are Credit, Absent, Derma, and Optical datasets. Due to the lack of space, we only present the results for the three datasets with the greatest number of class labels.

4   Experimental Results

This section presents the results of our experiments.

RQ1: How the performance of different classification algorithms changes when trained on anonymized datasets? The performance ratio for the considered classification algorithms (aggregated over the 10 datasets) is reported in Table 2. We can observe that the AB classifier shows the highest ratio for all performance metrics and anonymization techniques. The performance is even higher when AB is trained on 3-anonymous and 2-diverse datasets compared to when it is trained on the original dataset (the ratio is greater than 1). This performance improvement can also be observed for the kNN and SVM classifiers. The worst classifiers in terms of performance ratio are NB and BG.

Table 3. Average performance for the considered classifiers. The highest value in each row (representing the best performance) is highlighted in bold.

Anonymization	Metric	Classifier							
		DT	NB	kNN	SVM	RF	LR	AB	BG
Original	Accuracy	82.61	65.73	72.06	71.87	**85.73**	80.03	65.40	85.47
	Precision	82.60	81.45	68.39	71.72	83.74	77.25	58.89	**84.14**
	Recall	84.03	73.65	72.06	71.87	**85.73**	80.03	65.40	85.47
	F1 score	82.95	71.76	68.38	67.00	84.25	77.21	59.67	**84.41**
3-anonymity	Accuracy	73.03	59.23	71.10	**73.30**	73.28	72.10	67.61	72.04
	Precision	71.48	70.94	68.03	71.94	**72.65**	68.76	64.08	71.56
	Recall	73.03	59.23	71.10	**73.30**	73.28	72.10	67.61	72.04
	F1 score	71.38	58.80	68.01	71.79	**72.39**	68.77	62.96	71.27
2-diversity	Accuracy	64.27	56.71	65.28	66.04	64.75	**68.11**	64.36	63.80
	Precision	62.34	**68.63**	62.04	64.34	63.81	65.15	58.70	62.76
	Recall	64.27	56.71	65.28	66.04	64.75	**68.11**	64.36	63.80
	F1 score	62.47	56.30	62.18	64.21	63.82	**64.85**	60.12	62.81
0.2-closeness	Accuracy	50.97	37.99	49.74	51.44	50.41	**52.93**	51.76	50.41
	Precision	41.96	**46.12**	43.94	41.44	41.95	44.85	42.97	42.71
	Recall	50.97	37.99	49.74	51.44	50.41	**52.93**	51.76	50.41
	F1 score	45.72	36.20	45.89	45.60	45.57	46.14	**46.24**	45.74

Table 4. The best scored classifiers using the Holm methodology.

	Accuracy	Precision	Recall	F1-score
Original	BG	BG	BG	LR
3-anonymity	LR	RF	LR	LR
2-diversity	LR	NB	LR	LR
0.2-closeness	LR	NB	LR	LR

We used the Wilcoxon test to verify the statistical significance of these differences. Figure 2 depicts the heatmap of mutual comparison of classifier performance ratios in terms of p-values over accuracy results. The lower p-value (lighter color) shows more confidence in rejecting the null-hypothesis (*i.e.*, more different performance). It can be observed that the null-hypothesis of the Wilcoxon test is rejected in the mutual comparison of AB with DT, NB, RF, BG, and LR with high confidence (small p-value), *i.e.*, the AB classifier shows a different behavior compared to the other algorithms (except from SVM and kNN).

RQ2: Which classifiers are more affected by the employment of anonymization techniques? The average performance of classifiers is reported in Table 3. From the table, we can observe that 1) there is not a single classifier that outperforms all other classifiers for all performance metrics; 2) while for specific metrics and anonymization techniques, some classifier outperforms the others (highlighted values), in some cases the results are very close (*e.g.*, the accuracy of the SVM and RF on 3-anonymous datasets).

To get a better insight on these observations, we performed the Friedman test on average results. The null-hypothesis of this test was rejected in our experiments meaning that all classifiers are not equal in terms of performance and there is a significant difference. To determine where this difference comes from, in Tables 5, 6, 7, and 8 (in Appendix) we report respectively the Holm scores (simply score from now on) of classifiers with respect to accuracy, precision, recall, and F1-score. The higher scores represent higher performance for the associated classification algorithm and metric. Table 4 summarizes the best classifiers for each performance metric and anonymization technique according to the Holm scores. We can observe that the LR classifier outperforms the other classifiers in terms of accuracy, recall, and F1-score over anonymized datasets. However, in terms of precision, the RF and NB classifiers are the best scored ones over 3-anonymous and 2-diverse (and 0.2-close) datasets.

RQ3: Which dataset properties affect the performance of classifiers on anonymized datasets? To evaluate the impact of datasets on classifiers performance, we computed the accuracy, precision, recall, and F1-scores for each individual dataset averaged over all classifiers performance when trained on original and anonymized datasets. Figures 3a, 3b, 3c, and 3d show respectively the average accuracy, precision, recall, and F1-scores on all classifiers results. From these results, we can infer that the number of attributes has no direct impact on classifiers performance. The Bank and Optic datasets with an equal number of attributes show completely different impact on the classifiers' performance. The size of datasets also shows no direct impact on classifiers performance. It can be observed that the Credit and Absent, which have a comparable number of records, show different impact on classifiers performance. This can be seen for Network and Diabet datasets as well.

The number of class labels, differently from the previous properties, shows a considerable impact on classifiers performance on anonymized datasets. The datasets with two class labels, *i.e.*, Adult, Wine, Bank, Diabet, and Heart datasets, show better and more stable performance in both original and anonymized versions. On the other hand, the Credit and Absent datasets, which have multi class labels (239 and 17, respectively) result in poor performance in all scenarios.

To gain better insight on the impact of the number of class labels on classifiers performance, we compare the performance of binary classifiers (*i.e.*, classifiers trained on datasets which have two class values, namely Adult, Wine, Bank, Diabet, and Heart), and multi-class classifiers (*i.e.*, classifiers trained on datasets which have multiple class values, namely Credit, Absent, Derma, Network, and Optimal). Figure 4 shows the performance distribution of binary (blue) and multi-class (orange) classifiers when trained on the original data as well as when 3-anonymity, 2-diversity and 0.2-closeness are applied. Each box represents the distribution over five datasets (the blue ones over 2-labeled and the

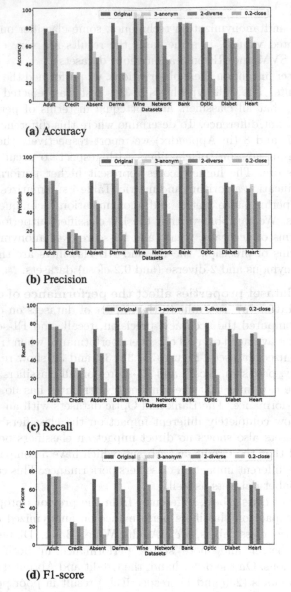

(a) Accuracy

(b) Precision

(c) Recall

(d) F1-score

Fig. 3. Average performance of the selected classifiers for each dataset.

orange ones over more-labeled datasets) and over classifiers' accuracy (*i.e.*, each box shows the average value computed over 5 accuracy values). We can observe that on the original datasets, the blue and orange boxes have close median values (*i.e.*, small difference). The difference in the distribution between the boxes (and median values) increases when anonymization is applied.

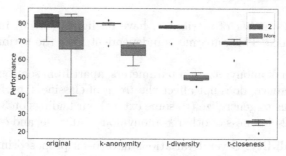

Fig. 4. Performance of the selected classifiers for datasets with two class labels vs. multi-class labels. (Color figure online)

RQ4: How the classifiers's performance is affected by changing the anonymization parameters? To evaluate whether a classifier' performance is affected by the values of k, ℓ, and t, we computed the performance metrics for all considered classifiers for different values of k, ℓ, and t on Credit, Absent, and Optic datasets. The results are reported in Figs. 5, 6, 7, 8, 9, 10, 11, 12, 13, 14, 15 and 16 (in Appendix). We observed that except from the kNN, NB, and RF classifiers showing a negligible difference for some values of k, ℓ, and t, the other classifiers preserve the performance trend when the values of k, ℓ, and t vary.

5 Discussion

In this work, we selected eight well-known classification algorithms and trained them over 10 datasets manipulated using three anonymization techniques, *i.e.*, k-anonymity, ℓ-diversity and t-closeness. We assessed how the employment of these anonymization techniques affect classifiers accuracy, precision, recall and F1-score, and investigated whether these effects depends on the chosen classification algorithm, dataset properties and anonymization parameters. We now discuss some interesting findings and report the threats to validity.

Findings: The performance of the considered classifiers when trained on anonymized datasets in comparison to when trained on the original datasets show that the AB classifier returns the most similar performance results between the original and anonymized datasets. The AB (outperformance) difference is significant compared to DT, NB, RF, LR, and BG, but it is not significant compared to the kNN and SVM classifiers. This result suggests the employment of AB, kNN, and SVM on anonymized data when these classifiers perform accurately when trained on the original data.

Our results also show that there is not a single classifier that significantly outperforms the other classifiers for all performance metrics. Nonetheless, we observe that LR is the best classifier in terms of accuracy, recall, and F1-score on anonymized datasets, whereas NB and RF are superior in terms of precision.

Among dataset properties, the number of class labels considerably affects the classifiers performance on anonymized dataset. The other properties, *i.e.*,

dataset size and number of attributes, have negligible (or no) impact on classifiers' performance. This outcome is independent from the performance metric
considered.

The variation of anonymization parameters, apart from some exceptions with
a negligible difference, does not affect the *trend* of classifiers' performance. This
outcome allows us to generalize (to some extent) our findings on 3-anonymity, 2-
diversity, and 0.2-closeness to other k-anonymous, ℓ-diverse, and t-close datasets.

Threats to Validity: Several variations have been proposed in the literature
for some classification algorithms. For instance, the polynomial and RBF kernel-
based SVM are two types of SVM classifiers, and the Bernoulli and Gaussian
are two types of Naïve Bayes classifiers. Moreover, each classification algorithm
has one or more configuration parameters, *e.g.*, the value of k in kNN classifier
or the number of trees in the AB classifier. The selection of other types of a
classification algorithm or tuning its configuration parameters might affect the
performance. Beside the selection of classification algorithms, the selection of
alternative datasets (*e.g.*, datasets with million records), the selection of other
anonymization algorithm (*e.g.*, Incognito instead of Mondrian) and anonymization parameters (*e.g.*, $k, \ell > 10$) might provide different results.

To mitigate the effect of aforementioned validity threats on our findings, we
have selected the classification algorithms and their configuration parameters as
suggested by the Scikit-learn library. These parameters have been tuned to their
highest performance (for the majority of cases) to provide a fair comparison
among classifiers. The datasets have been selected to meet the diverse requirements in terms of the dataset size, the number of attributes, and the number of
class labels. The selected anonymization algorithm (*i.e.*, Mondrian algorithm)
has shown higher performance compared to other algorithms [2].

While the control variables of this study were carefully chosen, we expect that
the selection of the alternative classification algorithm types (and parameter tuning), dataset, and anonymization technique, will not considerably affect (some
of) our findings. For instance, we expect that the anonymization process makes
the datasets linearly separable resulting in (generally) higher performance for the
LR classifier compared to other classifiers. Also, the anonymization approaches
tend to perform poorly when several records with multi-class labels are grouped
together. These claims and the other aforementioned findings of this study need
more work to investigate the results on a wider range of datasets, different types
of classifiers, other adjustments of the classification algorithms parameters (with
the use of cross-validation or greedy search for parameter adjustment), and different anonymization algorithms and approaches.

6   Related Work

Data anonymization has become a widely investigated research direction in an
effort to protect individuals' privacy when data is supposed to be released publicly. k-anonymity, which was proposed as the initial definition of anonymity
[21,25], has been extended to new additional constraints such as ℓ-diversity [15]

and t-closeness [12]. The proposed approaches have been optimized in terms of a generic measurement with no emphasis on the utility of anonymized data for classification. For instance, in [13], a novel anonymization technique, named *slicing* is proposed, which handles high-dimensional data and improves the data structure compared to preliminary generalization technique. Nergiz et al. [20] suggest a hybrid generalization technique which by data relocation provides a trade-off between utility and privacy. The design and application of anonymization techniques when data utility is critical for data classification has been investigated in several studies. Ye et al. [28] propose a new anonymization approach based on rough set theory, which measures data quality for accurate classifiers construction and guides the anonymization process through combining rough set theory and attribute value taxonomies. In [22] and [18], a trade-off between privacy and data utility is achieved by appropriate feature suppression to publish the anonymized dataset to be used for classification. The focus of these approaches is to create anonymized datasets in which the features effectively discriminate the class labels. However, the performance of a specific classifier over the anonymized datasets has not been sufficiently investigated. In [6], anonymization (k-anonymity) is embedded within the decision tree induction process, which provides better accuracy compared to the scenario in which data is first anonymized and then used for inducing the tree. A similar methodology has been proposed in [4] for embedding anonymization within the association rule mining algorithm. Mostly, the data perturbation for building a specific classification algorithm is performed with the use of differential privacy approach [7]. In this set of approaches, the perturbation process is task-specific dependent on the classification algorithm, *e.g.*, for constructing Naïve Bayes classifier.

Another research steam focuses on the effect of dataset features on data anonymization. In [19], for instance, novel methods are proposed to identify which features of documents need to change and how they must be changed to accomplish document anonymization. Brikke and Shmatikov [3] investigate whether generalization and suppression of quasi-identifier features offer any benefit over trivial sanitization which simply separates quasi-identifier features from sensitive ones. In [17] and [16], a series of experiments is conducted to study the effect of anonymization on classifiers accuracy by applying four different classifiers to the Adult dataset (before and after being perturbed).

While some work in the literature compares the impact of privacy in the context of classifier training, *e.g.*, over encrypted data [24] and under differential privacy [14], to the best of our knowledge, no prior work has provided a comparison of the performance achieved by different classifiers when trained on different datasets before and after being anonymized.

7   Conclusion

This paper investigate the performance achieved by classifiers when they are trained over anonymized datasets. Accordingly, ten benchmark datasets have been anonymized using k-anonymity, ℓ-diversity, and t-closeness approaches.

Then, eight well-known classifiers have been trained on these datasets, and their performance in terms of accuracy, precision, recall, and F1-score has been compared. Our experimental results show that depending on performance metric and dataset properties, one classifier might outperform the others.

In future work, we plan to provide a thorough comparison among a wider range of classifiers on a broader range of benchmark datasets with mixed types of attributes (*e.g.*, categorical). Moreover, we plan to evaluate the impact of classifiers' parameter configuration on classifiers performance trained on anonymized datasets.

Acknowledgement. This work has been supported by H2020 EU funded project SECREDAS [GA #783119].

Appendix

Tables 5, 6, 7, and 8 report respectively the Holm scores of classifiers with respect to accuracy, precision, recall, and F1-score. The higher scores show better performance results for the associated classification algorithm and associated metric.

Table 5. Classifier accuracy scores.

Anonymity	Classifier							
	DT	NB	kNN	SVM	RF	LR	AB	BG
Original	4.15	2.83	3.51	2.92	5.66	4.02	3.97	**5.80**
3-anonymity	4.06	3.33	3.79	3.97	4.34	**5.34**	4.47	3.56
2-diversity	3.10	4.70	4.79	4.15	3.38	**5.20**	4.11	3.42
0.2-closeness	3.61	3.33	3.79	4.93	3.38	**5.25**	4.93	3.65

Table 6. Classifier precision scores.

Anonymity	Classifier							
	DT	NB	kNN	SVM	RF	LR	AB	BG
Original	4.66	3.93	3.29	2.15	5.75	3.74	3.15	**6.21**
3-anonymity	3.83	5.11	2.88	3.65	**5.16**	4.11	3.65	4.47
2-diversity	2.92	**5.66**	4.43	3.61	4.02	4.75	4.02	3.47
0.2-closeness	3.79	**5.75**	3.97	3.01	3.01	5.39	3.93	4.02

Table 7. Classifier recall scores.

Anonymity	Classifier							
	DT	NB	kNN	SVM	RF	LR	AB	BG
Original	4.38	3.63	3.38	3.29	5.07	3.70	4.02	**5.39**
3-anonymity	4.52	3.33	3.79	3.97	4.34	**5.34**	4.47	3.56
2-diversity	3.10	4.70	4.79	4.15	3.38	**5.20**	4.11	3.42
0.2-closeness	3.61	3.33	3.79	4.93	3.38	**5.25**	4.93	3.65

Table 8. Classifier F1-score scores.

Anonymity	Classifier							
	DT	NB	kNN	SVM	RF	LR	AB	BG
Original	3.83	4.38	4.43	3.38	4.29	**5.11**	3.70	3.74
3-anonymity	4.02	3.88	3.97	3.56	4.70	**4.84**	4.02	3.88
2-diversity	2.83	4.47	4.83	3.65	4.11	**4.84**	4.20	3.93
0.2-closeness	4.02	4.47	4.56	3.51	3.83	**4.66**	4.52	3.29

Figures 5, 6, 7, 8, 9, 10, 11, 12, 13, 14, 15 and 16 show the classifiers performance trained on anonymized Credit, Absent, and Optic datasets for different values of k, ℓ, and t.

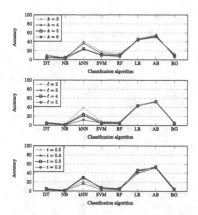

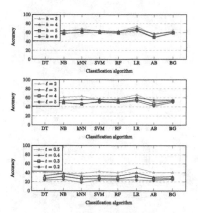

Fig. 5. Accuracy on Credit.

Fig. 6. Precision on Credit.

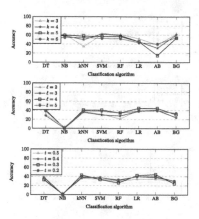

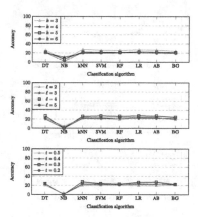

Fig. 7. Recall on Credit.

Fig. 8. F1-score on Credit.

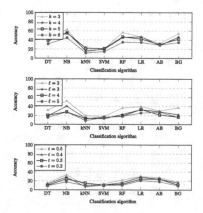

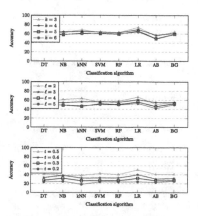

Fig. 9. Accuracy on Absent.

Fig. 10. Precision on Absent.

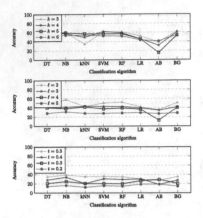

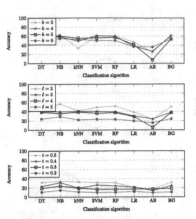

Fig. 11. Recall on Absent.

Fig. 12. F1-score on Absent.

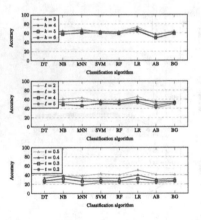

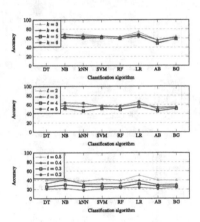

Fig. 13. Accuracy on Optic.

Fig. 14. Precision on Optic.

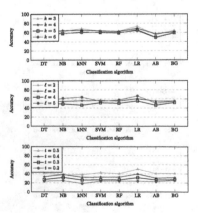

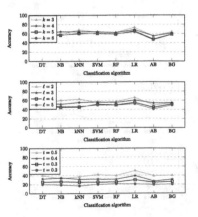

Fig. 15. Recall on Optic.

Fig. 16. F1-score on Optic.

References

1. Aggarwal, C.C.: Data Classification: Algorithms and Applications. Chapman and Hall CRC (2014)
2. Ayala-Rivera, V., McDonagh, P., Cerqueus, T., Murphy, L.: A systematic comparison and evaluation of k-anonymization algorithms for practitioners. Trans. Data Priv. **7**(3), 337–370 (2014)
3. Brickell, J., Shmatikov, V.: The cost of privacy: destruction of data-mining utility in anonymized data publishing. In: International Conference on Knowledge Discovery and Data Mining, pp. 70–78. ACM (2008)
4. Ciriani, V., di Vimercati, S.D.C., Foresti, S., Samarati, P.: k-anonymous data mining: a survey. In: Aggarwal, C.C., Yu, P.S. (eds.) Privacy-Preserving Data Mining: Models and Algorithms. ADBS, vol. 34, pp. 105–136. Springer, Boston (2008). https://doi.org/10.1007/978-0-387-70992-5_5
5. Demšar, J.: Statistical comparisons of classifiers over multiple data sets. J. Mach. Learn. Res. **7**, 1–30 (2006)
6. Friedman, A., Schuster, A., Wolff, R.: k-anonymous decision tree induction. In: Fürnkranz, J., Scheffer, T., Spiliopoulou, M. (eds.) PKDD 2006. LNCS (LNAI), vol. 4213, pp. 151–162. Springer, Heidelberg (2006). https://doi.org/10.1007/11871637_18
7. Gong, M., Xie, Y., Pan, K., Feng, K., Qin, A.: A survey on differentially private machine learning. IEEE Comp. Intell. Mag. **15**(2), 49–64 (2020)
8. Holm, S.: A simple sequentially rejective multiple test procedure. Scand. J. Stat. **6**(2), 65–70 (1979)
9. James, G., Witten, D., Hastie, T., Tibshirani, R.: An Introduction to Statistical Learning: With Applications in R. Springer, New York (2014). https://doi.org/10.1007/978-1-4614-7138-7
10. Khodaparast, F., Sheikhalishahi, M., Haghighi, H., Martinelli, F.: Privacy preserving random decision tree classification over horizontally and vertically partitioned data. In: Conference on Dependable, Autonomic and Secure Computing, pp. 600–607 (2018)
11. LeFevre, K., DeWitt, D.J., Ramakrishnan, R.: Mondrian multidimensional k-anonymity. In: International Conference on Data Engineering, p. 25 (2006)
12. Li, N., Li, T., Venkatasubramanian, S.: t-closeness: privacy beyond k-anonymity and l-diversity. In: 23rd International Conference on Data Engineering, pp. 106–115. IEEE (2007)
13. Li, T., Li, N., Zhang, J., Molloy, I.: Slicing: a new approach for privacy preserving data publishing. IEEE Trans. Knowl. Data Eng. **24**(3), 561–574 (2012)
14. Lopuhaä-Zwakenberg, M., Alishahi, M., Kivits, J., Klarenbeek, J., van der Velde, G.J., Zannone, N.: Comparing classifiers' performance under differential privacy. In: International Conference on Security and Cryptography (SECRYPT) (2021)
15. Machanavajjhala, A., Kifer, D., Gehrke, J., Venkitasubramaniam, M.: l-diversity: privacy beyond k-anonymity. ACM Trans. Knowl. Discov. Data **1**(1), 3-es (2007)
16. Malle, B., Kieseberg, P., Holzinger, A.: DO NOT DISTURB? Classifier behavior on perturbed datasets. In: Holzinger, A., Kieseberg, P., Tjoa, A.M., Weippl, E. (eds.) CD-MAKE 2017. LNCS, vol. 10410, pp. 155–173. Springer, Cham (2017). https://doi.org/10.1007/978-3-319-66808-6_11

17. Malle, B., Kieseberg, P., Weippl, E., Holzinger, A.: The right to be forgotten: towards machine learning on perturbed knowledge bases. In: Buccafurri, F., Holzinger, A., Kieseberg, P., Tjoa, A.M., Weippl, E. (eds.) CD-ARES 2016. LNCS, vol. 9817, pp. 251–266. Springer, Cham (2016). https://doi.org/10.1007/978-3-319-45507-5_17

18. Martinelli, F., Alishahi, M.S.: Distributed data anonymization. In: Conference on Dependable, Autonomic and Secure Computing (DASC), pp. 580–586 (2019)

19. McDonald, A.W.E., Afroz, S., Caliskan, A., Stolerman, A., Greenstadt, R.: Use fewer instances of the letter "i": toward writing style anonymization. In: Fischer-Hübner, S., Wright, M. (eds.) PETS 2012. LNCS, vol. 7384, pp. 299–318. Springer, Heidelberg (2012). https://doi.org/10.1007/978-3-642-31680-7_16

20. Nergiz, M.E., Gök, M.Z.: Hybrid k-anonymity. Comput. Secur. **44**, 51–63 (2014)

21. Samarati, P.: Protecting respondents' identities in microdata release. IEEE Trans. Knowl. Data Eng. **13**(6), 1010–1027 (2001)

22. Sheikhalishahi, M., Martinelli, F.: Privacy-utility feature selection as a privacy mechanism in collaborative data classification. In: Enabling Technologies: Infrastructure for Collaborative Enterprises, pp. 244–249 (2017)

23. Sheikhalishahi, M., Saracino, A., Martinelli, F., Marra, A.L.: Privacy preserving data sharing and analysis for edge-based architectures. Int. J. Inf. Secur. **1**(2), 1–23 (2021). https://doi.org/10.1007/s10207-021-00542-x

24. Sheikhalishahi, M., Zannone, N.: On the comparison of classifiers' construction over private inputs. In: International Conference on Trust, Security and Privacy in Computing and Communications, pp. 691–698 (2020)

25. Sweeney, L.: k-anonymity: a model for protecting privacy. Int. J. Uncertainty Fuzziness Knowl. Based Syst. **10**(05), 557–570 (2002)

26. Wilcoxon, F.: Individual comparisons by ranking methods. Biom. Bull. **1**, 60–83 (1945)

27. Yang, Q., Liu, Y., Chen, T., Tong, Y.: Federated machine learning: concept and applications. ACM Trans. Intell. Syst. Technol. **10**(2), 1–19 (2019)

28. Ye, M., Wu, X., Hu, X., Hu, D.: Anonymizing classification data using rough set theory. Knowl. Based Syst. **43**, 82–94 (2013)

A Rewarding Framework
for Crowdsourcing to Increase
Privacy Awareness

Ioannis Chrysakis[1,3](✉), Giorgos Flouris[1], Maria Makridaki[2],
Theodore Patkos[1], Yannis Roussakis[1], Georgios Samaritakis[1],
Nikoleta Tsampanaki[1], Elias Tzortzakakis[1], Elisjana Ymeralli[1],
Tom Seymoens[4], Anastasia Dimou[3], and Ruben Verborgh[3]

[1] FORTH, Institute of Computer Science, Heraklion, Greece
{hrysakis,fgeo,patkos,rousakis,samarita,tsaban,tzortzak,
ymeralli}@ics.forth.gr
[2] FORTH, PRAXI Network, Heraklion, Greece
makridaki@praxinetwork.gr
[3] IDLab, Department of Electronics and Information Systems,
Ugent, imec, Ghent, Belgium
{Ioannis.Chrysakis,Anastasia.Dimou,Ruben.Verborgh}@UGent.be
[4] imec-SMIT, Vrije Universiteit Brussel, Brussel, Belgium
Tom.Seymoens@vub.be

Abstract. Digital applications typically describe their privacy policy
in lengthy and vague documents (called PrPs), but these are rarely read
by users, who remain unaware of privacy risks associated with the use of
these digital applications. Thus, users need to become more aware of digital applications' policies and, thus, more confident about their choices.
To raise privacy awareness, we implemented the CAP-A portal, a crowdsourcing platform which aggregates knowledge as extracted from PrP
documents and motivates users in performing privacy-related tasks. The
Rewarding Framework is one of the most critical components of the platform. It enhances user motivation and engagement by combining features
from existing successful rewarding theories. In this work, we describe this
Rewarding Framework, and show how it supports users to increase their
privacy knowledge level by engaging them to perform privacy-related
tasks, such as annotating PrP documents in a crowdsourcing environment. The proposed Rewarding Framework was validated by pilots ran
in the frame of the European project CAP-A and by a user evaluation focused on its impact in terms of engagement and raising privacy
awareness. The results show that the Rewarding Framework improves
engagement and motivation, and increases users' privacy awareness.

Keywords: Data privacy · Privacy awareness · Privacy policies ·
GDPR · Crowdsourcing · Rewarding · Collective intelligence

© IFIP International Federation for Information Processing 2021
Published by Springer Nature Switzerland AG 2021
K. Barker and K. Ghazinour (Eds.): DBSec 2021, LNCS 12840, pp. 259–277, 2021.
https://doi.org/10.1007/978-3-030-81242-3_15

1   Introduction

Personal data is the hottest commodity in today's networked society [13,25]. On a daily basis, digital applications drive personal data use. These applications describe how they collect, control and process personal data in lengthy, vague and frequently changing privacy policy documents (PrP) [4][1]. Thus, it is hard for users to read, understand and follow the updates of the PrP documents.

Privacy awareness reflects the extent to which a user is informed about privacy practices and policies, and about how disclosed information is used [39]. In other words, it reflects how clearly users understand the manner at which their data are handled and processed by used applications. Privacy awareness (i) helps users understand the privacy implications of using digital applications [20], e.g., when accepting permissions in device sensors (Camera, GPS); (ii) makes users more thoughtful in relevant situations [6], e.g., when downloading an application or giving their consent to a service provider; and (iii) contributes in counteracting the privacy paradox [33], i.e., the observation that although users are concerned about their privacy in real life, they act differently in their digital life [22]; indeed, we argue that the privacy paradox is due to limited awareness, so improved privacy awareness mitigates the problem.

Achieving privacy-awareness is difficult [36], but better results are achieved if users join forces through a crowdsourcing approach [14,17]. For example, in [7,32,38], crowdsourcing has been employed to allow users to annotate PrPs to clarify privacy practices, and thus improve their privacy knowledge. Also, crowdsourcing could allow users to evaluate the privacy friendliness of apps, e.g., by identifying GDPR concepts[2] in PrP texts, so they are better informed about the use of their personal data by applications [19,30].

One of the most fundamental challenges in crowdsourcing platforms is recruiting and engaging users [27]. Without engagement and motivation, user participation is significantly lower and the platform's objective is not achieved [18]. Combining intrinsic (fun, autonomy, reputation) and extrinsic (money, learning, forcedness, implicitness, task autonomy) rewards in crowdsourcing motivates users [24,40] and increases user participation and engagement.

This paper presents a *Rewarding Framework (RF)* for crowdsourcing activities, used in tandem with a crowdsourcing application to improve privacy awareness based on the crowd's collective knowledge. Such novel combination of crowdsourcing and rewarding to raise privacy awareness has not been considered so far. To design this framework, we identify basic characteristics, such as rewarding features, components and gamification principles, which motivate users, increase participation and achieve a sustainable solution that addresses the data privacy problem. The proposed RF model is based on REWARD [10], a general-purpose ontology designed to represent a reward strategy. The RF is adopted by the

---

[1] https://www.varonis.com/blog/gdpr-privacy-policy/.
[2] https://eur-lex.europa.eu/eli/reg/2016/679/oj.

CAP-A portal[3], implemented and evaluated in the frame of the CAP-A European project[4], and supported by the established CAPrice Community[5].

2   Background and Related Work

In this section, we present related work on crowdsourcing with focus on privacy and existing gamification and rewarding features relevant to our approach.

2.1   Crowdsourcing and Privacy

By focusing on the problem of increasing data *privacy awareness* using crowdsourcing methods, we identify two main approaches enhanced by crowdsourcing: (i) *evaluation of applications* with respect to privacy based on user's opinion [1,2], and (ii) *annotation of Privacy Policy (PrP) documents* through participatory processes [7,32,38]. The former is highly relevant to ours as it focuses on mobile apps and use crowdsourcing tasks to monitor privacy goals. However, users' engagement is based exclusively on paid crowdsourcing activities (e.g., Mechanical Turk Human Intelligence Tasks[6]) and no other rewarding scheme, whereas ours is based on a combination of intrinsic and extrinsic rewards. In the latter, the design of crowdsourcing tasks is not straightforward due to privacy policies' vagueness. Users require more assistance and feedback to make successful contributions in privacy awareness. This needs to be considered when designing the crowdsourcing tasks included in the proposed Rewarding Framework (RF).

However, none of the aforementioned approaches applies a rewarding mechanism to maximize the crowd participation and improve the results. After all, the lack of participation and engagement is one of the most fundamental problems appearing in crowdsourcing [27]. Thus, ways to incentivize the crowd need to be identified, such as gamification schemes and rewarding methodologies. In this paper, we try to cover these needs by proposing a Rewarding Framework.

2.2   Gamification and Rewarding

Gamification refers to the application of game mechanics to a task that is not a game to increase user engagement, happiness or loyalty and constitutes a motivational driver to the success of a crowdsourcing technique [21]. McGonigal et al. [27] suggests a generic gamification scheme based on four principles: *goal, rules, voluntary participation and feedback*. This scheme is suitable in crowdsourcing; hence we follow and extend it in our framework.

Several gamification features were proposed [28], such as *rewards, points and tiers*. Except from gamification principles and features, a gamification approach

---

[3] https://www.cap-a.eu/portal.
[4] https://www.cap-a.eu.
[5] https://www.caprice-community.net.
[6] https://www.mturk.com/.

needs to be adjusted to the underlying crowdsourcing task's complexity [29]. Finnerty et al. [16] evaluated task complexity in crowdsourcing, and showed that a clearer and simpler design, with less demand on workers' attention, provides more accurate results. For this reason, we considered the principle of flexible task management in the design of our proposed crowdsourcing privacy tasks.

In addition to gamification, *rewarding* the users strengthens the members' commitment and increases users' motivation to participate in, and contribute to, any crowdsourcing activity [3]. The positive impact of rewards to encourage participation in open source communities or citizen science initiatives has been well-documented in the related literature [23,31].

Scekic et al. [35] identified different incentive mechanisms in different business environments used for social computing and crowdsourcing. The most relevant to our approach is the *Quota Systems and Discretionary Bonuses mechanism* [35]. In this mechanism, a number of performance metrics is set; when workers reach a threshold they earn a bonus. This is closer to the proposed RF, because users are rewarded according to their contributions and successful accomplishment of privacy-related crowdsourcing tasks.

3   The Rewarding Framework (RF)

First of all, in this section we present the relationship between CAP-A portal and RF. Then we describe the methodology that we used to design the RF. We analyze the critical rewarding features and rewarding components adopted by the RF, along with the gamification principles that we followed.

About the CAP-A Portal. The Rewarding Framework (RF) was implemented as a fundamental component of the CAP-A portal, a crowdsourcing platform aiming to raise privacy awareness in mobile applications [11,12]. In CAP-A portal we applied the proposed rewarding features and components of the RF to reward users' contribution in crowdsourcing privacy tasks. The CAP-A portal is available at: https://www.cap-a.eu/portal.

Methodology. Our framework (Fig. 1) is based on existing works that fit well with the crowdsourcing paradigm. The RF design included the identification of the most appropriate **rewarding features** from existing approaches [26,28,41] that fit well with the crowdsourcing paradigm. The final result is a mix of *platform and user-centric methodology* [40]: a platform for aggregating privacy-related information from several users, announcing results, promoting new crowdsourcing activities etc., and dedicated tasks for users, according to their expertise and preferences.

Before starting designing any Rewarding Framework, the audience needs to be defined. In our case, both users of digital products/services and developers or companies that offer these services are considered. However, the proposed RF is addressed to users only, so we focus on *user-related tasks and features*.

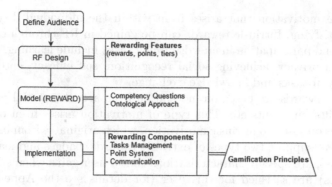

Fig. 1. The reward framework's design process

We described the **rewarding features** in a conceptual model and evaluated it through several competency questions [10]. We implemented the RF's conceptual model as an ontology, the REWARD ontology[7], because an ontology gives us flexibility in the design, e.g., agile adaptation of rewarding features and components. We considered critical **rewarding components** (e.g., Communication, Point System) according to state of the art rewarding approaches. Finally, we adapted **gamification principles** in our approach to contribute to the further engagement of users in executing privacy tasks. We analyze below the aforementioned elements that we adopted in RF: rewarding features, rewarding components and gamification principles.

Table 1. Privacy tasks that lead to rewarding

Task description	Task level
Complete user profile	1
Share installed applications through the CAP-A app	1
Declare favorite applications	1
Download and login to the CAP-A app	1
Claim and acquire ownership of an app	1
Add a privacy expectation/justification	2
Add a URL evidence	2
Vote on the credibility of a given URL	2
Edit/Create annotations on PrP documents	3
Identify GDPR aspects	4

Rewarding Features. Our RF considers the following rewarding features: (i) **rewards.** We support both *intrinsic and extrinsic rewards* [21]. Intrinsic rewards

---

[7] http://www.w3id.org/reward-ontology.

are based on motivation that arises from within the individual, because it is naturally satisfying. Intrinsic rewards can be gained in RF through community interaction and participation in crowdsourcing tasks include learning, improving skills in data privacy, achieving social recognition, self efficacy, entertainment through playful tasks and knowledge exchange.

Extrinsic rewards are based on motivation from actual material prizes, such as money, gifts, discounts etc. This type of motivation arises from outside the individual, as opposed to intrinsic motivation, which originates from inside of the individual. We support two types of extrinsic rewards, badges and leaderboards in RF. *Badges*, are special characterisations that a user gains when specific conditions (rules) are satisfied for this user (for details see the Appendix). The badges offer recognition in the community and allow participating in high-level crowdsourcing tasks, suggesting new ones, and having a more active role in the CAP-A portal. *Leaderboards* like in other similar approaches, are used to further motivate users [28]. They are a dedicated space in the CAP-A Portal that shows users with the highest points and applications with the highest ratings.

(ii) **points**. Our rewarding strategy was inspired by the *"Pay Per Performance"* and *"Quota Systems and Discretionary Bonuses"* [35] mechanisms. Thus, each user is rewarded with points for each completed task. The collection of points leads to the redemption of specified rewards which merit a pre-specified number of points. This policy motivates users to increase their privacy knowledge through the execution of privacy-tasks (Table 1), while the whole community benefits from the resulting aggregated knowledge.

(iii) **tiers**. As in many successful reputation point systems [41], earned points are used to rank users ranked into one of the following tiers: *Baby, Novice, Grown-Up, Enthusiast, Warrior, Expert, Guru, Royal*[8] (see details on the Appendix, Table 5). Tiers help in keeping crowd workers in the loop and motivating them to always try to accomplish new tasks [41].

Rewarding Components and Implementation. Our aim is to support simple and flexible tasks for users. This **Task Management** policy allows users to select the crowdsourcing tasks that best match their needs and, thus, it has positive effects for both users and the community [16]. Thus in RF, tasks are available to users based on their credentials and tier, as well as the tasks' difficulty. To determine which tasks are available to which user, we identified four *task levels* to facilitate task management, each of them representing a different level of difficulty and sophistication:

- **Task Level 1**: tasks that do not demand much effort or expertise;
- **Task Level 2**: tasks related to other users;
- **Task Level 3**: time-consuming and sophisticated tasks that possibly contain a lot of transactions and iterations, or that were initiated by other users;
- **Task Level 4**: high-quality tasks that need to be verified by a system admin.

---

[8] https://cap-a.eu/portal#info.

Higher-level tasks are initially locked and available only to users of higher-level tiers. This policy ensures high-quality feedback, avoiding the probable ad-hoc behaviour of first-time users. Each task follows the rules applied in each respective level, which denote the amount of points for each task level according to the point system, and the task's availability according to the user tier (see Appendix, Table 4).

Our RF includes a **Point System** to determine the details of the process of acquiring and redeeming points with rewards of the user's choice [41], following the practice of most rewarding systems where user tasks are associated with points [41]. The policy of the Point System ensures justice among users which comes from the equity theory: equity theory states that "people compare the ratios of their perceived outcomes to their inputs with the corresponding ratios of others"[34]. Due to lack of space we omit details regarding the Point System.

The **Communication** of the users' accomplishments and associated benefits contributes further to their engagement to the platform [28]. In our case, the Communication of results is performed through a specific notification mechanism, which informs users about their progress in the CAP-A portal (e.g., completed tasks, acquired points or other details with regards to the applied Point System). This mechanism also suggests steps for participating in more privacy tasks (and levelling-up), and announces the available rewards.

Gamification Principles. Our RF and its features (e.g., points, badges, leaderboards, tiers, etc.) are heavily influenced by existing common gamification theories. Gamification plays a critical role in any rewarding strategy, as it results to a more fruitful process.

We apply the generic approach of McGonigal et al. [27] which denotes that a gamification scheme should be based on: *common goal, rules of the game, feedback and voluntary participation*. In our case the *common goal* is raising privacy awareness. We extend this approach by setting as well *personal goals* for users following the expectancy theory [37]. This design option creates a wide gamification space for users and a clear relationship between tasks and rewards to contribute further to user's motivation.

Specifically, in the RF, gamification is supported by providing the following goals, which can be picked up by CAP-A users:

1. Level-up through community experience
2. Be included in the top20 list
3. Acquire knowledge on one's favourite applications
4. Express privacy concerns
5. Participate in the evaluation of applications
6. Improve the market towards building more transparent and privacy friendly applications

In our case the *rules of the game* are related to the accomplishment of tasks that cover several aspects of privacy focusing on PrP documents and are applied throughout the system.

Feedback is also a fundamental characteristic, because it makes goal achievement more realistic by showing gradual improvement and motivates users to participate further [21]. Thus, the CAP-A portal gives *feedback* in several milestones for the accomplishment of tasks, in tiers level ups, earning of badges, rewards availability and redemption.

Finally, voluntary participation requires that everyone who is participating knowingly and willingly accepts the goals, the rules, and the feedback [27]. For this reason, we pursue a transparent policy to users, so that they can clearly denote their goals, and understand the gamification rules and the offering feedback of the CAP-A portal (e.g., through supporting users' history, denoting next steps for rewarding).

4   Validation

The Rewarding Framework (RF) was validated through six pilots focused on specific app categories and ran in the frame of the CAP-A project (see Table 2). Each Pilot followed a pre-designed workflow scenario, including requests to perform crowdsourcing actions that lead to rewarding. In total, 108 users participated in the pilots and 141 users registered to the CAP-A portal.

Table 2. CAP-A Pilots overview

Pilot name	Duration	Participants	Apps category
Saferinternet4Kids	1/10–31/10	35	Games, Social Media & Communication
Bora	16/10–9/11	36	Business
REN	23/11–30/11	11	Conferencing
Devstaff	10/12 (Live)	8	Social media, Productivity
Praxi	16/12–23/12	5	Conferencing
Homodigitalis	15/10–15/12	13	All above categories

The CAP-A portal allows two main types of interaction with the user, both of which are included in the RF and give points to users upon successful completion: expressing privacy expectations, and adding annotations on PrP documents. Both activities were included in the pilot scenarios, and resulted in meaningful contributions in terms of populating our privacy repository with user expectations and annotations; analysing these contributions is out of the scope of the current paper, but the interested reader is referred to the CAP-A portal statistics (https://cap-a.eu/portal/#stats) for details.

Our validation showed that the CAP-A portal and the RF worked properly without facing any problems; moreover, useful feedback for further improvement was received, which was organised in three main categories, analysed below.

Overriding the Default Behavior of RF. During the pilots, the RF's default behaviour should be overridden in some cases for promotion and motivation purposes (e.g., to increase the points per task for a specific task and for a limited time period). Thus, the rewarding system should be flexible, parameterisable and adaptable. The implementation of our RF as an ontology helped us to easily address this demand; we added a *Boost Parameter* to the RF to support manual point adjustment. For example, the task of completing the user profile, important for the CAP-A portal's dashboard, was promoted for some of the last pilots.

Users Info - History of Activities. Users should be able to see their history of activities, e.g., for rewarding tasks. This functionality could help users monitor their past activities, and be further engaged in performing similar activities. This feature is related to the offered *"Feedback"* which is provided by the system to the users with regards to their accomplishments, and is a critical feature of reward systems [21,28]. To support this, we implemented in the CAP-A portal a *User's History* page, to clearly display past activities as a personal record dashboard.

UX Improvement Suggestions with Regards to RF. During the pilots, it was clear that more information about the steps that a user should follow to level-up tier was needed. We concluded that an interface encouraging a more stream-lined process would help users enjoy flexible navigation through the available tasks. This idea was implemented by redesigning the homepage of the CAP-A (to present alternative privacy tasks associated with the RF in a specific order).

5   Empirical Evaluation

After improving the CAP-A portal and the Reward Framework (RF) based on the feedback we received in the validation, we further performed an empirical evaluation with 11 participants who followed a specific workflow scenario [9]. The goal was to make a first assessment of how the system improvements can affect user engagement and privacy awareness. Our evaluation is an *exploratory study*, aimed at identifying critical success aspects of the RF and potential barriers or features that need to be further explored.

We considered both *objective* and *subjective metrics*. We evaluated engagement features, e.g., number of accomplished tasks compared to requested ones (objective metric), or users' opinion on the used features, e.g., points and tiers (subjective metric). We also evaluated privacy awareness with the *Privacy Awareness Index* [15], which measures the increase of privacy knowledge (objective) and the users' opinion on the acquired privacy knowledge (subjective).

We present below the *research questions (RQ)* and *hypotheses (H)* used to assess our targeted evaluation goals, as well as the evaluation setup, the methodology we followed to assess the impact of the RF, and the results of our study.

5.1 Research Questions and Hypotheses

We evaluated the impact of the RF along two dimensions: engagement of users, and raising of privacy awareness. Each dimension constitutes a different part in our evaluation. We also made a UX evaluation to ensure that the user interface is not affecting negatively our results regarding user engagement and privacy awareness.

Part 1. Engagement and Motivation. We assess engagement and motivation indicators of the RF based on the following research questions and hypotheses:

- RQ1.1: *Are the users engaged while participating in crowdsourcing activities (as enabled by the CAP-A portal) due to the RF?*
  H1.1: *Users' engagement and participation is increased because of the RF.*
- RQ1.2: *Does rewarding encourage the participation level of users in participating in privacy related tasks requiring interaction?*
  H1.2: *Users are motivated to participate in actions requiring their feedback due to the RF.*
- RQ1.3: *Does rewarding affect positively the performance of users in executing privacy-related tasks?*
  H1.3: *Users gaining rewards are strongly motivated to perform more crowdsourcing privacy tasks.*
- RQ1.4: *Does rewarding affect users' return in the portal?*
  H1.4: *Rewarding makes users willing to return to the portal to perform additional crowdsourcing activities.*

To validate the above hypotheses, we asked a set of questions [8] to capture users' opinion (subjective metrics). We also used objective metrics extracted from the users' interaction with the portal. For example, to validate the most generic hypothesis H1.1, we examined the *response rate of participation* in specific crowdsourcing tasks of the workflow [9], and the *total number of accepted invitations* to register to the CAP-A portal. We also considered metrics indicating the users' actual engagement with the portal, e.g., the actual users' *interaction with the invited apps* and *their total points*, which gives a sense of the amount of work performed, in addition to the work required by the workflow scenario.

Finally, additional metrics were used to measure the level of engagement in the defined scenarios: *number of declared expectations, number of favorite apps, number of annotations in PrP documents*, and *number of total examined apps*.

Part 2. Privacy Awareness. Common privacy awareness questions were asked before and after the use of the CAP-A portal to assess whether privacy awareness increased. Similarly to Part 1, we formulated a set of research questions and hypotheses to evaluate privacy awareness [8]:

- RQ2.1: *Does participation in the CAP-A portal improve privacy awareness?*
  H2.1: *Users who used the CAP-A portal improved their privacy awareness.*

- RQ2.2: *Does the RF have a positive impact in raising privacy awareness?*
  H2.2: *Users with less experience in rewarding tasks (low amount of points) are less privacy-aware.*
- RQ2.3: *Does the tier level up affect the increased privacy awareness of users?*
  H2.3: *Users that level-up tier increase their knowledge on privacy.*
- RQ2.4: *Does the RF encourage users to make privacy aware actions through the portal?*
  H2.4: *The RF motivates users to offer their privacy expectations, to improve the community's privacy awareness.*

5.2   Evaluation Setup

For the evaluation setup, following the approach of [15], we define the following elements: participant details, the goals of the evaluation, the applied methodology, and the final expected results.

Participant Details. For our evaluation process, we invited 15 participants from two different countries. The participants belonged to three different age groups and had different background and level of knowledge with respect to technology, mobile apps and privacy. By calculating the users' *Privacy Concerns Index* (PCI), following the approach of [5] we classified them into three categories to check the impact of the RF to different categories of users (see Users Classification subsection). All users who participated download and install mobile applications that handle their personal data, so it is interesting for them to learn more about their data utilization by service providers and developers.

Evaluation Goals. The goal of the evaluation is twofold: (i) assess the impact of the proposed RF in the engagement and motivation of users to participate in privacy related tasks; and, (ii) improve users' privacy awareness due to their participation in a crowdsourcing approach that applies this RF.

Evaluation Methodology. For our evaluation, we combined a Survey Part with an Experimental Part following [5].

The *Survey Part* aims to collect the participants' demographic characteristics, estimate their privacy concerns, and classify their concerns into different categories according to their common interests and acquired knowledge with regards to privacy. It consists of a *Pre-Questionaire* where we use the popular Likert scaling[9], as it is equidistant and well elaborated. After applying the Pre-Questionaire to the target audiences, *Pre-Activities* are used to explain the evaluation process of the experimental part, whereas *Post-Activities* are used to correlate the results from the experimental part (objective) with answers to subjective related questions for both engagement and awareness.

---

[9] https://conjointly.com/kb/likert-scaling/.

The Pre-Activities include a Pre-Questionaire to capture *Demographic* and *Background* information for users and calculate the *Privacy Concerns Index (PCI)* to classify them to different categories following the approach of [5].

The Post-Activities include a Post-Questionnaire to evaluate the success of the engagement and motivation strategy (Part 1), and privacy awareness methods used (Part 2), according to the evaluation's goals. To ensure that the applied User Interface does not get in the way of the goal of improving privacy awareness, we validated some common usability metrics[10] through a UX Questionaire, namely, *user's subjective satisfaction, success rate*, and *required time*.

The *Experimental Part* aims to assess the impact on users' behavior while interacting with the RF. It consists of a specific workflow scenario [9], in which we evaluate a set of parameters regarding user engagement and improvement of privacy awareness through interaction with the CAP-A portal and the applied RF. This part is also correlated with the respective Post-Questionnaire which is collected after this interaction to draw conclusions about user engagement and motivation and their level of privacy awareness.

Expected Results. Our expected results include:

1. a *users classification* according to their background and privacy expertise based on the Pre-Questionnaire;
2. insights about *user engagement and motivation* and comparisons on quantitative metrics that resulted from the user interaction with the RF;
3. insights regarding the RF impact on users' *privacy awareness level*;
4. *Correlation of results* of the Pre- and Post-Questionnaire.

5.3 Results

We present the users classification, results and conclusions for our evaluation.

Users Classification. 11 participants fulfilled the evaluation. Following the approach of [5], the categories and the respective classifications for users were:

- **Fundamentalists**, users really sensitive to privacy with high privacy concerns); 5 of our users were classified as fundamentalists.
- **Pragmatists**, users that care for privacy; 3 of our users were classified as pragmatists.
- **Unconcerned**, users that do not really have special concerns with regards to privacy; 3 of our users were classified as unconcerned.

Part 1. Engagement and Motivation. In total, 73% of users actively participated in the evaluation. We considered as active any user who was invited, registered to the CAP-A portal and completed at least one rewarding task of the defined scenario. Interestingly, most users have not participated in any privacy-awareness activity in the past according to the replies in the Pre-Questionaire,

---

[10] https://www.nngroup.com/articles/usability-metrics/.

and still accepted to participate in such a privacy activity; as a result, engaging them to the related crowdsourcing task becomes more challenging.

In the majority of the cases, and for all types of users, the participation rate exceeded the minimum number of 10 tasks specified in the scenario [9], revealing high engagement to the CAP-A portal. Each user accomplished on average 41 tasks, 13.5 tasks per user if we consider unique actions per app (Table 3).

Table 3. Completed tasks per user

User	Expectations (min 3)	Favorites (min 5)	Annotations (min 1)	Examined Apps (min 5)
User 1 (F)	30	5	0	5
User 2 (P)	18	8	3	8
User 3 (P)	3	5	1	5
User 4 (F)	53	8	3	8
User 5 (F)	30	5	0	5
User 6 (U)	24	9	4	10
User 7 (F)	28	5	0	7
User 8 (F)	49	5	2	5
User 9 (P)	41	5	1	5
User 10 (U)	26	8	2	8
User 11 (U)	42	20	1	22

(F): Fundamentalist; (P): Pragmatist; (U) unconcerned

Comparing these results, i.e. response rate of participation in completed tasks, with results of the *subjective* questions in Part 1 of the Post-Questionnaire (to validate research questions RQ1.1, RQ1.2 and RQ1.3), we can clearly draw positive results regarding user engagement. Specifically, most users agreed that the RF positively affects their activities in the CAP-A portal (Fig. 2(b)).

In addition, it seems that the RF motivates users in participating in privacy tasks requiring their interaction: 54% of the users agreed with this, 37% appeared neutral and only 9% disagreed (Fig. 2(c)). This argument is reinforced by the fact that the task that most users completed is the expression of expectations (Table 3). Thus, users preferred to express expectations than adding favorites which is a naive task. Thus, the interaction of users through participation in privacy tasks ultimately helps to improve their privacy knowledge. The applied rewarding features (e.g., points and tiers) motivates 72% of the users, namely user engagement is achieved through interaction with the RF (Fig. 2(d)).

More than half of the users had positive or neutral impression for the RF (72%). Specifically, 45% of users think that the RF is supportive and 18% believe it is essential, 9% are neutral, 18% think it is unnecessary and only 9% think that is disturbing. The results are displayed graphically in Fig. 2(a).

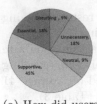   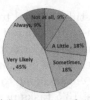

(a) How did users experience the RF?

(b) Does rewarding affect users activity in the CAP-A portal?

(c) The RF triggers users in tasks that require their feedback

(d) How much RF features motivate users to participate in more activities?

Fig. 2. Evaluating users' engagement and motivation

We further examined the users who appeared negative (1 user) or neutral (2 users). We noticed that these users performed 40, 40 and 52 tasks respectively, which is much more than the minimum total requested number of tasks (10 tasks). However, we also found that 2 of them did not participate in the annotation task. For this reason we included a dedicated question regarding the ease of annotation task in the UX evaluation below.

According to their replies in the Post-Questionnaire, 64% of users spent more than 10 min in the CAP-A portal, which is much more than the estimated 5 min necessary to complete the scenario. This shows engagement, and clearly demonstrates that the RF increases users' return to the portal (RQ 1.4).

Part 2. Privacy Awareness. In the second part of our evaluation, we compared the Privacy Awareness Index (PAI) for the same users before and after the execution of the workflow scenario. This experiment mainly assessed the research questions RQ2.1, RQ2.2 and RQ2.3. Our results show a very high increase on the level of privacy awareness for all users and all categories (the highest increase, 500%, appeared in User 9(P), see Fig. 3). For Unconcerned and Pragmatist users the increase was 288% and 286% on average, while for Fundamentalists we noticed a 151% increase. We also noticed a significant improvement in the level of privacy awareness for users who have children (246%). Thus, it

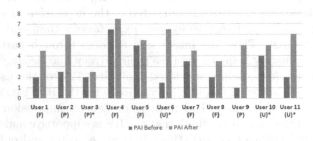

Fig. 3. Privacy Awareness Index (PAI), max:8, users who have kids marked with a star

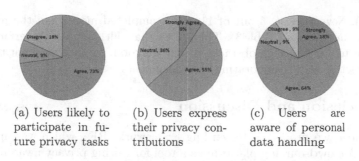

(a) Users likely to participate in future privacy tasks

(b) Users express their privacy contributions

(c) Users are aware of personal data handling

Fig. 4. Evaluating privacy awareness

is very encouraging that this category of users who were classified as the most unconcerned, exhibited considerable increase in their privacy awareness index.

Furthermore, the RF proved that it motivates users to contribute in privacy awareness actions through the CAP-A portal (RQ 2.4). This conclusion is drawn since most users (73%) agree that it is very possible to participate in a new privacy task in the future due to the RF (Fig. 4(a)).

Most users (55%) agree that due to the RF, they make their own privacy contributions such as expressing privacy expectations on device permissions (see Fig. 4(b)). The interaction of users with the RF helps them to become aware and knowledgeable about how personal data are handled by mobile applications and service providers; 82% of participants agreed with this claim (Fig. 4(c)).

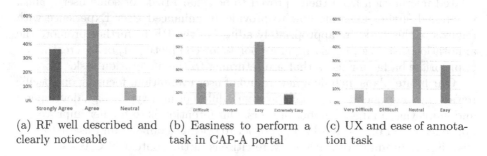

(a) RF well described and clearly noticeable

(b) Easiness to perform a task in CAP-A portal

(c) UX and ease of annotation task

Fig. 5. UX evaluation

UX Evaluation. Users did not experience any difficulty in understanding and identifying the main RF features. 91% agreed that the RF features (such as tasks, points and tiers) are well described, whereas 9% appeared neutral (Fig. 5(a)). The results of our User Experience (UX) evaluation showed that the CAP-A portal interface offers an easy and straightforward way to perform the main privacy tasks related to the RF. 64% or users found easy or extremely easy to perform a task in the portal (Fig. 5(b)). Adding annotations was the most difficult task: 9% of the users found it very difficult, 9% difficult, 55% neutral and 27% easy

(Fig. 5(c)). Nevertheless, 5 out of 11 users completed more than the minimum requested annotations (Table 3). This shows that, although the interface of the Annotator tool was not an obstacle, it still has room for improvement to offer a more attractive way of annotating PrP documents.

6   Conclusion and Discussion

In this paper, we propose a Rewarding Framework (RF), which was applied in the CAP-A crowdsourcing platform as a tool for raising privacy awareness. The validation and evaluation of the RF led to some interesting empirical findings, showed promising results regarding user engagement, and contributed in improving users' privacy awareness. To the best of our knowledge, rewarding has not been evaluated so far to raise privacy awareness.

Our study showed that it is very important for a rewarding framework to be agile (in terms of overriding its default behavior), supporting fair rewards according to the complexity of each task, and providing a transparent mechanism of offering feedback to users (showing their progress in personal goals, or how their contributions make an impact towards the common goal of the community).

It is also important for rewarding frameworks to provide an easy way to achieve initial goals, so that new users are not repelled. Equally important is to utilize appropriate terminology to refer to the RF features, such as the experience points earned or the achievements accomplished, so that the users feel that they are becoming domain experts, in some sense, as they become more involved.

In addition, dealing with annotating PrP documents and extracting privacy-related information from them, proved to be a hard task for some users. Thus, we should further examine how to provide an enhanced User Experience with regards to this task, by appropriately adjusting the RF to further promote the annotation task, e.g., by giving more points for completing it, or by creating an appropriate badge for users that are returning to the annotation task.

Our future plans include a new round of user evaluation, focusing on specific rewarding components (e.g., point system), different user types (e.g., developers) and employing a larger number of users. Our intention is to further improve our approach based on the feedback, in order to optimise our RF and use it to help sustain a community of privacy-aware citizens in the context of CAP-A.

Acknowledgement. This work has been supported by the CAP-A project which has received funding from the European Union's Horizon 2020 research and innovation programme under the NGI_TRUST grant agreement no 825618. The described research activities were also funded by Ghent University, imec, Flanders Innovation & Entrepreneurship (VLAIO). Ruben Verborgh is a postdoctoral fellow of the Research Foundation – Flanders (FWO).

Appendix: RF Implementation Details

We present below details regarding the rules that we applied for task levels, tiers and for the introduced badges in the RF implementation.

Table 4. Task levels

Task level	Earned points	Tier availability
1	$W_1 * C$ Points ($W_1 = 1$)	Baby to Royal
2	$W_2 * C$ Points ($W_2 = 1$)	Novice to Royal
3	$W_3 * C$ Points ($W_3 = 2$)	Grown-Up to Royal
4	$W_4 * C$ Points ($W_4 = 4$)	Expert to Royal

TaskLevelAdjustmentParameter $= W_i$ for $i > 0$ and
$i \leq$ Task Levels Number, C = Min Task LevelThreshold in
points

Table 5. Available tiers

Tier	Required points
Baby	Zero points
Novice	$V_1 * D$ Points ($V_1 = 5$)
Grown-up	$V_2 * D$ Points ($V_2 = 15$)
Enthusiast	$V_3 * D$ Points ($V_3 = 20$) points
Warrior	$V_4 * D$ Points ($V_4 = 50$) points
Expert	$V_5 * D$ Points ($V_5 = 100$)points
Guru	$V_6 * D$ Points ($V_6 = 500$) points
Royal	$V_7 * D$ Points ($V_7 = 1000$) points

TierLevelAdjustmentParam $= V_j$ for $j > 0$ and
$j \leq$ Tiers Number
D = FirstTierThreshold in points

Badges

- **Social/Buddy**: users who invited more than five friends to join the system or to accomplish a specific task within a month;
- **Super Star**: users who completed at least ten tasks in a month;
- **On Fire**: users who completed more than three tasks in the last week;
- **Ambassador**: users with high expertise on various tasks or on privacy issues; ambassadors are invited/suggested by the crowd or by other ambassadors.
- **Inactive**: users who joined the system but did not start/complete any task;
- **Sleepy**: users who completed at least one task but did not start a new one for the last three months.

References

1. Amini, S., Lin, J., Hong, J.I., Lindqvist, J., Zhang, J.: Mobile application evaluation using automation and crowdsourcing (2018). https://doi.org/10.1184/R1/6470255.v1
2. Amini, S.: Analyzing mobile app privacy using computation and crowdsourcing. Ph.D. thesis, Carnegie Mellon University (2014)
3. Antikainen, M.J., Vaataja, H.K.: Rewarding in open innovation communities-how to motivate members. Int. J. Entrepre. Innov. Manag. 11(4), 440–456 (2010)
4. Antón, A.I., Earp, J.B., He, Q., Stufflebeam, W., Bolchini, D., Jensen, C.: Financial privacy policies and the need for standardization. IEEE Secur. Privacy 2(2), 36–45 (2004)
5. Bergmann, M.: Testing privacy awareness. In: IFIP Summer School on the Future of Identity in the Information Society, pp. 237–253 (2008)
6. Bergram, K., Bezençon, V., Maingot, P., Gjerlufsen, T., Holzer, A.: Digital nudges for privacy awareness: from consent to informed consent? In: ECIS (2020)
7. Bhatia, J., Breaux, T.D., Schaub, F.: Mining privacy goals from privacy policies using hybridized task recomposition. ACM TOSEM 25(3), 1–24 (2016)
8. Chrysakis, I.: CAP-A rewarding framework evaluation - list of questions (2020). https://doi.org/10.6084/m9.figshare.13042772.v8
9. Chrysakis, I.: Introduction to CAP-A portal & rewarding evaluation scenario (2020). https://doi.org/10.6084/m9.figshare.13042787.v5
10. Chrysakis, I., Flouris, G., Patkos, T., Dimou, A., Verborgh, R.: REWARD: ontology for reward schemes. In: Harth, A., et al. (eds.) ESWC 2020. LNCS, vol. 12124, pp. 56–60. Springer, Cham (2020). https://doi.org/10.1007/978-3-030-62327-2_10
11. Chrysakis, I., et al.: CAP-A: a suite of tools for data privacy evaluation of mobile applications. In: JURIX (2020)
12. Chrysakis, I., et al.: Evaluating the data privacy of mobile applications through crowdsourcing. In: JURIX (2020)
13. Craig, T., Ludloff, M.E.: Privacy and Big Data: The Players, Regulators, and Stakeholders. O'Reilly Media, Inc., Sebastopol (2011)
14. Diamantopoulou, V., Androutsopoulou, A., Gritzalis, S., Charalabidis, Y.: An assessment of privacy preservation in crowdsourcing approaches: towards GDPR compliance. In: RCIS, pp. 1–9. IEEE (2018)
15. Fatima, R., Yasin, A., Liu, L., Wang, J., Afzal, W., Yasin, A.: Sharing information online rationally: An observation of user privacy concerns and awareness using serious game. J. Inf. Secur. Appl. 48, 102351 (2019)
16. Finnerty, A., Kucherbaev, P., Tranquillini, S., Convertino, G.: Keep it simple: reward and task design in crowdsourcing. In: CHItaly (2013)
17. Flouris, G., et al.: Towards a collective awareness platform for privacy concerns and expectations. In: ODBASE (2018)
18. Grobbink, E., Peach, K.: Combining crowds and machines (2020), https://www.nesta.org.uk/report/combining-crowds-and-machines/
19. Hatamian, M., Kitkowska, A., Korunovska, J., Kirrane, S.: 'It's shocking!': analysing the impact and reactions to the A3: android apps behaviour analyser. In: DBSec, pp. 198–215 (2018)
20. Kani-Zabihi, E., Helmhout, M.: Increasing service users' privacy awareness by introducing on-line interactive privacy features. In: Laud, P. (ed.) NordSec 2011. LNCS, vol. 7161, pp. 131–148. Springer, Heidelberg (2012). https://doi.org/10.1007/978-3-642-29615-4_10

21. Kavaliova, M., Virjee, F., Maehle, N., Kleppe, I.A.: Crowdsourcing innovation and product development: gamification as a motivational driver. Cogent Bus. Manag. **3**(1), 1128132 (2016)
22. Kokolakis, S.: Privacy attitudes and privacy behaviour: a review of current research on the privacy paradox phenomenon. Comput. Secur. **64**, 122–134 (2017)
23. Krishnamurthy, S., Ou, S., Tripathi, A.K.: Acceptance of monetary rewards in open source software development. Res. Policy **43**(4), 632–644 (2014)
24. Lee, T.Y., et al.: Experiments on motivational feedback for crowdsourced workers. In: 7th AAAI Conference on Weblogs and Social Media (2013)
25. Liang, F., Yu, W., An, D., Yang, Q., Fu, X., Zhao, W.: A survey on big data market: pricing, trading and protection. IEEE Access **6**, 15132–15154 (2018)
26. McCall, M., Voorhees, C.: The drivers of loyalty program success: an organizing framework and research agenda. Cornell Hosp. Q. **51**(1), 35–52 (2010)
27. McGonigal, J.: Reality Is Broken: Why Games Make Us Better and Bow They Can Change the World, Penguin, New York (2011)
28. Morschheuser, B., Hamari, J., Koivisto, J.: Gamification in crowdsourcing: a review. In: 49th Hawaii International Conference on System Sciences (2016)
29. Morschheuser, B., Hamari, J., Koivisto, J., Maedche, A.: Gamified crowdsourcing: conceptualization, literature review, and future agenda. Int. J. Hum. Comput. Stud. **106**, 26–43 (2017)
30. Nejad, N.M., Scerri, S., Lehmann, J.: KnIGHT: mapping privacy policies to GDPR. In: European Knowledge Acquisition Workshop, pp. 258–272 (2018)
31. Newman, G., Wiggins, A., Crall, A., Graham, E., Newman, S., Crowston, K.: The future of citizen science: emerging technologies and shifting paradigms. Front. Ecol. Environ. **10**(6), 298–304 (2012)
32. Oltramari, A., Piraviperumal, D., Schaub, F., Wilson, S., Cherivirala, S., Norton, T.B., Russell, N.C., Story, P., Reidenberg, J., Sadeh, N.: Privonto: a semantic framework for the analysis of privacy policies. Seman. Web **9**(2), 185–203 (2018)
33. Pötzsch, S.: Privacy awareness: a means to solve the privacy paradox? In: Matyáš, V., Fischer-Hübner, S., Cvrček, D., Švenda, P. (eds.) Privacy and Identity 2008. IAICT, vol. 298, pp. 226–236. Springer, Heidelberg (2009). https://doi.org/10.1007/978-3-642-03315-5_17
34. Samaha, S.A., Palmatier, R.W., Dant, R.P.: Poisoning relationships: perceived unfairness in channels of distribution. J. Market. **75**(3), 99–117 (2011)
35. Scekic, O., Truong, H.L., Dustdar, S.: Incentives and rewarding in social computing. Commun. ACM **56**(6), 72–82 (2013)
36. Solove, D.J.: The myth of the privacy paradox. Available at SSRN (2020)
37. Vroom, V.H.: Work and Motivation. Wiley, New York (1964)
38. Wilson, S., et al.: Crowdsourcing annotations for websites' privacy policies: can it really work? In: Proceedings of WWW-16, pp. 133–143 (2016)
39. Xu, H., Dinev, T., Smith, H.J., Hart, P.J.: Examining the formation of individual's privacy concerns: toward an integrative view. In: ICIS (2008)
40. Yang, D., Xue, G., Fang, X., Tang, J.: Crowdsourcing to smartphones: incentive mechanism design for mobile phone sensing. In: MobiCom (2012)
41. Zichermann, G., Cunningham, C.: Gamification by Design: Implementing Game Mechanics in Web and Mobile Apps. O'Reilly Media, Inc., Sebastopol (2011)

DUCE: Distributed Usage Control Enforcement for Private Data Sharing in Internet of Things

Na Shi[1]([✉]), Bo Tang[1], Ravi Sandhu[2], and Qi Li[3]

[1] Sichuan Changhong Electric Co., Ltd., Mianyang, China
{na.shi,bo.tang}@changhong.com
[2] Institute for Cyber Security, NSF Center for Security and Privacy Enhanced Cloud Computing, Department of Computer Science, University of Texas at San Antonio, San Antonio, USA
ravi.sandhu@utsa.edu
[3] Tsinghua University, Beijing, China
qli01@tsinghua.edu.cn

Abstract. The emerging Cloud-Enabled Internet of Things (CEIoT) is becoming increasingly popular since it enables end users to remotely interact with the connected devices, which collect real-world data and share with diverse cloud services. The shared data will often be sensitive as well as private. According to the General Data Protection Regulation (GDPR), the privacy issue should be addressed by the cloud services and subsequent data custodians. In this paper, we propose DUCE, an enforcement model for distributed usage control for data sharing in CEIoT. DUCE leverages both blockchain and Trusted Execution Environment (TEE) technologies to achieve reliable and continuous life-cycle enforcement for cross-domain data sharing scenarios. The core components of DUCE are distributed Policy Decision Points (PDPs) and Policy Enforcement Points (PEPs) to enable reliable execution of usage control policies without a centralized trusted authority. Policy administration is also distributed and controlled by the data owner, who can modify the rules anywhere anytime. The policy rules expressed in eXtensible Access Control Markup Language (XACML) are parsed into smart contracts to be executed on the blockchain service. A detailed explanation of the enforcement process is given for an example "delete-after-use" rule. A prototype system is implemented with an open-source permissioned blockchain system and evaluated on an experimental deployment. The results show reasonable performance and scalability overhead in comparison to OAuth 2.0. We believe additional cross-domain data usage control issues can also be addressed by DUCE.

Keywords: Cloud-enabled Internet of Things · Privacy · Usage control · Blockchain · Trusted execution environment.

© IFIP International Federation for Information Processing 2021
Published by Springer Nature Switzerland AG 2021
K. Barker and K. Ghazinour (Eds.): DBSec 2021, LNCS 12840, pp. 278–290, 2021.
https://doi.org/10.1007/978-3-030-81242-3_16

1 Introduction

The Internet of Things (IoT) extends the boundary of the familiar Internet by incorporating smart physical objects (things) embedded with sensors, actuators, software and communications hardware, for the purpose of connecting and exchanging data with other devices and systems. The continuing convergence of cloud computing and IoT has brought about the concept of Cloud-Enabled Internet of Things (CEIoT) [4–6,8] which is a new computing paradigm bringing together the complementary advantages of cloud computing and IoT. In this paradigm, a cloud computing service is used to provide convenient access to online applications and services, and a IoT service is used to enable sensing and control of the physical world, through which increasingly comprehensive data interactions facilitate "smarter" applications for end users.

Applications of CEIoT span a diverse set of consumer, industrial and professional scenarios. Data about healthcare collected and stored in a secure manner can be shared, to provide remote sensing detection services for elderly healthcare [22], to provide patients some facilities through telehealth [2], to promote medical research [9], etc. Contaminated water detected can be shared to prevent users and crops from outbreak of diseases [10,21]. As for transportation, smart parking service data can be shared and driving habits monitored to provide vehicle owners services such as warranty and insurance discounts for safe driving [12,13]. The proliferation of data sharing in CEIoT is an emerging trend with great potential and may even become the future of the Internet [11,28].

CEIoT presents significant privacy concerns especially in consumer-oriented applications [1,16,26]. Most shared data collected by user devices is sensitive and private. Despite the fact that access control mechanisms can be used to prevent the data from leakage during an access, the data shared to an external application is not subject to this restriction. Furthermore, the behaviors of the external application cannot be monitored or controlled once the data is shared, whereby the usage of the data may violate articles such as *"Rights to erasure"* in General Data Protection Regulation (GDPR). Thus appropriate privacy preserving mechanisms need to be developed to mitigate this risk and realize the true potential of such data sharing. In this paper, we propose a distributed usage control enforcement model, namely DUCE, to address the aforementioned privacy concerns in CEIoT.

The key contributions of this work are as follows. (i) A DUCE design overview is given with the system components including the distributed PDPs and PEPs. DUCE leverages permissioned blockchain technology to build a trusted relationship between data-sharing parties, whereby the rules and enforcement records are tamper-proof and visible to users. A Trusted Execution Environment (TEE) is used to ensure that the enforcement process of the rules and the usage of user data are trustworthy and controllable by users. (ii) The policy administration model of DUCE is also provided with a policy example of "delete-after-use" in XACML and the policy translation algorithm into Solidity language for smart contracts. (iii) A prototype system is implemented and deployed along with an

OAuth 2.0 benchmark system. The end-to-end delay and throughput are evaluated and analyzed to demonstrate the viability of DUCE.

Organization. Section 2 reviews essential technical concepts. Section 3 provides a typical user scenario, the problem statement and the design goals. DUCE is developed in Sect. 4. Section 5 discusses the experiment results. Related work is summarized in Sect. 6. Section 7 concludes the paper.

2   Background

Usage Control (UCON). Traditional access control models [27] deal with authorization as sole basis for access decisions and typically focus only on *server-side* controls. The UCON model, namely $UCON_{ABC}$ [24], enables mutability of subject and object attributes, as well as continuity of control on usage of digital resource, and focuses on both *server-side* and *client-side* controls. The basic access control decision in any access control model can be represented as a triple (s, o, r, c), in which s denotes the subject S exercising a right r for object O under conditions c. UCON comprises eight core components, as shown in Fig. 1 to resolve this question. There are three functional predicates that have to be evaluated for usage decisions. The authorizations denote specific rights that a subject may exercise, the obligations denote actions the subject must perform and the conditions denote criteria influenced only by system-wide conditions.

Cloud-Enabled Internet of Things (CEIoT) is a basic IoT three layers architecture [5,6]. Perception layer includes devices that can perceive and collect data. Middle layer, in which components have functions to transfer data, communication and provide data services. Application layer provides diverse applications to meet the needs of the society and users. IoT devices are typically resource-constrained while close to real data, while cloud computing can provide elastic scalable storage, computing, and analysis. Therefore, the current emerging and widely used architecture called CEIoT integrates the IoT and the cloud, wherein cloud service providers (CSPs) expand services and applications via Internet on the existing foundation based on the above basic IoT.

Distributed Ledger Technology (Blockchain) is a technology [18,19] linking records expressed as blocks on a chain through cryptography, initially deployed to address the double-spending and currency generation problems of the Bitcoin cryptocurrency [20]. Each block contains a cryptographic hash of the previous block, a timestamp, and transaction data usually expressed as a Merkel Tree. For use as a distributed ledger, nodes in a blockchain system are usually managed by a peer-to-peer network, and encouraged to follow protocols for communicating and validating new blocks by incentives. As a decentralized infrastructure and distributed computing paradigm with the characteristics of tamperproof, traceability, and joint maintenance by multiple parties, blockchain has considerable promise for the construction of future IoT systems. As an autonomous application program running in the isolated virtual machine on a blockchain system, the smart contract provides a novel mechanism that can autonomously manage and implement interaction-rules between related parties.

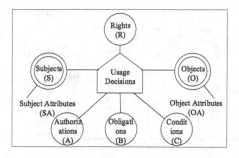
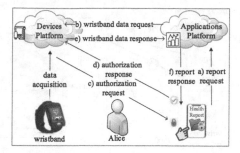

Fig. 1. UCON policy model [24]   **Fig. 2.** A private data-sharing scenario

3 Problem Statement and Design Goals

Problem Statement. In this paper, we address data sharing in CEIoT [23]. We assume a typical data usage scenario, as shown in Fig. 2. Suppose Alice acquires a wristband to collect the family's health data, such as heartbeat, exercise and sleep. Alice desires that the device platform of this wristband, can share her data with an application platform, such as a professional health institute, with her authorization. So Alice can obtain a health report after the data is processed by the institute. Presumably, some sensitive and private data is included in her data collection. As per article 17 of GDPR, user data has the right to be forgotten, i.e. *"Rights to erasure"*. Thereby, Alice expects that the institute can delete the data immediately after use, as well as not doing anything against her will during the usage, e.g., copying without her authorization or sharing to others directly in the current system. We note that any solution to such requirements must impose some perimeter restriction whereby all processing of the data takes place within this perimeter. In particular, so-called analog hole operations such as taking photos of display screens or manually copying data are beyond the scope of purely technical solutions.

To satisfy Alice's expectations, the wristband uploads data sporadically to its platform, viz. Devices Platform. Subsequently, a health report request is initiated by Alice via the application button on her smart phone, and the action triggers a professional institute, viz. Applications Platform to send a data request to the devices platform, in steps a) and b). After receiving a data request, the devices platform indicates to Alice that authorization is required through a visualized and unambiguous view in step c). Then, the interface is redirected back to the application with permissions, as indicated in step d). Successful authorization by Alice allows the data to be communicated to applications platform by devices platform in step e). Finally, the data is used to compute a health report which is delivered to Alice by the applications platform in step f).

Design Goals. In the above scenario, Alice wants a professional analysis report, which is a task that a devices platform or a general data storage party cannot fulfill. Thus the data needs to be shared with a third party such as a health institution. Moreover, Alice wants continued control of data usage wherein a

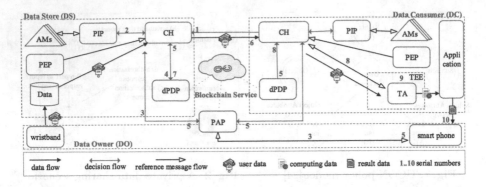

Fig. 3. System overview of DUCE, a trusted and distributed enforcement model.

"delete-after-use" rule is defined. However, in a distributed architecture, once data is shared, users lose control of the data. Due to mutual untrustworthy relationship between participants, users cannot be sure whether the applications platform follows the rules, and the applications platform cannot prove to users that they did not break the rules and breach privacy.

Therefore, to prevent privacy compromises, a trusted relationship should be established between mutually untrustworthy participants to keep data usage completely visible and in absolute control of users. This motivation drives our goals as follows, and inspires us to design a privacy-preserving distributed usage control enforcement model for data sharing in CEIoT. We recognize the following derivative goals. **Privacy Preserving** requires that the shared user data and the keys in authorization used to decrypt this data should be protected. **Integrity Protection** requires that the policy defined by users and enforcement records should not be tampered with. **Traceability** requires that violations must be able to be traced through enforcement records, and are visible to users.

4   The DUCE Model for Cloud-Enabled IoT

In this section, we present an overview of DUCE and its various components, and develop its enforcement process and an administration model. A system overview of DUCE is given in Fig. 3. A blockchain service is leveraged to construct a trusted and distributed architecture, in which data-sharing participants including data stores, data consumers and data owners are orchestrated and the data usage control rules can be enforced with administration and visibility by data owners via distributed PDPs and PEPs. Moreover, the policy defined by data owners is not only an authorized foundation to share data for a data store, but also a rule constraint for use of data by a data consumer. DUCE connects Data Stores (DSs), Data Consumers (DCs) and Data Owners (DOs) via a Blockchain Service. The system components of DUCE are discussed below.

Policy Enforcement Point (PEP) is a distributed component coupled with protected resources (i.e. the user data stored in a data store), which can intercept usage requests initiated by accessing subjects to trigger a decision through assessing the access request via available attributes, and finally enforce the result returned by an allow or deny decision. In DUCE, the PEP is a distributed engine used to enforce usage requests and perform specific decisions for user data. Additionally, the PEPs incorporate a Context Handler (CH) which plays the role of coordinator in an entire usage policy decision-making process and manages workflow by interacting with all other components.

Distributed Policy Decision Point (dPDP) is an adjudicator who makes decisions including allow or deny. It returns the decision to PEPs after the parameters including policy, usage requests, and current available attributes are evaluated. It is an essential aspect of DUCE.

Policy Administration Point (PAP) is a component responsible for management, storage and retrieval service during the evaluation process of usage requests. Meanwhile, PAP can also help decision-makers to define and modify policy, or perform other more complex and related policy management actions.

Attribute Manager (AM) is a component in charge of managing usage, retrieval and update of subject, object and environmental attributes. In DUCE, subject attributes mainly refer to general, authorization, obligation and condition attributes. The object attributes mainly refer to attributes, such as times to use and unique identifiers. The environmental attributes mainly refer to attribute values that are only effected by administrative operations in systems. AMs are not confined to an authorization service, but can be extended to local services, cloud services, or other services in different management domains.

Policy Information Point (PIP) is an interactive interface between diversified AMs, which provides attribute retrieval and update services, whereas attribute sets required for evaluation are collected by different AMs and protocols.

Data Owner (DO) provides PAP and ownership service to IoT devices and user data. DO is responsible for administrating usage policy and can control the entire enforcement by interacting with context handlers in PEPs.

Data Store (DS) provides a hosting service for user data. As a PEP, the data store translates the policy defined by a data owner into a form that can be evaluated by dPDPs, and then determines whether to call the PIP to perform data-sharing based on the distributed evaluation results. Whether or not the data is shared, the data store needs to send a notification to the user through CH. Formally, DS=<CH,PEP,dPDP,PIP,AMs,Data>, where the Data denotes an object. More precisely, we denote CH in DS as CH_{DS}, thus $CH_{DS}=<H_{pol},H_{oat},H_{obj}, H_{not}>$, where H_{pol} receives policy defined in XACML from DO, and translates policy into a form that can be evaluated in a dPDP. H_{oat} updates object attributes. H_{obj} handles user data, and H_{not} sends notifications to data owners.

Data Consumer (DC) is a data consumer who enforces data applications services according to policies defined by DO. If a usage request is allowed,

DC receives user data and keys through CH. After a data-sharing process is completed, a notification is sent to the data owner through CH. Formally, DC=<CH,PEP,d-PDP,PIP,AMs,TA,Application>, where the Application denotes a subject, viz. data requester, and also performs computing functions by using the data from the TA that excluding sensitive information, then returns a result showing on the smart phone of DO. More precisely, we denote CH in DC as CH_{DC}, thus $CH_{DC}=<H_{pol},H_{sat},H_{sub},H_{not}>$, where H_{pol} receives policy and enforce a distributed evaluation directly through the dPDP. H_{sat} updates subject attributes. H_{sub} sends data requests, and H_{not} sends notifications to data owners.

Trusted Agent (TA) is an agent of a TEE belonging to DC, which interacts with an external untrusted environment. In a data-sharing process, the user data received through sharing actions is stored by PIP, and the key is directly delivered to TA for storage in the TEE. It is worth noting that the entire process of providing specific application services and distributed policy evaluation is completed in TEE, and interactions required with the external environment also is completed by TA. If a violation occurs in distributed evaluation processes, CH or TA is triggered to send alerts and notifications to the DO.

Blockchain Service is a service provided by blockchain technology to build a trust relationship among DO, DS, and DC, in charge of providing distributed services including policy decision, policy enforcement, and policy administration.

Next, we illustrate **the enforcement process of DUCE** as follows.

Initialization Phase refers to an initial preparation of DUCE, including service, communication, and data preparation. First of all, the distributed policy-decision services, i.e., nodes of a permissioned blockchain, and the enforcement environments need to be prepared by DS and DC. Then, the communication ability namely CH with data owners need to be prepared in DS and DC, such as P2P network or an internal protocol of DS or DC. Next, a TEE needs to be provided in DC, in which TA can interact with DS and DO. Finally, the user data uploaded by devices should be prepared and retrievable in DS. In this paper, to protect user privacy, we default that the device data is encrypted for storage in DS and can only be decrypted with an authorization of data owners.

Enforcement Phase is divided into the following four segments.

Authorization. After the initialization is completed, the application service that DO wants triggers DC to initiate a data request to DS in step 1. After the request is received, CH retrieves relevant information of authorization on blockchain by executing smart contracts in step 2. There is no authorization information related to this DC since it is in initial status. We assume that authorization in DUCE is instant, i.e. permission will be automatically revoked if the relevant operation is not performed within a limited time, so DC should request authorization every time before an access. Then, DS initiates an authorization request through CH to ask DO for authorizations. After receiving the request through a smart phone, the user authorizes with a defined policy, namely rule in 3. The policy is received and translated into a smart contract, and issued on blockchain by CH in 4. Simultaneously, CH knows the authorization through synchronization in 5.

Operation. The policy-related information on blockchain, <subject,object, right> is used to make a decision to access DC to data by dPDPs. The data is shared to DC by DS through the CH in 6. After receiving the data, DC performs related operations by TA in TEE in 9. The relevant records of operations in 7 are uploaded to blockchain for storage and evaluation subsequently in 8 by CH.

Evaluation. The records stored in an *Operation* are analyzed by executing smart contracts, either make a decision to update the subject, object, and environmental attributes by dPDPs via CH in 8 or enforce a revocation.

Notification. This segment is enforced by CH in PEP and TA. Once the operation the user follows is completed, or related evaluation is triggered, the TA or the CH should notify the user regardless of the conditions met in 5. Definitely, TA should send a notification to users when all the above segments are completed in 10. Thus to be completely visible and controllable to the user.

Policy 1 Usage Control
<Policy PolicyID="UCONPolicy">
<Rule Effect="Permit" RuleID="usage-data-consumer-rule">
<Target> <AllOf>
<Match
MatchID="urn:oasis:names:tc:xacml:1.0:function:date-greater-than">
<AttributeValue
DataType="http://www.w3.org/2001/XMLSchema#date">2021-02-08
</AttributeValue>
<AttributeDesignator
AttributeId="urn:oasis:names:tc:xacml:1.0:resource:data-collected-date"
Category="urn:oasis:names:tc:xacml:3.0:attribute-category:
user-wristband-data"
DataType="http://www.w3.org/2001/XMLSchema#date"
Issuer="ID_{DO}"
MustDeleteAfterUse="true"
MustMeetSystemCondition="true" />
</Match>
</AllOf> </Target>
</Rule>
</Policy>

Algorithm 2 UCON Policy Translation

1: **procedure** TRANSLATE(xa, sc)   ▷ translate a XACML file into a smart contract
2:    $rule \leftarrow xa.Rule$
3:    $s \leftarrow rule.Target$
4:    $res' \leftarrow retrieve(rule.\{Category, AttributeID, AttributeValue, Issuer\})$
5:    **while** $res \in res'$ do traversed   ▷ traverse res to find the data
6:       **if** ($res.AttributeValue \in rule.MatchID$) **then**
7:          $o \leftarrow res.AttributeValue$
8:       $b \leftarrow rule.MustDeleteAfterUse$
9:       $c \leftarrow rule.MustMeetSystemCondition$
10:      $r \leftarrow rule.Effect$ ▷ parse xacml file to object successfully
11:      $sc \leftarrow constructSC()$ ▷ begin to construct a smart contract to load object
12:      $uconManager \leftarrow uconManagerContract(rule.Issuer)$
13:      **if** ($uconManager.AttributeValue \in o$) **then**
14:         **if** ($r=="Permit"$ && $b=="ture"$ && $c=="true"$) **then**
15:            $uconManager.Permit \leftarrow "true"$
16:         **else**
17:            $uconManager.Permit \leftarrow "false"$
18:      $sc \leftarrow uconManager$
19: **return** sc ▷ translate XACML file into a smart contract successfully

Administrating the Enforcement Model. In particular, we translate the policy so that we can evaluate the policy by executing smart contracts. We use XACML [3], the most popular expression policy language, to help data owners to define policy, as shown in Policy 1, "UCONPolicy" is used to identify the policy, which is briefly described as the "Permit" permission is granted to the subject who obeys the two conditions "delete after use" and "use and delete in TEE" by the data owner (Issuer), the subject is allowed to use the user data in "user-wristband-data" category and the date is after "2021-02-08".

Moreover, smart contract is a core component that supports trusted and distributed policy decision-making services of blockchain technology, it ensures functionality and security of policy through automated execution and evaluation. We translate the policy into smart contracts, that are issued on Blockchain Service through CH by DS and waits execution triggers for data usage decision-making, as shown in Algorithm 2.

5   Performance Evaluation

Implementation. We prototype the DUCE authentication and authorization service built upon SpringFramework. In particular, we utilize smart contracts enabled in FISCO BCOS[1] to realize blockchain based authorization in DUCE.

---

[1] http://www.fisco-bcos.org.

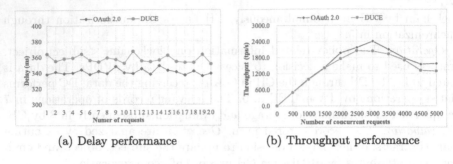

(a) Delay performance   (b) Throughput performance

Fig. 4. Comparison of delay and throughput between DUCE and OAuth 2.0.

We store the accessToken on the blockchain through the CRUD feature of FISCO BCOS, and use Solidity to realize the accessToken authentication service. We replace the authentication service logic of OAuth 2.0 as the DUCE service, which ensures that the authentication process in DUCE is tamper-proof. A user can use an acccessToken stored in the blockchain to get authorized.

Experiment Setup. Our prototype is deployed on the Alibaba Cloud Elastic Compute Service. Also, we use the default OAuth module as a baseline, to implement our OAuth authorization service. We use MySQL database to store the user identifier information, and the Redis cache mechanism to cache accessToken to reduce the delay of the OAuth authorization. We also deploy the FISCO BCOS blockchain service of DUCE on the same cloud server.

Results. Based on the above implementation and setup, we run the project and define three metrics to evaluate performance.

Delay, the time required for communication messages transmitting from one network end to another, including transmission, propagation, processing, and queuing delay. Since the processing and queuing delay are mainly determined by the communication message size, in DUCE, we focus on the transmission and propagation delay, namely end-to-end transfer delay.

Throughput, the maximum request number that the system can handle per unit time, We focus on authorization and authentication throughput in DUCE.

To demonstrate the effectiveness of DUCE, we first use Postman to test the transfer delay, as shown in Fig. 4 (a). Then, we use JMeter to test the throughput of DUCE, as shown in Fig. 4 (b).

Discussion. According to the above experimental results, we find that the realization of authentication and authorization by using blockchain services increases the delay and decreases the throughput. In the experiment, the additional blockchain services requires more time (i.e., 350 ms in the OAuth 2.0 system and 370 ms in the DUCE) to process end-to-end communication than the OAuth 2.0 system. The choice of OAuth 2.0 may be limited to the experimental configuration, and the throughput performance is around 2400 tps/s. In

DUCE, the selection of different blockchain may result in different throughput, and the throughput reaches about 2000 tps/s in FISCO BCOS. Therefore, as the circumstance that there is a same ratio of the peak throughput (y-axis) to number of concurrent requests (x-axis) in DUCE and OAuth 2.0, i.e., both are 80%, the decrease of about 17% (less than 20%) is within the acceptable range. In other words, we demonstrate that compared to the existing widely used solution namely OAuth, DUCE does not introduce excessive overhead, while preventing user privacy from being compromised.

6 Related Work

Privacy preserving of static data refers to a protection of static storage data, methods include the access control mechanism, the encrypted storage and the anonymization of sensitive information. Both academic researchers and industry cloud service providers, such as Microsoft, Amazon, Google, have deployed CEIoT platforms and novel access control models. Google [11] developed GCP-IoTAC, a fine-grained access control model based on attribute extensions, and demonstrated two main use cases which are more privacy-conscious of IoT. Fernández et al. [9] designed a data collection and data sharing model based on the DataBank architecture and implemented it on an open-source platform Privasee. Liu et al. [16] proposed BC-SABE, a blockchain-assisted mechanism with effective revocation and decryption functions based on attribute-based encryption. Xu et al. [30] designed the Key Compromise Resilient Signature (KCRS) system. To protect IoT device data, an authentication framework based on a decentralized ledger namely DIoTA [29] is proposed. Patil et al. [25] used the concept of anonymous tokenization to make up for the shortcomings of current communication technology that cannot protect the anonymity of users.

Privacy preserving of dynamic data refers to prevention of privacy leakage due to improper data usage during data-sharing transmission and computing, the main prevention methods include Federated Learning, Homomorphic Encryption, and Trusted Execution Environment. In order to balance utility and privacy, Ramesh et al. [26] proposed a framework namely proxy re-ciphering as a service that using Fully Homomorphic Encryption and Chameleon Hash to customize the solution to ensure long-term computing with privacy-preserving of device data. Federated Learning can be used to train a global machine-learning model using data distributed across multiple sites without data movement. Choudhury et al. [7] proposed a grammatical method, different from differential privacy, that can support privacy-preserving at the defense level while maximizing the effectiveness of the model. Zhang et al. [31] proposed a system solution called BatchCrypt for the cross-silo federated learning system, which can ensure update of the local gradient is concealed when is aggregated. Zhang et al. [32] designed Cerberus by combining blockchain technology provided distributed data storage and TEE for state maintenance, data storage and off-chain computing in a computing scenario outsourced to edge nodes.

Moreover, Lazouski et al. [15] designed U-XACML to express UCON, implementing a prototype system for evaluation. Marra et al. [14] proposed a realization of usage control in a smart home use case. Ma et al. [17] proposed BlockBDM, a decentralized trust management scheme for IoT big data.

7   Conclusion

Utilizing blockchain or DLT to build a trust relationship between participants in a data-sharing scenario to prevent user privacy leakage is one of the most popular methods. Whereas the IoT device data contains sensitive or private information, combining two new computing models, cloud computing and IoT, can provide users with efficient services with privacy-preserving. To address the problem that applications service or data consumer violates articles such as *"Right to erasure"* in GDPR and leads to user privacy disclosure, we propose DUCE, a trusted and distributed enforcement architecture. In DUCE, blockchain is used to enforce distributed usage control policy to make decisions by distributed PDPs and PEPs, the policy is defined in XACML and translated into smart contracts for automatic execution and evaluation. Utilizing a TEE to limit obligations and conditions, we demonstrate the enforcement process of DUCE, and conducted functional and performance evaluations by comparing our prototype with OAuth 2.0 system. However, DUCE integrates and relies on TEE, thus the protection of user data depends on the security strength of cryptography and TEE. In future work, we devote to research more secure and trusted enforcement models, and figure out methods for encrypted data protection.

References

1. Almolhis, N., Alashjaee, A., Duraibi, S., Alqahtani, F., Moussa, A.: The security issues in INT-cloud: a review. In: 2020 16th IEEE International Colloquium on Signal Processing & Its Applications (CSPA), pp. 191–196. IEEE (2020)
2. Alzahrani, B., Irshad, A., Alsubhi, K., Albeshri, A.: A secure and efficient remote patient-monitoring authentication protocol for cloud-Iot. Int. J. Commun. Syst. **33**(11), e4423 (2020)
3. Anderson, A., et al.: eXtensible access control markup language (XACML) version 1.0. OASIS (2003)
4. Bhatt, S., Patwa, F., Sandhu, R.: An access control framework for cloud-enabled wearable internet of things. In: 2017 IEEE 3rd International Conference on Collaboration and Internet Computing (CIC), pp. 328–338. IEEE (2017)
5. Bhatt, S., Sandhu, R.: ABAC-CC: Attribute-based access control and communication control for internet of things. In: Proceedings of the 25th ACM Symposium on Access Control Models and Technologies, pp. 203–212 (2020)
6. Chen, R., et al.: Trust-based service management for mobile cloud IoT systems. IEEE Trans. Netw. Serv. Manag. **16**(1), 246–263 (2018)
7. Choudhury, O., et al.: Anonymizing data for privacy-preserving federated learning. arXiv preprint arXiv:2002.09096 (2020)
8. De Donno, M., Tange, K.,.: Foundations and evolution of modern computing paradigms: cloud, IoT, edge, and fog. IEEE Access **7**, 150936–150948 (2019)

9. Fernández, M., Franch Tapia, A., Jaimunk, J., et al.: A data access model for privacy-preserving cloud-IoT architectures. In: Proceedings of the 25th ACM Symposium on Access Control Models and Technologies, pp. 191–202 (2020)
10. Foughali, K., Fathallah, K., Frihida, A.: Using cloud IoT for disease prevention in precision agriculture. Procedia Comput. Sci. **130**, 575–582 (2018)
11. Gupta, D., et al.: Access control model for google cloud IoT. In: (BigDataSecurity), (HPSC) and (IDS). pp. 198–208. IEEE (2020)
12. He, W., Yan, G., Xu, L.: Developing vehicular data cloud services in the IoT environment. IEEE Trans. Ind. Inform. **10**(2), 1587–1595 (2014)
13. Kianoush, S., et al.: A cloud-IoT platform for passive radio sensing: challenges and application case studies. IEEE Internet Things J. **5**(5), 3624–3636 (2018)
14. La Marra, A., Martinelli, F., Mori, P., Saracino, A.: Implementing usage control in internet of things: a smart home use case. In: 2017 IEEE Trustcom/BigDataSE/ICESS, pp. 1056–1063. IEEE (2017)
15. Lazouski, A., Martinelli, F., Mori, P.: A prototype for enforcing usage control policies based on XACML. In: Fischer-Hübner, S., Katsikas, S., Quirchmayr, G. (eds.) TrustBus 2012. LNCS, vol. 7449, pp. 79–92. Springer, Heidelberg (2012). https://doi.org/10.1007/978-3-642-32287-7_7
16. Liu, S., Yu, J., et al.: Bc-SABE: Blockchain-aided searchable attribute-based encryption for cloud-IoT. IEEE Internet J. **7**(9), 7851–7867 (2020)
17. Ma, Z., et al.: Blockchain-enabled decentralized trust management and secure usage control of IoT big data. IEEE Internet Things J. **7**(5), 4000–4015 (2019)
18. Di Francesco Maesa, D., Mori, P., Ricci, L.: Blockchain based access control. In: Chen, L.Y., Reiser, H.P. (eds.) DAIS 2017. LNCS, vol. 10320, pp. 206–220. Springer, Cham (2017). https://doi.org/10.1007/978-3-319-59665-5_15
19. Maesa, D., Mori, P., Ricci, L.: A blockchain based approach for the definition of auditable access control systems. Comput. Secur. **84**, 93–119 (2019)
20. Nakamoto, S.: Bitcoin: a peer-to-peer electronic cash system. Tech. Report (2019)
21. Nandakumar, L., et al.: Real time water contamination monitor using cloud, IOT and embedded platforms. In: 2020 International Conference on Smart Electronics and Communication (ICOSEC), pp. 854–858. IEEE (2020)
22. Neagu, G., et al.: A cloud-IoT based sensing service for health monitoring. In: 2017 E-Health and Bioengineering Conference (EHB), pp. 53–56. IEEE (2017)
23. Ouaddah A., Elkalam, A.A., Ouahman, A.A.: Towards a novel privacy-preserving access control model based on blockchain technology in IoT. In: Europe and MENA Cooperation Advances in Information and Communication Technologies. p. 520 (2017)
24. Park, J., Sandhu, R.: The uconabc usage control model. ACM Trans. Inf. Syst. Secur. (TISSEC) **7**(1), 128–174 (2004)
25. Patil, S., Joshi, S., Patil, D.: Enhanced privacy preservation using anonymization in IoT-enabled smart homes. In: Satapathy, S.C., Bhateja, V., Mohanty, J.R., Udgata, S.K. (eds.) Smart Intelligent Computing and Applications. SIST, vol. 159, pp. 439–454. Springer, Singapore (2020). https://doi.org/10.1007/978-981-13-9282-5_42
26. Ramesh, S., et al.: An efficient framework for privacy-preserving computations on encrypted IoT data. IEEE Internet Things J. **7**(9), 8700–8708 (2020)
27. Sandhu, R., Samarati, P.: Access control: principle and practice. IEEE Commun. Mag. **32**(9), 40–48 (1994)
28. Stergiou, C., Psannis, K., Kim, B., Gupta, B.: Secure integration of IoT and cloud computing. Fut. Gen. Comput. Syst. **78**, 964–975 (2018)
29. Xu, L., Chen, L., Gao, Z., et al.: Diota: decentralized-ledger-based framework for data authenticity protection in IoT systems. IEEE Network **34**(1), 38–46 (2020)

30. Xu, L., et al.: KCRS: a blockchain-based key compromise resilient signature system. In: Zheng, Z., Dai, H.-N., Tang, M., Chen, X. (eds.) BlockSys 2019. CCIS, vol. 1156, pp. 226–239. Springer, Singapore (2020). https://doi.org/10.1007/978-981-15-2777-7_19

31. Zhang, C., Li, S., Xia, J., Wang, W., Yan, F., Liu, Y.: BatchCrypt: efficient homomorphic encryption for cross-silo federated learning. In: 2020 USENIX Annual Technical Conference (USENIX ATC 2020), pp. 493–506 (2020)

32. Zhang, D., Fan, L.: Cerberus: privacy-preserving computation in edge computing. In: IEEE INFOCOM 2020-IEEE Conference on Computer Communications Workshops (INFOCOM WKSHPS), pp. 43–49. IEEE (2020)

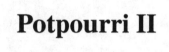

Potpourri II

A Digital Twin-Based Cyber Range
for SOC Analysts

Manfred Vielberth[1]([✉])[iD], Magdalena Glas[1][iD], Marietheres Dietz[1][iD],
Stylianos Karagiannis[2][iD], Emmanouil Magkos[2][iD], and Günther Pernul[1][iD]

[1] Chair of Information Systems, University of Regensburg, Regensburg, Germany
{manfred.vielberth,magdalena.glas,marietheres.dietz,
guenther.pernul}@ur.de
[2] Department of Informatics, Ionian University, Corfu, Greece
{skaragiannis,emagos}@ionio.gr

Abstract. Security Operations Centers (SOCs) provide a holistic view
of a company's security operations. While aiming to harness this potential, companies are lacking sufficiently skilled cybersecurity analysts. One
approach to meet this demand is to create a cyber range to equip potential analysts with the skills required. The digital twin paradigm offers
great benefit by providing a realistic virtual environment to create a
cyber range. However, to the best of our knowledge, tapping this potential to train SOC analysts has not been attempted yet. To address this
research gap, a concept of a digital twin-based cyber range for SOC
analysts is proposed and implemented. As part of the virtual training
environment, several attacks against an industrial system are simulated.
Being provided with a SIEM system that displays the real-time log data,
the trainees solve increasingly complex tasks in which they have to detect
the attacks performed against the system. Thereby, they learn how to
interact with a SIEM system and create rules that correlate events aiming
to detect security incidents. To evaluate the implemented cyber range, a
comprehensive user study demonstrates a significant increase of knowledge within SIEM-related topics among the participants. Additionally,
it indicates that the cyber range was subjectively perceived as a positive
learning experience by the participants.

Keywords: Cyber range · Security operations center · Digital twin

1 Introduction

As cyber-attacks become increasingly sophisticated and use more and more
points of attack, it is essential to establish a holistic view of organizations' security. As a recently published report [2] indicates, organizations are becoming better at detecting and mitigating direct attacks. However, more advanced attacks
are on the rise, targeting the victim indirectly through weak spots in the business ecosystem or the supply chain. Over the recent years, Security Operations
Centers (SOCs) have emerged to address this problem by providing a holistic

© IFIP International Federation for Information Processing 2021
Published by Springer Nature Switzerland AG 2021
K. Barker and K. Ghazinour (Eds.): DBSec 2021, LNCS 12840, pp. 293–311, 2021.
https://doi.org/10.1007/978-3-030-81242-3_17

view of organizations' cybersecurity. However, this has increased the demand for security personnel, making it difficult to find enough well-trained analysts for SOCs. This is worsened by the so-called "alert burnout", since an analyst's daily work can be quite tedious and tiring. According to a SANS survey [23], the key to low attrition rates is to invest more in analysts' training. Therefore, it is crucial to create a means to train analysts as quickly and effectively as possible, considering that the requirements can vary from company to company. To create a suitable training environment, cyber ranges can be used to train analysts by simulating realistic scenarios without disrupting business operations. To be as close as possible to the specifics of the company, the integration of a digital twin is a promising option. Thereby, the relevant section of the company infrastructure for which the experts are to be trained can be mirrored, creating a training environment that barely differs from the company's real environment.

The contribution of this paper is twofold. First, we examine which components of a digital twin can be used for cyber ranges. Based on this, a cyber range for SOC analysts is designed and prototypically implemented. To show that the proposed concept offers advantages for the training of security analysts, it is evaluated through an extensive empirical user study.

The remainder of this paper is structured as follows. Section 2 provides the foundation of the conducted research. In Sect. 3, the digital twin's potential for cyber ranges is outlined along with the current research gap. Based on that, Sect. 4 proposes a concept for a digital twin-based cyber range, including a scenario and learning concept and concludes with a description of the prototypical implementation of the concept. Section 5 covers the evaluation of the concept in the form of a comprehensive user study by presenting the methodology and the results of the evaluation. Finally, the work is concluded in Sect. 6.

2   Background and Related Work

2.1   Cyber Range

As conventional training methods that only focus on transferring theoretical knowledge do not meet the demand for practical knowledge and skills within the cybersecurity domain, cyber ranges have gained attention over the past years [32]. Generally, cyber ranges are virtual environments, which are used for cybersecurity training [28]. As the name indicates, the expression is derived from shooting range, as both provide an environment in which people can be trained without harming or interfering with the environment for which they are educated. Application areas range from public settings such as military defense and intelligence, academic and educational, to commercial purposes driven by the industry [29].

The idea of using cyber ranges to train specialists in attack detection and in cybersecurity in general is not entirely new. For example, the Austrian Institute of Technology recently introduced a cyber range of industrial control systems [20], not only targeting education, but also serving as a platform for conducting research and development by testing new approaches and methods. This is only one example in this context. For a deeper insight into related approaches, we would like to refer to two extensive literature reviews which provide a good overview of preliminary work [29, 32]. Although some works in this area exist, to the best of our knowledge, no approach combines the digital twin's potential with the concept of cyber ranges for educating SOC analysts to date. Additionally, the effect of the approaches on the obtained knowledge has not gained sufficient attention in previous works.

According to Yamin et al. [32], a cyber range can be described by following a taxonomy with six domains. However, the description of a cyber range does not necessarily have to consider all domains, but instead, can focus on selected ones. As this paper applies the taxonomy for describing the developed cyber range, the six domains are elaborated briefly in the following:

Scenario: A scenario defines the storyline and context of a training exercise performed in a cyber range. It supports the purpose of the training, such as education, experimenting, or testing. Thereby, it is allocated to a domain (e.g., networking, critical infrastructure, or IoT). Additionally, a scenario can either be static or dynamic. A dynamic scenario means that changes are made during the exercise, for example, by simulating infrastructure components.

Environment: The environment presents the topology in which the scenario is executed. This includes the underlying technology used to build a system model (simulation, emulation, hardware, or hybrid).

Teaming: Teaming describes which teams are part of the scenario. The most important teams are a red team with the goal to exploit vulnerabilities of the system, and a blue team with the task to defend the system against attacks. Teams can also be autonomous if specific technologies automate them.

Learning: The learning domain covers explanatory elements of a scenario such as texts, images, or video clips used for initial knowledge transfer.

Monitoring: Participants' actions can be monitored in real-time during an exercise by using appropriate tools.

Management: This domain covers how management tasks, such as role and resource allocation, are performed. It also comprises interfaces for controlling the scenario or the environment during the exercise.

Furthermore, it is worth mentioning in this context that the term can be narrowed down further. Kavallieratos et al. [17] define a cyber-physical range as a testbed that enables the testing of the security posture of cyber-physical systems. The cyber range presented in this paper can be assigned to this class.

2.2 Security Operations Center (SOC)

The term Security Operations Center has been around in research for more than a decade. However, attention has significantly increased in the last three to five years as SOCs have emerged as a central pivotal point for security operations in practice [30]. The SOC represents an organizational aspect of an enterprise's security strategy. It combines processes, technologies, and people [21, 27] to manage and enhance an organization's overall security posture. This goal can usually not be accomplished by a single entity or system, but rather by a complex structure. It creates situational awareness, mitigates the exposed risks, and helps to fulfill regulatory requirements [19]. Additionally, a SOC provides governance and compliance as a framework in which people operate and to which processes and technologies are tailored. A central role within a SOC is taken by security analysts. Using appropriate tools, they can attempt to detect security incidents, then analyze them and react appropriately. Therefore, the success of a SOC depends to a large extent on the skills and training of the analysts. Within a SOC, a SIEM system is usually used as the central tool [31]. A SIEM aims to collect security-relevant data (usually log data) in a central location and analyze it in a correlated manner to detect security incidents. For this purpose, SIEM systems use detection rules that are usually created by analysts, in most cases in JSON or XML format. These fulfill the purpose of triggering an alert if defined conditions within the log data apply.

2.3 Digital Twin

The digital twin refers to a concept that differs in meaning depending on its application area [22]. In general, a digital twin can be defined as a virtual representation of any real-world asset (e.g., system or process). The digital twin accompanies its real-world asset's lifecycle, which may range from phases like idea/planning over operation to decommissioning [6]. The digital twin gathers data about its real-world twin during these phases and enriches the data with semantics [3]. This way, the twin is able to represent its counterpart in-depth and provides a solid basis for simulations and further analytical measures.

Especially in cybersecurity, the digital twin holds several benefits [26]. It can support lifecycle security [11], including the security-by-design paradigm by offering simulations and system testing, in which the security level of the asset can be assessed. Moreover, digital forensics may profit from the vast data and documentary capabilities of a digital twin [7].

3   Investigating the Potential of the Digital Twin for Building Cyber Ranges

In order to extract what digital twins offer for cyber ranges, we must first regard the foundation of digital twin deployment in cybersecurity. According to [11], the digital twin is required to provide sufficient fidelity for security measures that rely on its data. A digital twin offering this characteristic can then be successfully implemented for cybersecurity. This definition presents the prerequisite for combining digital twins and cyber ranges. Currently, one work conceptually proposes to utilize a digital twin as a cyber range [4]. However, an implementation has not been realized to date. In their approach, the digital twin is merely applied as cyber range with the purpose of security training, while other purposes are not considered. However, the digital twin originally serves completely different purposes, such as monitoring and controlling its counterparts' operation [6]. Thus, in this paper, we propose to use the digital twin as a valuable input to create a cyber range rather than turning it into one. In this matter, we investigate which digital twin characteristics can provide valuable input for cyber ranges in the following. The core parts included in a digital twin represent (a) *data, enhanced with semantic technologies*, (b) *analysis, simulation and other intelligent services* as well as (c) *access control and interfaces* [6].

Data of the digital twin's real-world counterpart is produced along its lifecycle, stored in the digital twin and given context by adding *semantics* [3]. This data supports high-fidelity modeling of the counterpart to virtually represent the real-world system. Added semantics offer better comprehension and modeling of the connection and the context of the system's components. This can prove to be an essential input for creating cyber ranges as well. To maximize the training potential of cyber ranges, the virtually represented system and related security incidents should resemble reality as close as possible. This way, security analysts can be trained in a highly realistic environment. However, not all data held in a digital twin may be relevant for building cyber ranges. The virtual system, used to build a cyber range might represent only a part of a complex real-world system, e.g., by focusing on the network level. In this case, the physics-related data of the system might not be of interest. Moreover, the resulting data of digital twin analyses (like predictive maintenance) typically are not relevant. In general, only a subset of digital twin data is required for creating the cyber range – depending on the complexity level, granularity, and the part of the system being represented.

Analysis, simulation, and other intelligent services represent operation modes of a digital twin. According to [7], three modes can be used for security purposes as well: analysis, simulation, and replication. Table 1 summarizes these modes and their potential benefits for building cyber ranges. Each operation mode relies on digital twin data and has already been tackled in terms of security in some works (see Table 1). **Analysis** usually takes historical/state data of the physical counterpart into account to apply analytical measures such as anomaly detection, pattern recognition, etc. For cyber ranges, this data has to be virtually reproduced (moderate effort). However, there is no virtual system

Table 1. Digital twin security operation modes and their potential for cyber ranges.

Operation mode	Required data	Related work	Benefit	Effort
Analysis	Historical/state data	[24]	Low	Moderate
Simulation	Specification data (for emulation)	[8, 10]	High	Moderate
Replication	Specification data (for emulation), historical/state data (stimuli)	[9, 13]	Moderate	High

that can be explored by security analysts (low benefit). **Simulation**, in contrast, requires only specification data to build the emulation. On top of the emulation, different (security) scenarios can be applied to a virtual system to create a simulation, where the security analyst in training can not only see produced data of the virtual system but also interact with the system (high benefit). Moreover, the simulations can be taken from digital twins and directly used in or tailored to the cyber range (moderate effort). **Replication**, on the other hand, requires high effort to be used for cyber ranges as it relies on integrating not only specification data to build the emulation, but also on current state data of the physical counterpart to defer the stimuli changing the systems state. However, it only provides moderate benefit as the system is always in synchronization with its real-world counterpart and alternative scenarios (e.g., security incidents and countermeasures) cannot be tested.

Other important parts of digital twins are *access control and interfaces* (e.g., implemented in [25]). Although control mechanisms in digital twins for accessing their data and analytic capabilities represent no relevant input for cyber ranges, interfaces might be used to transmit data from the digital twin.

To conclude, some parts of the digital twin offer benefits for building cyber ranges. Especially the operation mode simulation can be used to create a virtual environment close to reality. Such a system simulation model can be directly transferred from the digital twin into the cyber range and – if necessary – customized to meet the cyber range's needs. The interfaces part of the digital twin might help to transfer the model, while additional data might help to create simulation scenarios or to get an overview of the system that is virtually represented. Overall, the simulation capability of the digital twin presents a valuable input for cyber ranges and will be concentrated on in the following.

4   A Cyber Range for SOC Analysts

To create the cyber range, it is first necessary to define the *learning objectives*, the *target group*, and the *requirements*. In the case of our cyber range, the analyst in training - hereafter referred to as the trainee - should be introduced to the tasks of a SOC analyst and learn how to work with a SIEM system. In the process, he or she should acquire the following skills and know-how:

S1: Knowledge of how selected incidents or attacks on the industrial system work.
S2: Manual detection of anomalies or incidents by analyzing log data with a SIEM system.
S3: Create both syntactically correct and semantically appropriate rules to detect the incidents.

The target group are individuals who want to achieve skills in security analytics within a SOC – for example, because they want to work as analysts in a SOC in the future. They are assumed to have basic cybersecurity skills but have never worked with a SIEM system or in a SOC. Even though incident response often lies within an analyst's responsibilities, this will not be considered in this cyber range as, in our opinion, it is too complex to start with and would create too steep a learning curve. However, this could be addressed in future work.

One requirement for the cyber range is to run it entirely virtual in order for the trainees to take part in the cyber range remotely without physical presence. This allows the trainees to take part without too much effort. Additionally, it facilitates the evaluation with an international user study. Furthermore, since the user study is to take place in times when – due to COVID-19 restrictions – face-to-face contact should be kept to a minimum, conducting it in a classroom setting is not an option.

In the following, first the general concept of the cyber range is presented. Based on this, the scenario is described with the help of which the user should acquire the skills outlined above. Subsequently, the prototypical implementation of the cyber range is elaborated upon, with a brief description of the technologies used.

4.1   Cyber Range Concept

Our cyber range consists of five main building blocks (compare Fig. 1): A *virtual environment*, a *SOC*, a *management and monitoring* unit, a *learning management system*, and the *digital twin*, which lies outside of the cyber range. Thus, it represents a security analytics service [12] combined with cyber range specific components. In the following, these building blocks are explained in more detail.

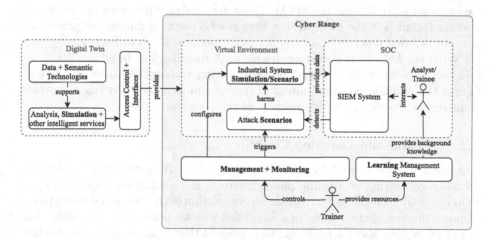

Fig. 1. Basic concept of the digital twin-based cyber range for SOC analysts.

As investigated in Sect. 3, the **digital twin** provides a system *simulation* model used to create the *virtual environment*. The simulation model is supported by specification data, enabling a realistic simulation of the physical counterpart with which the trainee can dynamically interact, like with the real system within the organization. The data for the simulation is provided through respective *interfaces* and protected by *access control* capabilities.

The **virtual environment** implements and reflects the scenario of the cyber range through the simulation. For this purpose, an *industrial system* is simulated on the one hand and simulated *attacks* harming the industrial system are carried out on the other. Thereby, the planned training *scenario* is reproduced, guiding the trainee through several training units similar to a playbook, elaborated in more detail in Sect. 4.2. In the process, the simulated industrial system produces log data documenting its operation and providing traces pointing to the attack scenarios.

Within the **SOC** building block, a *SIEM system* is provided, which provides the actual point of interaction with the trainee. The SIEM represents the system for which an analyst is trained, and ideally is also a system in practical use in the trainee's organization. This ensures that the trainee learns to work with a system that is as close to the real SIEM as possible or even identical to it. The log data of the industrial system is fed into the SIEM. In the first step, the trainee interacts with the SIEM to analyze and manually detect the simulated attacks based on the available data. In the next step, the trainee can use this to create correlation rules in the SIEM, which detect attacks automatically.

The **learning management system (LMS)** provides additional learning material for the trainee and introduces the scenario. This information can be presented in various forms, such as videos or simple textual descriptions. In our case, an introduction to the functioning of SIEM systems and the structure of SIEM rules is provided. In addition, hints on the attacks are given to make it easier to get started using the SIEM. These materials are prepared by the trainer and are included in the LMS so that they can be accessed during the procedure. A more detailed description of the prepared media is given in Sect. 4.2.

With the help of the **management and monitoring** building block, the trainer can oversee the trainees' progress during training. Additionally, it configures the simulation of the industrial system and automatically triggers attack simulations depending on the progress of the training.

4.2 Scenario and Learning Concept

The scenario represented by the cyber range is an Industrial Control System (ICS)-based setting of a filling plant. Thereto, the simulation from the digital twin is used, which enables a realistic representation of the industrial filling plant. Figure 2 illustrates the setting in a simplified way for better understanding. The filling plant consists of a tank containing liquid that is to be filled into bottles. The tank is equipped with a sensor measuring the liquid level at regular intervals. To control how much liquid is bottled, the system includes a motoric valve that can be opened and closed. The flow-level sensor is being used to check how much liquid flows through the pipe towards the bottle at any given time. The level of the bottle itself is monitored with another sensor. Each sensor and the actuator is controlled by one of the three Programmable Logic Controllers (PLCs) connected through a switch via Ethernet, which store the sensor data and communicate via Ethernet/IP. The interface between the employees and the industrial plant is realized with the help of a Human-Machine Interface (HMI). This allows an employee to read the measured and logged sensor values and intervene in the plant's operation. Within the scenario, it is assumed that an attacker has gained direct access to the network of the industrial plant. This allows him or her to carry out various attacks, which can then be detected in the SIEM.

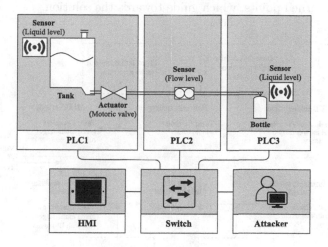

Fig. 2. ICS Scenario of the cyber range.

As shown in Fig. 3 the trainee is guided through the scenario by several learning materials provided by the LMS. Each step within the scenario is accompanied by a task that the trainee must complete.

The scenario is designed to slowly introduce to rule creation by requiring the trainee to solve increasingly elaborate tasks. It starts with a general introduction, where only simple questions about the events captured by the SIEM have to be answered. Once the first step is complete, increasingly complex attacks are simulated one after another, which the trainee must first detect manually (S2). Then he or she is required to create rules (S3) that automatically detect these attacks. The rules to be created also increase in complexity. In order not to over-tax the trainee, large parts of the rule are initially given, and the trainee only has to add certain parts. Then, starting with the scenario step "log file manipulation", the trainee has to create the whole rules themselves. The complexity of the rules to be learned can be divided into three difficulty levels: Starting with very simple rules for which only one condition must be met, to multi-stage rules that build on each other and for which several conditions must be met, to rules that also query an IP address range.

The LMS provides various media to support the trainee's learning between each scenario step. These are either explanatory texts or videos that convey knowledge for the subsequent step in the scenario. In each case, the simulated attack is briefly presented from the attacker's point of view (S1) to provide guidance on what the trainee must look for in the SIEM. It also explains how to use the SIEM and how rules are structured. Gamification elements are used to motivate the trainee during the training session. The trainee receives points for each task he or she solves and can use them to move up levels. If a task is answered incorrectly, the trainee can correct the answer, but points are lost to prevent solutions from simply being guessed. If the trainee is stuck, hints can be bought with earned points, which guide towards the solution.

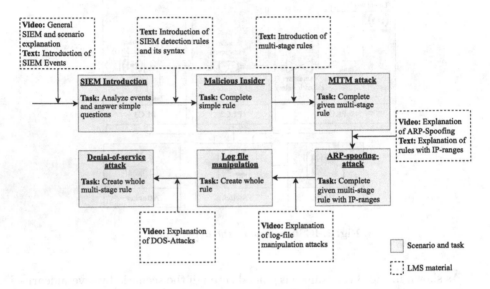

Fig. 3. Learning concept for the cyber range.

4.3   Prototypical Implementation

The overall architecture of the cyber range is shown in Fig. 4. To simulate the industrial system, the digital twin's simulation component is transferred to the cyber range to create a realistic virtual environment. The simulation is realized with MiniCPS[1], an academic framework for simulating cyber-physical systems which builds upon Mininet[2]. To monitor the network traffic, a firewall captures the TCP-traffic within the network and detects certain abnormalities such as ambiguous responses to ARP-requests. The firewall functionalities are implemented with scapy[3]. The PLCs and the HMI produce system logs on the main functions of the filling process and the firewall monitoring, which are stored as log files in a common logs directory.

---

[1] https://github.com/scy-phy/minicps.
[2] http://mininet.org/.
[3] https://scapy.net/.

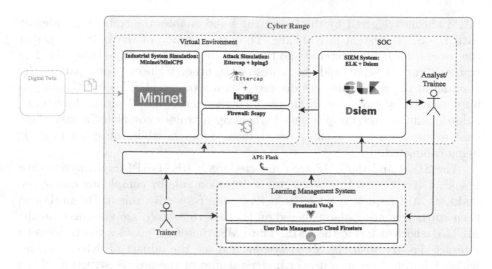

Fig. 4. Architecture of the prototypical implementation.

As described in Sect. 4.2, the attacker performs various attacks against the network components. To implement the attacks, the network tools Ettercap[4] (for the ARP-Spoofing/Man-In-The-Middle-Attack) and hping3[5] (for the Denial-Of-Service-Attack) are used. The Log-File-Manipulation-Attack is performed by simply deleting the log file in which the system logs of PLC1 are stored. For the filling plant simulation to produce consistent system logs over the cyber range's lifetime, the attacks are automated and repeated periodically. The open-source tool Dsiem[6] is the implemented SIEM system of the cyber range. It builds upon Elasticsearch, Logstash, Filebeat, and Kibana. With Logstash and Filebeat, the aforementioned log data is parsed and normalized as so-called SIEM events, which are then forwarded to Dsiem. Dsiem correlates SIEM events with prede-fined rules to generate SIEM alarms. Finally, these SIEM events and alarms are transferred to Elasticsearch and visualized in Kibana. The virtual environment and the SIEM system are realized as a microservice-infrastructure separated from the LMS and with each component being deployed in a docker container. This modular architecture facilitates reusing the infrastructure for future work and enables its extension as well as the replacement of one or more of the com-ponents.

The LMS is realized with the JavaScript framework Vue.js[7]. A screenshot of the user interface of the cyber range is presented in the Appendix (Fig. 6). One section of the LMS displays a Kibana-based SIEM dashboard for Dsiem. It visualizes the SIEM events produced by the digital twin-based simulation and the

---

[4] https://www.ettercap-project.org/.

[5] http://www.hping.org/.

[6] https://www.dsiem.org/.

[7] https://vuejs.org/.

SIEM alarms triggered by the Dsiem rules and enables the trainees to interact with the SIEM system in real-time. The other section of the LMS consists of the provided learning material and the tasks the trainees need to complete. The trainee's current score, and the scores of the other trainees taking part in the training at the same time, are displayed on a scoreboard. This functionality is implemented by storing each trainee's current score in a Realtime Firestore[8]. Additionally, a timestamp is saved whenever a trainee completes a task. This enables the trainer to monitor the trainees' progress while the cyber range is being conducted.

The SIEM and the LMS are connected via a REST-API implemented with Flask[9]. Every time a trainee creates a detection rule by completing one of the tasks, an API request is set off to activate the respective rule in Dsiem. Dsiem then starts triggering alarms based on the new rule which are visualized on the SIEM dashboard inside the LMS. The LMS, therefore, enables the trainees to interact directly with the SIEM system and see the impact of detection rules without having to gain a deeper understanding of the project structure of the SIEM system beforehand. Furthermore, the Flask API provides functions for the trainer to interact with the microservice architecture of the digital twin-based simulation and the SIEM system. These functions can be used to start and stop the infrastructure and reset single components in case any technical issues occur while the cyber range training is being conducted. The source code of the project, together with further documentation, is available on GitHub[10].

5   User Study Evaluation

5.1   Method

To measure the effectiveness of the cyber range, it is necessary to evaluate whether it leads to an improvement of the participants' knowledge or skill level. Since a cyber range in our case is similar to a serious game according to the definition of Girard et al. [15], methods from this context can be applied to measure the effectiveness. Besides qualitative methods [16], it is possible to quantitatively evaluate this by measuring the participants' skills and knowledge before and after the training [15]. In the present case, to the best of our knowledge, a comparable system targeting the training of analysts within a SOC does not exist. Therefore it is not possible to evaluate the increase of performance of participants of the cyber range training against participants of a control group in order to compare it to a similar training concept. Instead, it is more suitable to use a one group pre-test/post-test design proposed by Hauge et al. [16] to show whether or not an increase in knowledge has been achieved. Therefore, two assessment questionnaires are constructed consisting of 13 multiple-choice questions (Q1–Q13) for evaluating the learning outcomes of the cyber range. These aim at testing the

---

[8] https://firebase.google.com/docs/firestore.

[9] https://flask.palletsprojects.com/en/1.1.x/.

[10] https://github.com/DigitalTwinSocCyberrange.

knowledge of the participants, whereby four answer options are given for each question. These questionnaires are disseminated before and after the training to measure the improvement of the participant's knowledge.

As the cyber range concept should not only lead to an increase of knowledge but also provide a positive learning experience, the training aims to attract the participant's attention and provide a high level of engagement. Metrics for measuring the engagement levels of the participants are provided by Keller's ARCS model of motivational design [18] which has been used in the past to evaluate security and privacy educational approaches before [14]. It focuses on the intrinsic attributes enhancing motivation, and includes metrics that relate to Attention, Relevance, Confidence, and Satisfaction. The ARCS model can be extended by an extra metric for perceived learning, which measures the subjective impression of whether learning has occurred [1,5]. This part of the evaluation was implemented by constructing a feedback questionnaire based on the ARCS model, extended by the perceived learning condition. Thereto, the participants can indicate the degree of agreement to 16 statements, with a Likert scale ranging from 1 to 5 ("completely disagree" to "fully agree") after the training.

Participants. Participants were recruited in cybersecurity-related courses at both the University of Regensburg (Germany) and Ionian University (Greece). This ensures that all participants have at least a basic knowledge of cybersecurity, reflecting the target group of the cyber range. In total $n = 44$ test persons participated in the study: 22 German students and 22 Greek students, whereby 12 were female and 32 male. 24 students were undergraduate and 20 were postgraduate students.

Procedure. The study was conducted entirely online over several video conferencing sessions. For each session, 10 virtual machines with one cyber range each were available, limiting the simultaneous number of participants to 10. The user study was divided into three phases. After a short welcome and introduction to the cyber range at the start of the session, the participants were asked to complete the first questionnaire to record their previous knowledge. In the second phase, they were asked to open the cyber range and complete the training contained within. Participation was not time-limited, but most of the participants completed all tasks after a maximum of 2 h. After having completed the second phase, the test persons were asked to fill in the two remaining questionnaires in the third phase, which tested their knowledge afterwards and assessed their motivation during the training. During the execution of the cyber range, we ensured that the trainer intervened as little as possible in the test persons' performance of the tasks in order to avoid influencing them and their results.

5.2   Results

To show that the participants of the study achieved a learning effect, the results from the assessment of the pre-, and post-knowledge are analyzed in the following. The study's questions can be divided into three classes: General knowledge about cybersecurity attacks, general knowledge about SIEM, and specific knowledge about the structure and functionality of SIEM detection rules. Figure 5 shows the results of the pre-, and post-test. Thereto, the mean percentage of correctly answered questions in both test runs is visualized. The dashed lines indicate the mean in the respective knowledge classes.

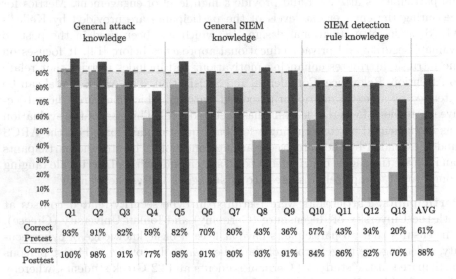

Fig. 5. Comparison of test persons' knowledge (measured by percentage of correct answers for questions Q1 to Q13) before and after participation in the cyber range. Grouped by knowledge classes, with the dashed lines visualizing the mean of each class.

A paired t-test was conducted to examine the increase in knowledge overall and across the individual knowledge classes[11]. It shows that the mean of correctly answered questions significantly increased by 26.92% ($t = -12.472, SD = 0.143191, p < 0.001$). In the first class, "general attack knowledge" the mean increase is smaller (10.23%) and less significant ($t = -3.448, SD = 0.196763, p = 0.0013$). This is, however, expected because the test persons possess a certain level of pre-knowledge in cybersecurity and therefore about simple attacks. Thus, the increase from an already high level is smaller. Within the class "general SIEM knowledge", an increase of 28.18% is observed ($t = -7.398, SD = 0.252681, p < 0.001$). Based on the pre-test, it could be determined that some pre-knowledge was already present within this class. However, a significant increase could still be achieved. Within the "SIEM detection rule knowledge" class, a significant increase of 42.05% is indicated ($t = -8.417, SD = 0.331368, p < 0.001$).

Since an increase in knowledge does not necessarily show that the cyber range was a positive experience for the participants, it is necessary to evaluate the results from the feedback survey. The aggregated results can be found in

---

[11] The SPSS output of the t-test can be found in Fig. 7 in the appendix.

Table 2. The results indicate that the cyber range was, in general, received quite well by the test persons. Both the mean and the median are at least 4 for all conditions on a scale of 1 to 5 (where a higher value indicates the participants' agreement).

Table 2. Results of the feedback questionnaire.

Condition	Mean	Median	Standard deviation
Attention	4.395	5	0.753
Relevance	4.352	4	0.724
Confidence	4.090	4	0.778
Satisfaction	4.284	4	0.738
Perceived learning	4.460	5	0.602

To ensure a high standard of reproducibility and reusability, the anonymized data of all the results and the used questionnaires are available as a public data set[12].

5.3 Discussion

Overall, the results of the user study reveal that an increase in knowledge could be achieved among the participants. Although the increase in general knowledge about attacks (S1) was quite small, a significant increase in knowledge about attack detection using a SIEM system (S2 and S3) is shown – leading to the conclusion that the previously defined goals are achieved. Taking into consideration the results of the evaluation, in the following, we discuss some details we found to be particularly noteworthy.

Within the cyber range, the participants were able to score points by solving the tasks provided as described in Sect. 4.2. The score of a participant thereby indicates to what extent he or she was able to solve the tasks without requiring many attempts to provide the correct solution. While this score was not explicitly used for evaluating the effectiveness of the cyber range, we find it worth examining - especially for participants with particularly high or low increase in knowledge. Five participants showed a notably large increase in knowledge in the assessment questionnaire from 50% or less to more than 90% after participating in the cyber range. The score results of these participants vary from 43 to 100 out of 101 possible points. This shows that though initially failing some tasks of the cyber range, a participant can still gain a large increase of knowledge. In contrast, three participants did not present any improvement in the pre-, and post-assessment. These participants achieved comparably low scores ranging

---

[12] https://github.com/DigitalTwinSocCyberrange/userStudy.

from 28 to 33 points. This indicates difficulties in engaging with the overall app-roach. However, it is noteworthy that these participants still provided positive feedback on the cyber range.

Considering the results of the feedback survey, a noticeable aspect is a some-what lower result for Confidence compared to the other values. This is also confirmed by some participants' oral feedback, who told us that they were some-what overstrained at the beginning. In our estimation, this was mainly due to an information overload, as they were confronted with both the SIEM and the LMS. In the future, the cyber range could be adapted so that trainees are not shown all information from the start, but only selected content that is then gradually expanded. The value for perceived learning also sticks out, indicating whether the participants themselves assess whether they learned something during the procedure. With a value of 4.460, it is slightly higher than the others. This confirms the result from comparing the pre-, and post-test, as the participants themselves also have the impression of having gained knowledge.

6   Conclusions

This work demonstrates how cyber ranges can be utilized for training security analysts in a SOC. It shows that cyber ranges are suitable for the acquisition of general knowledge about SIEM as well as for specific training on how to create SIEM rules. The provided cyber range concept builds upon the simula-tion component of a digital twin of an industrial filling plant. This ensures that the analysts are trained based on a realistic scenario. To show the increase in knowledge and the perceived learning experience, the concept is implemented and evaluated in an international study among both Greek and German partic-ipants. To the best of our knowledge, this is the first cyber range to utilize the potential of a digital twin, specifically targeting the training of SOC analysts.

Like any other research effort, this paper contains limitations. Since, to our knowledge, no approach with the same objective exists, it was not possible to compare the knowledge gains. However, we were able to show that a cyber range is, in general, suitable for imparting knowledge. Nonetheless, we did not concen-trate on an evaluation comparing our cyber range to other concepts.

In summary, this work provides a new approach to train SOC analysts. By proposing security training, it addresses the current problem of the increasing demand for security analysts personnel, which will continue to grow. Further-more, the attack detection training of SOC analysts is only one of many possible applications of the presented cyber range. Among many other possibilities, it could also be used for penetration testing of industrial plants or incident response exercises in future research.

Appendix

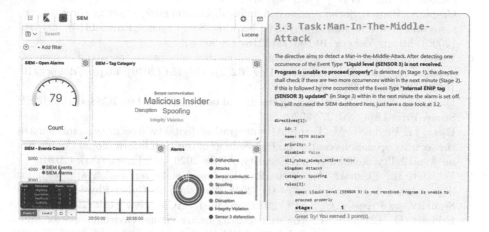

Fig. 6. Screenshot of the cyber range interface: SIEM dashboard and LMS

Paired Samples Statistics

		Mean	N	Std. Deviation	Std. Err. Mean
Pair 1	Correct_pre	,60839	44	,139608	,021047
	Correct_post	,87762	44	,148677	,022414
Pair 2	Correct_pre_att	,81250	44	,187742	,028303
	Correct_post_att	,91477	44	,142069	,021418
Pair 3	Correct_pre_SIEM	,62273	44	,280252	,042250
	Correct_post_SIEM	,90455	44	,146199	,022040
Pair 4	Correct_pre_rule	,38636	44	,255489	,038516
	Correct_post_rule	,80682	44	,240298	,036226

Paired Samples Test

		Paired Differences							
					95% Conf. Interval of the Diff.				
		Mean	Std. Deviation	Std. Err. Mean	Lower	Upper	t	df	Sig. (2-tailed)
Pair 1	Correct_pre - Correct_post	-,269231	,143191	,021587	-,312765	-,225697	-12,472	43	,000
Pair 2	Correct_pre_att - Correct_post_att	-,102273	,196763	,029663	-,162094	-,042451	-3,448	43	,001
Pair 3	Correct_pre_SIEM - Correct_post_SIEM	-,281818	,252681	,038093	-,358640	-,204996	-7,398	43	,000
Pair 4	Correct_pre_rule - Correct_post_rule	-,420455	,331368	,049956	-,521199	-,319710	-8,417	43	,000

Fig. 7. SPSS output of the t-test

References

1. Barzilai, S., Blau, I.: Scaffolding game-based learning: impact on learning achievements, perceived learning, and game experiences. Comput. Educ. **70**, 65–79 (2014)
2. Bissel, K., Lasalle, R., Dal Cin, P.: Third annual state of cyber resilience report. Accenture (2020)
3. Boschert, S., Heinrich, C., Rosen, R.: Next generation digital twin. In: Proceedings of the 12th International Symposium on Tools and Methods of Competitive Engineering, TMCE 2018, pp. 209–217 (2018)

4. Bécue, A., et al.: CyberFactory1 – securing the industry 4.0 with cyber-ranges and digital twins. In: 2018 14th IEEE International Workshop on Factory Communication Systems (WFCS), pp. 1–4 (2018)
5. Caspi, A., Blau, I.: Social presence in online discussion groups: testing three conceptions and their relations to perceived learning. Soc. Psychol. Educ. **11**(3), 323–346 (2008). https://doi.org/10.1007/s11218-008-9054-2
6. Dietz, M., Pernul, G.: Digital twin: empowering enterprises towards a system-of-systems approach. Bus. Inf. Syst. Eng. **62**(2), 179–184 (2019). https://doi.org/10.1007/s12599-019-00624-0
7. Dietz, M., Pernul, G.: Unleashing the digital twin's potential for ICS security. IEEE Secur. Priv. **18**(4), 20–27 (2020)
8. Dietz, M., Vielberth, M., Pernul, G.: Integrating digital twin security simulations in the security operations center. In: Proceedings of the 15th International Conference on Availability, Reliability and Security, ARES 2020. ACM, New York (2020)
9. Eckhart, M., Ekelhart, A.: A specification-based state replication approach for digital twins. In: Proceedings of the 2018 Workshop on Cyber-Physical Systems Security and PrivaCy, CPS-SPC 2018, pp. 36–47. ACM, New York (2018)
10. Eckhart, M., Ekelhart, A.: Towards security-aware virtual environments for digital twins. In: Proceedings of the 4th ACM Workshop on Cyber-Physical System Security (CPSS 2018), pp. 61–72 (2018)
11. Eckhart, M., Ekelhart, A.: Digital twins for cyber-physical systems security: state of the art and outlook. In: Security and Quality in Cyber-Physical Systems Engineering, pp. 383–412. Springer, Cham (2019). https://doi.org/10.1007/978-3-030-25312-7_14
12. Empl, P., Pernul, G.: A flexible security analytics service for the industrial IoT. In: Proceedings of the 2021 ACM Workshop on Secure and Trustworthy Cyber-Physical Systems, pp. 23–32. ACM, New York (2021)
13. Gehrmann, C., Gunnarsson, M.: A digital twin based industrial automation and control system security architecture. IEEE Trans. Ind. Inf. **16**, 669–680 (2020)
14. Giannakas, F., Papasalouros, A., Kambourakis, G., Gritzalis, S.: A comprehensive cybersecurity learning platform for elementary education. Inf. Secur. J. **28**(3), 81–106 (2019)
15. Girard, C., Ecalle, J., Magnan, A.: Serious games as new educational tools: how effective are they? A meta-analysis of recent studies. J. Comput. Assist. Learn. **29**(3), 207–219 (2013)
16. Hauge, J.B., et al.: Study design and data gathering guide for serious games' evaluation. In: Tennyson, R., Connolly, T.M., Hainey, T., Boyle, E., Baxter, G., Moreno-Ger, P. (eds.) Psychology, Pedagogy, and Assessment in Serious Games. Advances in Game-Based Learning, pp. 394–419. IGI Global (2014)
17. Kavallieratos, G., Katsikas, S.K., Gkioulos, V.: Towards a cyber-physical range. In: Proceedings of the 5th on Cyber-Physical System Security Workshop - CPSS 2019, pp. 25–34. ACM Press, New York (2019)
18. Keller, J.M.: Development and use of the ARCS model of instructional design. J. Instr. Dev. **10**(3), 2–10 (1987). https://doi.org/10.1007/BF02905780
19. Kelley, D., Moritz, R.: Best practices for building a security operations center. Inf. Syst. Secur. **14**(6), 27–32 (2006)
20. Leitner, M., et al.: AIT cyber range: flexible cyber security environment for exercises, training and research. In: Proceedings of the European Interdisciplinary Cybersecurity Conference, pp. 1–6 (2020)

21. Madani, A., Rezayi, S., Gharaee, H.: Log management comprehensive architecture in Security Operation Center (SOC). In: 2011 International Conference on Computational Aspects of Social Networks (CASoN), pp. 284–289. IEEE (2011)
22. Negri, E., Fumagalli, L., Macchi, M.: A review of the roles of digital twin in CPS-based production systems. Procedia Manuf. **11**, 939–948 (2017)
23. Pescatore, J., Filkins, B.: Closing the critical skills gap for modern and effective security operations centers (SOCs). SANS Institute (2020)
24. Pokhrel, A., Katta, V., Colomo-Palacios, R.: Digital twin for cybersecurity incident prediction: a multivocal literature review. In: Proceedings of the IEEE/ACM 42nd International Conference on Software Engineering Workshops, ICSEW 2020, pp. 671–678. ACM, New York (2020)
25. Putz, B., Dietz, M., Empl, P., Pernul, G.: EtherTwin: blockchain-based secure digital twin information management. Inf. Process. Manag. **58**(1), 102425 (2021)
26. Rubio, J.E., Roman, R., Lopez, J.: Analysis of cybersecurity threats in industry 4.0: the case of intrusion detection. In: D'Agostino, G., Scala, A. (eds.) Critical Information Infrastructures Security. LNCS, vol. 10707, pp. 119–130. Springer, Cham (2018). https://doi.org/10.1007/978-3-319-99843-5_11
27. Schinagl, S., Schoon, K., Paans, R.: A framework for designing a security operations centre (SOC). In: 2015 48th Hawaii International Conference on System Sciences, pp. 2253–2262. IEEE (2015)
28. Tian, Z., et al.: A real-time correlation of host-level events in cyber range service for smart campus. IEEE Access **6**, 35355–35364 (2018)
29. Ukwandu, E., et al.: A review of cyber-ranges and test-beds: current and future trends. Sensors **20**(24), 7148 (2020)
30. Vielberth, M., Bohm, F., Fichtinger, I., Pernul, G.: Security operations center: a systematic study and open challenges. IEEE Access **8**, 227756–227779 (2020)
31. Vielberth, M., Pernul, G.: A security information and event management pattern. In: 12th Latin American Conference on Pattern Languages of Programs (SugarLoafPLoP 2018), pp. 1–12. The Hillside Group (2018)
32. Yamin, M.M., Katt, B., Gkioulos, V.: Cyber ranges and security testbeds: scenarios, functions, tools and architecture. Comput. Secur. **88**, 101636 (2020)

The *tkl*-Score for Data-Sharing Misuseability

Kalvin Eng[✉] and Eleni Stroulia

University of Alberta, Edmonton, AB T6G 2R3, Canada
{kalvin.eng,stroulia}@ualberta.ca

Abstract. Estimating the potential for data misuse is essential for all data-sharing decisions. This work presents *tkl*-Score which extends the state of the art M-Score and L-Severity measures. The new proposed measure is sensitive to the increased misuse potential when records are more identifiable in a source table with l-Distinguishing Factor and also when sensitive attributes are less granular in a source table with l-Distinguishing Factor and t-Distinguishing Factor; in contrast, the earlier M-Score and L-Severity only account for record identifiability in a source table with k-Distinguishing Factor. *tkl*-Score is shown to better characterize the risk of releasing records compared to M-Score and L-Severity due to accounting for sensitive attribute granularity.

Keywords: Data misuse · Data sharing · Data privacy

1 Introduction

As data-collection software becomes more ubiquitous, due to mobile apps, social platforms, and the internet of things (IoT), the privacy and sensitivity of collected data is becoming an increasing concern to data stakeholders. This ever-increasing amount of devices and apps leads to a plethora of data being collected. Organizations who wish to share their data need to be more aware of the consequences that can affect the different stakeholders of the data being shared. Unnecessary disclosures of sensitive information can lead to severe consequences for the subjects of the data and legal repercussions to the organizations themselves. For example, the personal information of over 50 million Facebook users were unintentionally exposed to Cambridge Analytica and used for political gain leading to backlash against Facebook [1].

The objective of our work is to develop a misusability score that when given a dataset to be shared, the score accurately estimates the risk of potential misuse of this data and helps inform the decision making of the owner organization. The rest of this paper is organized as follows. In Sect. 2, we review the current state of misuseability metrics and highlight their drawbacks that we wish to address. Based on these drawbacks, we introduce *tkl*-Score and its derivative *tkl*-Score$_{max}$ in Sect. 3 which augments current misuseability metrics. In Sect. 4, we perform a comparative analysis between *tkl*-Score and previous misuseability

© IFIP International Federation for Information Processing 2021
Published by Springer Nature Switzerland AG 2021
K. Barker and K. Ghazinour (Eds.): DBSec 2021, LNCS 12840, pp. 312–324, 2021.
https://doi.org/10.1007/978-3-030-81242-3_18

scores highlighting why previous scores should be augmented. Finally in Sect. 5, we summarize our work and introduce future avenues of exploration.

2 Related Work on Misuseability Scores

We base our work on M-Score and L-Severity which are two key metrics for estimating the potential of data misuse. Both are calculated using weights derived from a "source table" to help estimate the risk of potential misuseability in a subset of a source table called a "published table". The scores also rely on two types of attributes. *Quasi-identifiers* which are attributes that, when combined with other information, may partially reveal an individual's identity. For example, when two datasets have "city of residence" and "gender" columns, their join based on these quasi-identifier attributes may reveal identifying information such as an individual's name present only in one of the datasets. As well, there are *sensitive attributes* which convey information that should not be exposed publicly (e.g. one's health condition), because the release of this information has the potential to harm an individual such as damaging their reputation.

2.1 M-Score

M-Score [2] is the first metric designed specifically for identifying the potential negative impacts of a dataset release. It is a score for tabular data that quantifies the ability of a user to maliciously exploit exposed data and takes into account the anonymity of individuals in a dataset as well as the sensitivity of the data attribute values. The process of calculating the M-Score for a published table consists of two steps: (1) eliciting weights for sensitive attribute values, and (2) calculating the M-Score for the published table.

(1) Eliciting Weights of Sensitive Attribute Values. There are several methods for eliciting sensitive value weights from domain experts, but the authors of M-Score argue that the Analytic Hierarchy Process [5] elicits the best results for discretized data.

(2) M-Score of the Published Table. Given the weights of the sensitive attribute values, a source table, and a published table (i.e. a data subset of the source table), M-Score can be calculated. To begin, each record of a published table is given a record score as follows:

$$RS_{M_r} = \frac{\min(1, \sum_{A_{S_i} \in r} \text{weight}(A_{S_i}))}{DF_{k_r}} \qquad (1)$$

A record score RS_{M_r} of the rth record of a published table is the sum of each ith sensitive attribute value weight of the record minimized to 1 divided by the k-Distinguishing Factor DF_{k_r} of the rth record.

The k-Distinguishing Factor is a measure dependent on comparing the published table to the source table. It quantifies how easily an individual can be identified, based on the distinctiveness (or uniqueness) of records in a "lookup table".

The "lookup table" is collection of records related to a population that can identify an individual. Since such a collection is not easily acquired, the "lookup table" is approximated to be the source table. It is based on the k-anonymity measure [6] that groups rows with similar quasi-identifier values into "equivalence classes", where an equivalence class is a set of records that have the same values for quasi-identifiers attributes [3].

The k-Distinguishing Factor of a record in a published table is the size of the equivalence class in the source table that contains the record in the published table. If there are no quasi-identifiers to form an equivalence class, then the k-Distinguishing Factor of a record is the size of the published table. The k-Distinguishing Factor is meant to account for how distinguishable an exposed record is when it is published from the source table, and helps to differentiate the records by their identifiability—when the k-Distinguishing Factor is smaller, the record is more identifiable and the record score therefore becomes larger.

With all record scores of a published table calculated, the M-Score of a published table can be computed based on the maximal record score and the number of records, n, in the published table.

$$M\text{-Score} = n^{\frac{1}{x}} \times \max_{0 \leq r \leq n} \left(\frac{\min(1, \sum_{A_{S_i} \in} \text{weight}(A_{S_i}))}{DF_{k_r}} \right) \tag{2}$$

where $x \geq 1$ is a parameter for the importance of the amount of records, A_{S_i} is the ith sensitive attribute value of a record r, and DF_{k_r} is the k-Distinguishing Factor of a record r.

Using the definition above, the M-Score of the published table can be computed by multiplying the highest individual record score among the n records weighted with a power $\frac{1}{x}$. The x of $n^{\frac{1}{x}}$ is a parameter for specifying the importance of the quantity of records in a published table. If $x = 1$ then the amount of records is given more importance compared to the sensitivity of data. If $x \to \infty$, then $n^{\frac{1}{\infty}} \approx 1$ which means that we would like to know the highest individual record score of M-Score. The parameter x can be assigned any value where $x \geq 1$ with a trade-off between the importance of the highest individual record score to the importance of the amount of records.

Drawbacks. M-Score is approximative in nature as it takes the maximum record score of a published table for its score calculation. For example, consider a source table where 99 records of 100 records have a record score of 0.0001 with the remaining record having a record score of 1. A published table with nine records with score 0.0001 and one record with score 1 will result in the same M-Score, i.e., 10, as a published table with nine records of 1 and one of 0.0001.

The approximation in M-Score leads to issues when attempting to identify the "percentage of severity" a published table takes from the source table score. If we consider the example from the previous paragraph, where the published table has a score of 10, and the source table has a score of 100, we can say that the published table makes up 10% of the severity of the source table: $10/100 = 0.1$.

However, this is not the case when we compare the sum of the actual values of the record scores in each of the tables: $1.0009/1.0099 = 0.991$.

It can also be difficult to decide on the parameter x to model the trade-off between the importance of the number of records or the maximum record score in a published table. If we would like to approximate the severity based on the amount of records, then x can be set as 1. If we assume that releasing any single maximum record of a published table is the maximum severity, then x can be set as $x \to \infty$. However, to decide on a value between 1 and $x \to \infty$, that represents the trade-off between these two factors, an ad hoc decision would need to be made.

The last drawback to note is that the k-Distinguishing Factor in M-Score only accounts for identity disclosure attacks. The authors of M-Score suggest that measures such as l-diversity can be used to account for attribute disclosure attacks, but provide no method to do so.

2.2   *L*-Severity

Building on the work of M-Score, Vavilis et al. [7] designed L-Severity, a misuse-ability score aiming to address the approximative nature of the M-Score calculation. The process of calculating L-Severity can be divided into three steps: (1) developing a data model of the sensitive attributes in a source table, (2) eliciting weights for sensitive attribute values from the data model, and (3) calculating the overall L-Severity for the published table.

(1) Developing the Data Model. The data model in L-Severity is designed to represent the hierarchy of concepts surrounding the sensitive attributes in a source table. Each sensitive attribute of a source table is represented as a node and can fall under more general nodes assigned by a domain expert. As well, the concept of "inference relationships" which link sensitive attributes is also introduced to highlight how sensitive attributes may be related.

(2) Eliciting Sensitive Attribute Value Weights using the Data Model. Using the data model, domain experts assign "sensitivity values" to highlight the importance of nodes in the data model as well as "inference values" to quantify the importance between any relationships of attributes in the data model. Once all values have been assigned, they can be used to calculate the sensitive attribute value weights needed for L-Severity.

(3) *L*-Severity of the Published Table. To determine the L-Severity of a published table, the record scores are summed together. A record score can be calculated as follows:

$$RS_{L_r} = \frac{\sum_{A_{S_i} \in r} \text{weight}(A_{S_i})}{DF_{k_r}} \tag{3}$$

A record score RS_{L_r} of the rth record of a published table is the sum of each ith sensitive attribute value weight of the record divided by the k-Distinguishing Factor DF_{k_r} of the rth record. Then, given a published table T, the L-Severity of the table is:

$$L\text{-Severity} = \sum_{r \in T} \left(\frac{\sum_{A_{s_i} \in r} \text{weight}(A_{s_i})}{DF_{k_r}} \right) \tag{4}$$

where r is each record of the published table, A_{S_i} is the ith sensitive attribute value of a record r, and DF_{k_r} is the k-Distinguishing Factor of a record r.

Drawbacks. Although L-Severity addresses the approximative nature of M-Score by summing record scores, L-Severity cannot account for the case that assumes releasing the maximum record score of a published table is the maximum severity. Since L-Severity does not have a x parameter like M-Score, it cannot control the trade-off between the importance of the number of records and the maximum severity to calculate its score. Instead, it only accounts for the case that calculates severity based on the number of published records.

L-Severity also suffers the same drawback as M-Score, failing to consider anonymity measures like l-diversity and t-closeness to account for attribute disclosure. Only identity disclosure is accounted for with k-Distinguishing Factor. The authors of L-Severity suggest that attribute disclosure measures can be integrated into misuseability scoring, but provide no method to do so.

3   Calculating tkl-Score

Expanding on these earlier measures, we propose tkl-Score which incorporates the privacy preserving data publishing metrics: l-diversity as l-Distinguishing Factor, and t-closeness as t-Distinguishing Factor to account for attribute disclosure attacks in addition to identity disclosure attacks with k-Distinguishing Factor. It is important to also account for attribute disclosure as it may be difficult to be certain of an individual's identity, but attributes relating to an individual can still be disclosed when similar information about identities are grouped together.

To demonstrate how l-Distinguishing Factor and t-Distinguishing Factor are used in tkl-Score, we introduce the record score of tkl-Score which is defined as:

$$RS_{tkl_r} = \frac{DF_{t_r} + \sum_{A_{S_i} \in r} \text{weight}(A_{S_i})}{DF_{l_r}} \tag{5}$$

where r is a record, DF_{l_r} is the l-Distinguishing Factor of a record, DF_{t_r} is the t-Distinguishing Factor of a record, and $\text{weight}(A_{S_i})$ is the weight of the ith sensitive attribute value of a record.

It should be noted that l-Distinguishing Factor and t-Distinguishing Factor, like k-Distinguishing Factor, aims to quantify the identity and attribute uniqueness of records in a "lookup table" that contains all records related to a population. However, since it is difficult to obtain all records related to a population, the "lookup table" is approximated to be the source table of a published table.

3.1   *l*-Distinguishing Factor

l-diversity [4] is the measure used to determine how distinguishable an individual is based on attribute frequency in an equivalence class (a group of records with common quasi-identifier attribute values). In this paper, we use the *l*-diversity definition where every equivalence class in a table contains at least *l* distinct values for a sensitive attribute in order to be *l*-diverse. An interesting property of *l*-diversity is that *l* will always be ≤ *k* of *k*-anonymity.

Proof. Let *k* be the *k*-anonymity of a dataset which is the smallest equivalence class with size *k* rows. Assume that the *l*-diversity of the smallest equivalence class has $k < l$. Based on the assumption, let $l = k + 1$. Then based on the definition of *l*-diversity, there must be $k + 1$ unique attribute values. However, this is a contradiction as the size of the equivalence class must be $k + 1$ to have $k+1$ unique attribute values. Now assume that the dataset has multiple sensitive attributes and therefore multi-attribute *l*-diversity [4] is used to create different grouping combinations. The largest equivalence class of the combinations will still be at most be the size of the original equivalence class that is matched with only quasi-identifiers. I.e. as more attributes need to be matched to form a grouping of records, the size of the groupings either remains the same as the attribute values are all the same, or the size of the grouping becomes smaller when the attribute values are different. Therefore, $l \leq k$.

Recall that *k*-Distinguishing Factor is used to determine how distinguishable a record is in a source table and is a divisor in the score equations of *M*-Score and *L*-Severity. A large *k*-Distinguishing Factor implies a lower risk of identifiability, as more records are required to uniquely identify an individual. Hence, we wish to capture the maximal severity by minimizing *k*-Distinguishing Factor.

Since $l \leq k$, the *k*-anonymity metric for identity disclosure attacks of *k*-Distinguishing Factor will always be accounted for when using *l*-Distinguishing Factor. Therefore in *tkl*-Score, *l*-Distinguishing Factor replaces the *k*-Distinguishing Factor factor used in *M*-Score and *L*-Severity.

The definition of *l*-Distinguishing Factor uses the following definition of multi-attribute *l*-diversity: Let T be a table with nonsensitive attributes $Q_1, ..., Q_{m_1}$ and sensitive attributes $S_1, ..., S_{m_2}$. If for all iterations $i = 1...m_2$, the table T is *l*-diverse when S_i is treated as the sole sensitive attribute and $\{Q_1, ..., Q_{m_1}, S_1, ..., S_{i-1}, S_{i+1}, ..., S_{m_2}\}$ is treated as the "quasi-identifiers" to form "equivalence classes" [4].

The *l*-Distinguishing Factor of a record in a published table is the minimal multi-attribute *l*-diversity [4] "equivalence class" in the source table that contains the record in the published table. If there are no quasi-identifiers to form an equivalence class in the source table, then the minimal multi-attribute *l*-diversity of the published table (forming equivalence classes based on the sensitive attributes only) is the *l*-Distinguishing Factor of a record.

3.2 t-Distinguishing Factor

t-closeness [3] provides a measurement for the similarity between the attribute value distribution of an equivalence class and the attribute value distribution of an entire table. The similarity of distributions for sensitive attribute values helps to determine the true diversity of sensitive attributes globally. In comparison, l-diversity only considers the diversity of sensitive attributes within an equivalence class, which means sensitive attributes may still be prone to attribute disclosure if all the unique attributes are in a single equivalence class. As a result, t-closeness and l-diversity are two different anonymity measures that should be used together as a way to measure the anonymity of sensitive attributes in a table.

The objective of integrating t-closeness into tkl-Score is to increase the relative severity of an exposure as the value of the t-closeness of the released records increases. This is because the higher the t-closeness, the higher the likelihood of an attribute disclosure attack as the distributions are less similar and sensitive attributes are easier to discern. Therefore, t is added to the record score.

We assume to have a record score function $\frac{S}{DF}$ modeled after Eq. 3 where $1 \geq S \geq 0$ that represents the sum of sensitive attribute value weights of a record, and $DF \geq 1$ that represents the l-Distinguishing Factor. The addition of t to the numerator of the function to calculate a record score produces a linear translation of the score by $\frac{t}{DF}$ on the y-axis when visualized on a plot. In the best case the record score is minimally reduced when $t \approx 0$ (less risk because the attribute distributions are similar throughout the table) and in the worst case when $t = 1$ the score increases by $\frac{t}{DF}$ (more risk since equivalence class attribute distributions are not similar to the global attribute distribution). As a result, the record store will either be maintained or increased by a proportional t factor to indicate the severity of a record release more finely.

The t-Distinguishing Factor of a record in a published table is the maximal t-closeness of the equivalence class in the source table that contains this record. If there are no quasi-identifiers to form an equivalence class in the source table, then the t-Distinguishing Factor of a record is 0 as there are no equivalence classes to compare the distribution of sensitive attribute values. This is consistent with the limitation of k-anonymity which would also be 0 when there are no quasi-identifiers to form an equivalence class and indicates that there are no attributes that can be referenced to identify a person.

3.3 tkl-Score

To calculate the tkl-Score of a published table, every record in the table is scored and summed together using sensitive attribute value weights. To obtain sensitive attribute value weights for tkl-Score, the same methods used to derive weights in M-Score and L-Severity can be used. Equation (6) defined below uses the record score defined in Eq. (5).

Given a published table T, the *tkl*-Score is calculated as:

$$tkl\text{-Score} = \sum_{r \in T} \left(\frac{DF_{t_r} + \sum_{A_{S_i} \in r} \text{weight}(A_{S_i})}{DF_{l_r}} \right) \tag{6}$$

where for each record r, the total sum of each ith sensitive attribute value weight A_{S_i} is summed with the t-Distinguishing Factor DF_{t_r} of r and divided with the l-Distinguishing Factor DF_{l_r} of r.

We also introduce *tkl*-Score$_{\max}$ which is a score that is modeled after M-Score $(x \to \infty)$ to signify that releasing any one maximum record score is of maximum severity. The difference of *tkl*-Score$_{\max}$ and M-Score $(x \to \infty)$ is that it also accounts for attribute disclosure because it incorporates l-Distinguishing Factor and t-Distinguishing Factor, instead of only k-Distinguishing Factor.

Given a table with n records, the table's *tkl*-Score$_{\max}$ is then:

$$tkl\text{-Score}_{\max} = \max_{0 \le r \le n} \left(\frac{DF_{t_r} + \sum_{A_{S_i} \in r} \text{weight}(A_{S_i})}{DF_{l_r}} \right) \tag{7}$$

where for each record r, the total sum of each ith sensitive attribute value weight A_{S_i} is summed with the t-Distinguishing Factor DF_{t_r} of r and divided with the l-Distinguishing Factor DF_{l_r} of r.

4   Comparative Analysis

To compare *tkl*-Score against its predecessors M-Score and L-Severity, three alternative normalization assumptions can be considered: (1) the maximum severity is when a complete source table is released, (2) the maximum sever- ity is when any one record with the maximum record score in the source table is released, and (3) the maximum severity is when a score reaches the theoreti- cal maximum score determined by bounding the sum of sensitive attribute value weights and distinguishing factors. With assumption (1) the severity is related to the number of released records and therefore *tkl*-Score, L-Severity, and M-Score $(x = 1)$ are compared. With assumption (2) the severity is related to the maxi- mum record of a source table, and therefore *tkl*-Score$_{\max}$ and M-Score $(x \to \infty)$ are compared. For assumption (3) any misuseability score can be used, but it is best used with *tkl*-Score$_{\max}$ and M-Score $(x \to \infty)$ to compare their severity estimates.

Harel et al. [2] suggest that M-Score can be normalized under assumption (1) by taking the M-Score of the published table and dividing by the M-Score of the source table. This form of normalization assumes that releasing a complete source table is the maximum severity—when a subset of a source table is published, it will take a percentage of the source table score.

In this section, we illustrate how t-Distinguishing Factor and l-Distinguishing Factor in *tkl*-Score affects the row scores of two different tables using assumption (1) for normalizing the misuseability scores to the $[0, 1]$ range for comparison. Figure 1 illustrates how the misuseability scores can score rows differently based on attributes in a table.

4.1   A Case for t-Distinguishing Factor

The t-Distinguishing Factor is used to determine a record's distinctiveness based on the similarity of the distribution of sensitive attributes in the equivalence class of the source table containing this record and the distribution of sensitive attributes in the whole table. Recall that the higher the t value, the higher the severity a record would be as the distribution of the values of the sensitive attributes and the whole table are more distinct and therefore easier to differentiate.

To illustrate how t-Distinguishing Factor helps to distinguish records, the record scores for the source Table 1a are calculated in Table 1d. From the normalized scores in Table 1b, we can see that: (i) tkl-Score is reduced from row 5 to row 1, while M-Score ($x = 1$) and L-Severity maintains consistency for rows 0, 1, and 5; and (ii) the tkl-Score is reduced from row 2 to row 4, while M-Score ($x = 1$) and L-Severity are increased. These observations can be visualized in Fig. 1a.

Note that the M-Score (regardless of x parameter) and L-Severity calculated for each record seen in Table 1d are the same because there is only a single record for the misuseability score calculation. However, if we were to calculate the misuseability scores for a larger subset of published records, M-Score ($x = 1$), M-Score ($x \to \infty$) and L-Severity will produce different results.

The difference between (i) and (ii) stems from the equivalence class of rows 4 and 5 in Table 1a. The distribution of the *Initial Diagnosis* values in this equivalence has a lower t value compared to the other equivalence classes of the table as "HIV" and "Migraine" occur elsewhere in the table. In contrast, the equivalence class with rows 0 and 1, and the equivalence class with rows 2 and 3 have a unique value that does not appear in any other equivalence class leading to a higher t value. Therefore, rows 4 and 5 have a lower t-Distinguishing Factor than the other records, and as a result, rows 4 and 5 are considered less severe by tkl-Score than the other records in the table that have the same l-Distinguishing Factor and sum of sensitive attribute value weights as seen in Table 1d.

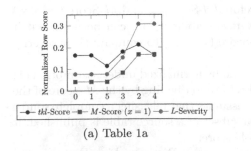

(a) Table 1a

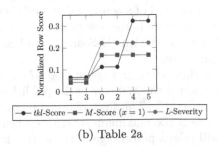

(b) Table 2a

Fig. 1. Per row scores normalized against the source table score for tkl-Score, M-Score ($x = 1$), and L-Severity of the records in Table 1a and Table 2a. It is sorted ascending by L-Severity.

Intuitively, observing the attribute values of Table 1a, we can see that either a female lawyer from Edmonton or a male lawyer from Edmonton has "HIV". Likewise, either a female lawyer from Edmonton or a female lawyer from Calgary has a "Migraine". Since "HIV" and "Migraine" must be distinguished from two distinct equivalence classes, it is less likely that these sensitive attributes will be inferred as opposed to "Flu" and "Hypertension" which occur only in a single equivalence class.

4.2 A Case for *l*-Distinguishing Factor

The *l*-Distinguishing Factor is used to determine a record's distinctiveness based on the size of the equivalence class that contains this record in the source table and also the values of sensitive attributes in the equivalence class that contains this record in the source table. Recall that *l*-Distinguishing Factor is part of the denominator of a record score (Eq. 5), and therefore the smaller the *l*-Distinguishing Factor the more the potential implications of releasing a record. So if there are more unique sensitive attribute values in an equivalence

Table 1. A case for using *t*-Distinguishing Factor.

id	Job	City	Gender	Initial Diagnosis
0	Lawyer	Calgary	Female	Flu
1	Lawyer	Calgary	Female	Migraine
2	Lawyer	Edmonton	Male	HIV
3	Lawyer	Edmonton	Male	Hypertension
4	Lawyer	Edmonton	Female	HIV
5	Lawyer	Edmonton	Female	Migraine

(a) The sensitive attribute of this source table is *Initial Diagnosis*; the quasi-identifiers are *Job*, *City*, and *Gender*.

Row	*tkl*-Score	M-Score ($x = 1$)	L-Severity
0	0.1642	0.0417	0.0769
1	0.1642	0.0417	0.0769
5	0.1149	0.0417	0.0769
3	0.1804	0.0833	0.1538
2	0.2128	0.1667	0.3077
4	0.1635	0.1667	0.3077

(b) Misuseability scores for each row of Table 1a normalized against the source table score rounded to four decimal places.

Disease	Medication	Age	Initial Diagnosis
$W(Flu) = 0.0864$	$W(Antibiotics) = 0.10432$	$W(30+) = 0.08$	$W(Migraine) = 0.05472$
$W(H1N1) = 0.3456$	$W(Paracetamol) = 0.10432$	$W(< 30) = 0.08$	$W(Flu) = 0.05472$
$W(Hypertension) = 0.3456$	$W(ARV) = 0.10432$		$W(Hypertension) = 0.10944$
$W(HIV) = 0.432$	$W(Tamiflu) = 0.10432$		$W(HIV) = 0.21888$
	$W(Statin) = 0.10432$		

(c) Sensitive Attribute Value Weights

Row	*tkl*-Score	M-Score	L-Severity	DF_t	DF_k	DF_l	Weights
0	0.27736	0.02736	0.02736	0.50000	2	2	0.05472
1	0.27736	0.02736	0.02736	0.50000	2	2	0.05472
5	0.19403	0.02736	0.02736	0.33333	2	2	0.05472
3	0.30472	0.05472	0.05472	0.50000	2	2	0.10944
2	0.35944	0.10944	0.10944	0.50000	2	2	0.21888
4	0.27611	0.10944	0.10944	0.33333	2	2	0.21888

(d) For each row of Table 1a: the raw misuseability scores, distinguishing factors, and sum of sensitive attribute value weights (Table 1c) rounded to five decimal places.

class, the harder it will be to link specific sensitive attributes to the identities of an equivalence class.

To illustrate how l-Distinguishing Factor can distinguish records, the record scores for the source Table 2a are calculated in Table 2c. From the normalized scores in Table 2b, we can see that M-Score and L-Severity maintain consistent scores for rows 0, 2, 4, and 5 while tkl-Score has greater scores in rows 4-and-5 compared to rows 0-and-2. This observation can be visualized in Fig. 1b.

We can account for the discrepancies between rows 4-and-5 and rows 0-and-2 in tkl-Score by observing the corresponding l-Distinguishing Factor and t-Distinguishing Factor values. In Table 2c, we see that the t-Distinguishing Factor values of rows 4-and-5 are twice as much as the t-Distinguishing Factor values of rows 0-and-2. This increases tkl-Score additively, but does not explain why the tkl-Score of rows 4-and-5 are more than double rows 0-and-2. To explain this multiplicative effect, we examine the l-Distinguishing Factor values of the rows.

Table 2. A case for using l-Distinguishing Factor

id	Job	City	Gender	Initial Diagnosis
0	Lawyer	Calgary	Female	HIV
1	Lawyer	Calgary	Female	Flu
2	Lawyer	Edmonton	Male	HIV
3	Lawyer	Edmonton	Male	Flu
4	Lawyer	Edmonton	Female	HIV
5	Lawyer	Edmonton	Female	HIV

(a) The sensitive attribute of this source table is *Initial Diagnosis*; the quasi-identifiers are *Job, City,* and *Gender.*

Row	tkl-Score	M-Score ($x = 1$)	L-Severity
1	0.0647	0.0417	0.0556
3	0.0647	0.0417	0.0556
0	0.1126	0.1667	0.2222
2	0.1126	0.1667	0.2222
4	0.3227	0.1667	0.2222
5	0.3227	0.1667	0.2222

(b) Misuseability scores for each row of Table 2a normalized against the source table rounded to four decimal places.

Row	tkl-Score	M-Score	L-Severity	DF_t	DF_k	DF_l	Weights
1	0.11069	0.02736	0.02736	0.16667	2	2	0.05472
3	0.11069	0.02736	0.02736	0.16667	2	2	0.05472
0	0.19277	0.10944	0.10944	0.16667	2	2	0.21888
2	0.19277	0.10944	0.10944	0.16667	2	2	0.21888
4	0.55221	0.10944	0.10944	0.33333	2	1	0.21888
5	0.55221	0.10944	0.10944	0.33333	2	1	0.21888

(c) For each row of Table 2a: the raw misuseability scores, distinguishing factors, and sum of sensitive attribute value weights (Table 1c) rounded to five decimal places.

From the equivalence class containing rows 4-and-5 as seen in Table 2a, it can be observed that "HIV" is the only unique sensitive attribute in this equivalence class. In every other equivalence class, there are two unique values. This means that if we knew a female lawyer was from Edmonton, we can deduce that they have HIV if the table were to be released, as opposed to having to distinguish between two different sensitive attribute values with the other equivalence class groupings. Because the equivalence class containing rows 4-and-5 has only "HIV" as the sensitive attribute value, the l-Distinguishing Factor for rows 4-and-5 is

1. As a result, the record scores of rows 4-and-5 are not reduced by a factor of 2 like rows 0-and-2.

4.3   Limitations

The basis of misuseability scores are the sensitive attribute value weights defined by domain experts. Therefore the "misuseability" of data is entirely dependent on how attributes are classified. This drawback is no different than M-Score or L-Severity, but our work shows how misuseability using pre-existing classifications can be more finely determined when integrating additional anonymity measures.

As well, misuseability scores normalized against one source table should not be compared to a score normalized against a different source table. The reason is that source tables can have different sensitivities of attributes, vary in size, and have different attribute values meaning that the records may have different distinguishing factors. However, misuseability scores are beneficial when deciding the misuseability of a subset of records in a source table that have been/will be shared (i.e. which release of a subset may be more risky than another release).

5   Conclusion

In this paper, we extend previous misuseabiltiy scoring by incorporating new distinguishing factors to account for sensitive attribute distinctiveness. We demonstrate how l-Distinguishing Factor and t-Distinguishing Factor in *tkl*-Score can account for data misuse scenarios not detected by previous misuseability scores. Two cases are presented to demonstrate how t-Distinguishing Factor and l-Distinguishing Factor can distinguish records better than k-Distinguishing Factor and make record scores more granular. Because of these new distinguishing factors, *tkl*-Score and *tkl*-Score$_{max}$ are better at characterizing the severity of records compared to L-Severity and M-Score.

Misuseability scoring enables comparisons between different datasets concerning the severity of a release by quantifying the sensitivity of data. Our *tkl*-Score is presented as an improvement to existing misuseability scoring by accounting for attribute disclosure attacks in addition to identity disclosure attacks.

Future work includes investigating systematic methodologies for determining sensitive attribute weights to address the limitation of misuseability scores being dependent on the classification of attributes. As well, applications of *tkl*-Score and *tkl*-Score$_{max}$ should also be investigated for how they might be used to quantify the sensitivity of records leaked in a data breach and indicate the extent of a breach. Furthermore, *tkl*-Score and *tkl*-Score$_{max}$ could be integrated with data-loss prevention systems to monitor for any anomalies in user behaviour when accessing database data by calculating a score each time data is accessed by a user and identifying any extreme scores. *tkl*-Score and *tkl*-Score$_{max}$ could also be used as part of a risk-assessment process to determine, based on a score, which sensitive records could cause issues when released.

References

1. Graham-Harrison, E., Cadwalladr, C.: Revealed: 50 million Facebook profiles harvested for Cambridge Analytica in major data breach. The Guardian (2018). https://www.theguardian.com/news/2018/mar/17/cambridge-analytica-facebook-influence-us-election
2. Harel, A., Shabtai, A., Rokach, L., Elovici, Y.: M-score: a misuseability weight measure. IEEE Trans. Dependable Secure Comput. 9(3), 414–428 (2012)
3. Li, N., Li, T., Venkatasubramanian, S.: t-closeness: privacy beyond k-anonymity and l-diversity. In: 2007 IEEE 23rd International Conference on Data Engineering, pp. 106–115. IEEE (2007)
4. Machanavajjhala, A., Kifer, D., Gehrke, J., Venkitasubramaniam, M.: l-diversity: privacy beyond k-anonymity. ACM Trans. Knowl. Discov. Data (TKDD) 1(1), 3-es (2007)
5. Saaty, T.L.: A scaling method for priorities in hierarchical structures. J. Math. Psychol. 15(3), 234–281 (1977)
6. Sweeney, L.: k-anonymity: a model for protecting privacy. Int. J. Uncertainty Fuzziness Knowl. Based Syst. 10(05), 557–570 (2002)
7. Vavilis, S., Petković, M., Zannone, N.: A severity-based quantification of data leakages in database systems. J. Comput. Secur. 24(3), 321–345 (2016)

Automated Risk Assessment and What-if Analysis of OpenID Connect and OAuth 2.0 Deployments

Salimeh Dashti[1,2]([⊠])(iD), Amir Sharif[1,2]([⊠])(iD), Roberto Carbone[1]([⊠])(iD), and Silvio Ranise[1,3]([⊠])(iD)

[1] Fondazione Bruno Kessler, Trento, Italy
{sdashti,asharif,carbone,ranise}@fbk.eu
[2] DIBRIS, University of Genova, Genova, Italy
[3] Department of Mathematics, University of Trento, Trento, Italy

Abstract. The introduction of the Payment Service Directive (PSD2) has accelerated financial services and open banking growth. Deploying appropriate identity management solutions is crucial. This implies the adoption of secure protocols for authentication and authorization, such as OpenID Connect and OAuth 2.0. The PSD2 also requires the application of the General Data Protection Regulation (GDPR) when transactions involve personal data. In turn, the GDPR mandates a Data Protection Impact Assessment (DPIA) for assessing risks posed to data subjects' rights and freedom. This is a time-consuming and challenging task requiring heterogeneous skills that include the knowledge of best practices for deploying protocols, security mechanisms adopted by available identity management providers, and the capability to perform careful what-if analysis of the possible alternatives. To assist users in this task, we propose a methodology based on the formalization of the what-if analysis as an optimization problem that available tools can solve. The formalization is derived from the OAuth 2.0 and OpenID connects standards, security best practices to mitigate threats, and thorough the evaluation of 19 identity management providers to check their supported features concerning the identified set of features for OAuth/OIDC solutions. We apply the methodology to assist controllers and identify the most appropriate security setup to drive the process of making financial services compliant with the PSD2.

Keywords: Digital identity · PSD2 · OAuth 2.0 · OIDC · GDPR · DPIA

1 Introduction

The growing importance of financial services and the regulatory push by the PSD2 to share user's financial data held by banks with third-party services have made the market more competitive. Although that brings a range of economic opportunities, it comes together with risks such as loss or theft of personal data, data protection violations, etc. One of the key points to make these services trustworthy—as required by PSD2—is to deploy an appropriate identity management solution. OAuth 2.0 (OAuth) and OpenID

© IFIP International Federation for Information Processing 2021
Published by Springer Nature Switzerland AG 2021
K. Barker and K. Ghazinour (Eds.): DBSec 2021, LNCS 12840, pp. 325–337, 2021.
https://doi.org/10.1007/978-3-030-81242-3_19

Connect (OIDC) are two widely used solutions among other identity management solutions. PSD2 states that where personal data is processed—as in the authorization and authentication process—relevant security requirements laid down in the General Data Protection Regulation (GDPR) should be met.

OAuth/OIDC solutions provide a secure and frictionless process [8,19,20]. However, wrong implementation choices of these solutions may result in data breaches that impact the rights and freedoms of data subjects in a large scale. A recent example is the internet bank account takeover of +1M users without user interaction due to an implementation flaw within their OAuth solution [7]. GDPR requires conducting Data Protection Impact Assessment (DPIA) to identify and evaluate risks to data subjects' rights and freedoms where the processing involves a large amount of personal data and affects a large number of data subjects (recital 75). That is true for OAuth/OIDC solutions.

Conducting a DPIA-compliant risk assessment for OAuth/OIDC solutions requires to: (1) assess the risk for both IdMP's deployment and its integration within web applications (hereafter clients); (2) face a maze of documents and guidelines to perform a comprehensive and flawless risk assessment which is a challenging task for non-security experts; (3) be aware of which Best Current Practices (BCPs) to follow that meets the requirements of their clients, e.g., for PSD2 open banking, they need to consider Financial-grade API (FAPI) [16] instead of OAuth/OIDC Core documents [4,17]; and (4) be aware of DPIA requirements to conduct a risk analysis. Meeting such requirements is a daunting task whose burden, according to the GDPR, is on the shoulder of the (data) controller. We propose a methodology to conduct a DPIA-compliant risk assessment for OAuth/OIDC solutions to assist controllers. It is designed to address the requirements of OAuth/OIDC-based financial services and any OAuth/OIDC-based client processing either a special category of personal data or common personal data. Our main contributions are:

- demystifying the maze of OAuth/OIDC documents to create a reference model that characterizes: (i) secure IdMPs deployment and their integration, and (ii) privacy-preserving components to meet DPIA requirements;
- formalizing a what-if DPIA-compliance risk analysis as an optimization problem, using the introduced reference model;
- proposing a methodology to solve the optimization problem and assist controllers in modeling and evaluating risks.

Paper Structure. The remainder of this paper is organized as follows. Section 2 introduces the background concepts used in the paper. Section 3 articulates the problem our methodology solves and our OAuth/OIDC security and privacy reference model. Section 4 presents our proposed methodology. Section 5 concludes the paper and provide some insights for future work.

2   Background

This section discusses the related concepts that make the paper self-contained.

Access Delegation: OAuth Standard. The OAuth authorization framework enables a third-party client to obtain access to a resource server, either on behalf of the resource owner or on its behalf [4]. The client redirects the resource owner through the browser into the IdMP authorization server, where the resource owner performs the authentication. After the successful authentication of the resource owner, the authorization server issues an *access_token* which clients use to access the resource owners' resources in the resource server.

Single Sign-On Login: OIDC Standard. OIDC is an authentication layer developed on top of the OAuth standard, that adds two main features: `id_token` and *userInfo* endpoint. An `id_token` enables the client to verify that the received token is issued for its previous token request. It is a structured JSON token [6] that contains information about: token issuer, subject, and the audience (the intended client); all signed by the OIDC provider (IdP). An *userInfo* endpoint is to obtain identity-related attributes concerning the users (e.g., the email and address).

Privacy Goals. Security goals (confidentiality, integrity, availability, i.e. the CIA) need to be complemented with further privacy goals [3] to evaluate the impact on all aspects of privacy and data protection. They are: data unlinkability, data minimization, purpose specification, transparency, and intervenability. Recent research efforts have come up with these goals [1,3,13,14] to provide an interdisciplinary standard model to assess the consequences of a complex IT systems concerning privacy and data protection [3]. As the CIA are well-known, we discuss only the last three. *Data unlinkability* refers to hiding the link between two or more actions, identities, and pieces of information [22]. *Data minimization* requires avoiding unnecessary data to achieve the determined purpose, that is, *purpose determination*. The mentioned privacy goals are requested by article 6.4.e and 32.1.a. *Transparency* requires data processing to be understandable and reconstructable by concerned individual [1] (article 5.1.a and 12.2). *Intervenability* requires that intervention (for the individual whose data are processed) is possible concerning all ongoing or planned privacy-relevant data processing [1] (article 12.2). To comply with the GDPR, controllers need to address data processing principles (article 5). Beside, *accountability*, *data accuracy* and *storage limitation*, the rest overlap with the privacy goals.

Data Protection Impact Assessment. GDPR requires controllers to carry out a Data Protection Impact Assessment (DPIA). Among others, when the processing involves a large amount of personal data and affects many data subjects. The controller is the natural or legal person, public authority, agency, or other body that determines the purposes and means of processing personal data (article 4). DPIA is a risk-based approach to data protection. Article 35.7 articulates the steps to conduct it; which overlap with risk management steps (e.g., ISO 3100026).

3   Problem Definition and Reference Model

Analyzing risks to the rights and freedoms of data subjects is an important step of a DPIA. Conducting risk analysis for OAuth/OIDC solutions—that are the backbone of clients—is critical but complex. That is due to: many choices of IdMPs with various configuration options, various implementation patterns, many security implications and guidelines documents, and lack of guideline to meet privacy goals (see Sect. 2.3) using OAuth/OIDC components. Therefore, we propose a tool-based methodology to perform a DPIA-compliant risk analysis that requires: to consider the security and privacy level required by the client in question, e.g., for open banking, they need to consider FAPI instead of OAuth/OIDC Core documents; and to meet the privacy goals. As such, we identify the following problem:

P_{risk} *security and privacy risk assessment and what-if analysis, taking into account the risk propagation.*

Below, we introduce a reference model of OAuth/OIDC solutions (Sect. 3.1). We briefly describe how the usage of recommended components by the OAuth WG and OIDF can help to achieve the proper level of security and privacy (Sect. 3.2). Finally, we formalize the identified problem (Sect. 3.3).

3.1   OAuth/OIDC Reference Model

To build the reference model, we studied OAuth/OIDC security documents (e.g., [4,17]) to extract the components that protect OAuth/OIDC deployment, and satisfy privacy and security goals (introduced in Sect. 2.3). We call these components *atomic features* (*af*), listed in Table 1. Each *af* represents either: (*i*) OAuth/OIDC parameter name (e.g., nonce), (*ii*) OAuth/OIDC parameter value (e.g., code), (*iii*) OAuth/OIDC functionality aspects (e.g., full redirect_uri), or (*iv*) client specific implementation-related tasks (e.g., Id token validation). A detailed definition for each *af* in the reported categories is provided in our companion website[1]. *af*s are to be set/implemented either: (*i*) in Authorization/Token requests; (*ii*) in the IdMP developer console, set by controllers; (*iii*) parameters set in the Header; or (*iv*) checks implemented by the controller.

Table 1 represents that by column *Deployment Place*. For example, controller needs to set *subject identifier type*[2], that could be either pairwise or public, in the IdMP developer console. Note that, due to the page limit and simplifying the table, we only consider the *Authorization Code* flow, and interested readers can refer to our companion website (see footnote 1) for the completed reference model.

The OAuth Working Group (WG) and the OpenID Foundation (OIDF) recommends different *af*s—with different levels of contributes (*protection level*)—to achieve a common goal for varied use-case scenarios. We group together such *af*s, and call them *composed feature* (*cf*). A client can implement one *af* from a *cf*, per request. For example, request, request_uri and query are grouped together under *cf Request*. They are introduced to meet goal *access token confidentiality*, by protecting authorization request against the threat *obtain code* (see Table 1). While *af* request and request_uri pass OAuth/OIDC requests in a signed and optionally encrypted single, self-contained parameter manner, the *af* query passes it directly in the URL. Thus, the

---

[1] https://sites.google.com/fbk.eu/oidc-dpia.

[2] A locally unique and never reassigned identifier within the Issuer for the user, which is intended to be consumed by the Client.

Table 1. OAuth/OIDC reference model

Deployment place	Atomic features	Threat (T)	Goal (G)	Consecutive T	Consecutive G	PL	LL
		Security Feature					
	code					3	
	token					1	
	client_credentials	Obtain AT	PD Conf	-	-	3	3
	password					1	
	hybrid					3	
	nonce	Obtain AT , Session misuse	PD Conf	-	-	5	5
	state	Session misuse	PD Conf	-	-	5	5
Authorization request	request					5	
	request_uri	Obtain code	AT Conf	Obtain AT	PD Conf	5	5
	query					1	
	form_post					5	
	fragment	Obtain code	AT Conf	Obtain AT	PD Conf	1	5
	query					1	
	Code challenge	Obtain AT , Session misuse	PD Conf	-	-	5	5
	plain	Obtain AT , Session misuse	PD Conf	-	-	1	5
	s256					5	
	mtls					3	
	client_secret_jwt					3	
Token request	private_key_jwt	Obtain AT	PD Conf	-	-	3	3
	client_secret_basic					2	
	cleint_secret_post					2	
	code_verifier	Obtain AT , Session misuse	PD Conf	-	-	5	5
Authorization request / Token request	full redirect_uri[idmp]	Obtain code	AT Conf	Obtain AT	PD Conf	3	3
	pattern redirect_uri[idmp]					1	
Header	binding IdMP metadata[cl]	Obtain AT	PD Conf	-	-	2	2
Console	referrer[cl]	Obtain code	AT Conf	Obtain AT	PD Conf	3	3
	distinct redirect_uri[cl]	Obtain code	AT Conf	Obtain AT	PD Conf	5	5
	open redirect validation[cl]	Obtain code	AT Conf	Obtain AT	PD Conf	5	5
Client check	state validation[cl]	Obtain code	PD Conf	-	-	5	5
	storing IdMP metadata[cl]	Obtain code	AT Conf	Obtain AT	PD Conf	5	5
	issuer validation[cl]	Obtain code	AT Conf	Obtain AT	PD Conf	5	5
	id token validation[cl]	Impersonation	PD Conf	-	-	5	5
		Privacy Features					
	claims	Comp data mini	Data mini	-	-	5	5
	scope	Comp data mini	Data mini	-	-	5	5
	purpose	Comp. Pur spec, Trans	Pur Spec, Trans	-	-	5	
Authorization request	verified_claims	Comp. PD accuracy	PD accuracy	-	-	5	5
	vot	Impersonation, Comp PD accuracy	PD accu, PD Conf	-	-	5	
	acr					5	5
	login					3	
	select_account	Comp transparency	Transparency	-	-	3	3
	consent					3	
Console	pairwise[cl]	Linkability	Unlinkability	-	-	5	
	public[cl]					1	5

LEGEND. idmp: idmp only feature, cl: client only feature, comp: compromise, PD: personal data, AT: access token, trans: transparency
data mini: data minimization, conf: confidentiality, pur spec: purpose specification, PL: protection level, LL: likelihood level

first two provide a higher protection level as they provide request integrity and confidentiality. We have used this reasoning to assign protection level to the *af*s. Thus, the protection level for the discussed *af*s are 5, 5, and 1, respectively. This is an important consideration, as it allows controllers to make an informed decision on the IdMP s/he chooses (not all IdMPs support all the *af*s), or/and the *af*s they decide to implement. When an *af* is not implemented, the protection level against its related threat(s) will decrease and adds to the likelihood level of the threat(s) to pose. Such that, the likelihood level of *af*s are equal to their protection level. While in case of *cf*s, the likelihood level will be the maximum value among its *af*s Protection/likelihood levels range from 1 to 5. Controllers can modify the likelihood and protection level if needed. Please find the details about the evaluation of the protection level of all *af*s in our website (see footnote 1).

Table 2. Composed features.

Composed Feature	Related Atomic Features
Response type	{code,token,client_credentials,password,hybrid}
Request	{request,request_uri,query}
Response mode	{form_post,fragment,query}
Code challenge method	{plain,SHA256}
Client authentication	{mTLS,client_secret_jwt,private_key_jwt,client_secret_basic,client_secret_post}
Redirect uri	{full redirect_uri,pattern redirect_uri}
Identity proofing	{vot,acr}
Prompt	{login,select_account,consent}
Subject type	{public,pairwise}

Each *af/cf* is introduced to protect OAuth/OIDC deployment against a threat to satisfy privacy/security goals. A threat could expose the system to another threat, which itself compromises another goal. We call them *consecutive threats* and *consecutive goal*. They are as likely to raise as their main threat. For example, as represented in Table 1, not implementing *af* request leads to threat *obtain code*, which itself let the attacker to *obtain access token*. They relate to goal *access token confidentiality*, and *personal data confidentiality*, respectively.

3.2 Best Current Practice Specification

To assist clients and IdMPs in achieving appropriate security and privacy levels based on their operating domain, the OAuth WG and the OIDF have published a set of BCPs in [4,9,10,16,17]. Depending on the domain the BCPs mandate to use some optional *af*s; and for *cf* to use an *af* over the others. For example, [16] requires using optional *af* state.

The OAuth WG and the OIDF do not provide any specific privacy considerations to meet privacy goals (See Sect. 2.3). However, to comply with the DPIA requirements, controllers need to address them. We have systematically studied the following documents [5,12,21] to provide easy-to-implement privacy recommendations for controllers based on the reported *af*s in Table 1. For the sake of brevity, we omit the full explanation and only give a couple of examples. As reported in Table 2 *cf* Response type comprises of *af*s code, token, client, credentials, password, and hybrid. However, the BCPs enforces the usage of *af* code among the other *af*s because it does not return the *access_token* in the front channel and it can be protected by *af*s code challenge and code_verifier. Concerning privacy, controllers can achieve privacy goal *purpose specification* by using *af* purpose to state the purpose of asking each individual claim.

3.3 Problem P_{risk} Definition

In this section, we formalise the problem, namely P_{risk}, considering the reference model reported in Table 1.

Let $\mathcal{AF}$ be the set of *af*s associated with an OAuth/OIDC deployment shown in the second column of Table 1. We split the set $\mathcal{AF}$ in three disjoint subsets: $\mathcal{AF} = \mathcal{AF}_{common} \cup \mathcal{AF}_{idmp} \cup \mathcal{AF}_{cl}$. The set $\mathcal{AF}_{common}$ includes the *af*s that the client cannot implement unless the IdMP supports them, like nonce.

An IdMP can support more than one af from a cf, while the client can implement only one for a given request. The set $\mathcal{AF}_{idmp}$ includes afs that IdMPs need to enforce (marked with the "idmp" superscript in Table 1) and clients need only to adopt, like pattern redirect_uri. The set $\mathcal{AF}_{cl}$ includes the afs that clients need to implement and checks that they need to perform (marked with the "cl" superscript in Table 1), like issuer validation.

Let $support_{idmp}$ be a Boolean mapping from the set $\mathcal{AF}_{common} \cup \mathcal{AF}_{idmp}$ for $idmp$ ranging over a given set of IdMPs. An af in $support_{idmp}$ maps to a true value iff $idmp$ supports the af. Let $implement_{cl}$ be a Boolean mapping from the set $\mathcal{AF}_{common} \cup \mathcal{AF}_{cl}$ for a given client cl. An atomic feature af maps to a true value iff cl implements af.

Notice, for $af \in \mathcal{AF}_{common}$, we say that $implement_{cl}$ is <u>constrained</u> by $support_{idmp}$, that is, $implement_{cl}$ is strictly dependent on $support_{idmp}$. Other words, $af \in \mathcal{AF}_{common}$ can map to true only if it maps to true in $support_{idmp}$. Indeed, controllers can decide whether to implement some atomic features among the one supported by $idmp$.

Let $C\mathcal{F}^* \subset \mathbf{P}(\mathcal{AF})$ (where $\mathbf{P}$ stands for the power set) be a set of sets of afs, including the cfs (see Table 2) as well as a set $\{af\}$ for each atomic feature $af \in \mathcal{AF}$ that does not belong to any composed feature in Table 2. Thus, $C\mathcal{F}^* = \{\ldots, \{\text{nonce}\}, \{\text{state}\}, \{\text{request}, \text{request\_uri}, \text{query}\}, \ldots\}$. Note that all the sets $S \in C\mathcal{F}^*$ are pairwise disjoint and $\bigcup_{S \in C\mathcal{F}^*} S = \mathcal{AF}$, that is $C\mathcal{F}^*$ is a partition of $\mathcal{AF}$. Let $\mathcal{T}$ be the set of threats and consecutive threats, listed in columns 3 and 5 of Table 1. Let $\mathcal{G}$ be the set of goals and consecutive goals, listed in columns 4 and 6 of Table 1. We thus define the following mappings and relations:

- Let p be a mapping from $\mathcal{AF}$ to the set $\{1, \ldots, 5\}$ of protection levels. The definition of this mapping can be obtained by considering the features in column 2 and the corresponding protection level in column 7 of Table 1.
- Let the likelihood mapping ℓ be a mapping from $C\mathcal{F}^*$ to the set $\{1, \ldots, 5\}$ of likelihood levels. The definition of this mapping can be obtained by considering the (sets of) atomic features in column 2 and the corresponding likelihood level in column 8 of Table 1.
- Let i be a mapping from the set $\mathcal{T}$ of threats to the set $\{1, \ldots, 5\}$ of impact levels. The definition of this mapping is decided by the controller and depends on the particular scenario in which the OAuth/OIDC solution is deployed.
- Let $CF2T \subseteq C\mathcal{F}^* \times \mathcal{T}$ be a relation between each composed feature $cf \in C\mathcal{F}^*$ and its related threat. The pairs in this relation can be defined by reading the elements reported in columns 2 and 3 of Table 1.
- Let $T2G \subseteq \mathcal{T} \times \mathcal{G}$ be a relation between each threat and the goal compromised by the threat itself. The pairs in this relation can be defined by reading the elements reported both in columns 3 and 4, and 5 and 6 of Table 1. Indeed, the relation between a threat and the goal is independent of the fact that the threat is consecutive to another threat or not.
- Let $T2CT \subseteq \mathcal{T} \times \mathcal{T}$ be a relation between a threat and its consecutive threat. The pairs in this relation can be defined by reading the elements reported in columns 3 and 5 of Table 1. We also use the notation $CT(t)$ to indicate the set of threats consecutive to the threat t, and $CT(T) \triangleq \bigcup_{t \in T}(CT(t))$.

The problem P_{risk} consists in finding the *idmp* and mapping *implement_{cl}* such that

$$min_{idmp,implement_{cl}} \mathcal{R}(p,\ell,i,support_{idmp},implement_{cl},CF2T,T2G,T2CT) \quad (1)$$

subject to $idmp \in$ IdMPs and $cl \in ClAdm$, where IdMPs is a set of IdMPs that support certain features and *ClAdm* is the set of admissible mappings for a given client. The definition of the set IdMPs and the *ClAdm* derive from external considerations performed by the controller of the client. For instance, the choice of an IdMP can be affected by the costs of IdMPs' services or supported features. Similarly, the controller can consider that some features—among the ones constrained (as defined above) by the selected *idmp*—will take longer to implement or charge more in terms of costs. Thus, the controller can further constrain the admissible *implement_{cl}* mappings accordingly.

$\mathcal{R}$ is thus an operator that returns the risk level given the selected *idmp*, the configuration of the client *cl*, the protection and likelihood mappings and the impact level while considering how the risk propagates among the various components by using relations *CF2T*, *T2G*, and *T2CT*.

As a final remark, note that the problem P_{risk} is solved by considering $p, l, i, CF2T$, *T2G*, and *T2CT* (obtained from Table 1), while *support_{idmp}* and *implement_{cl}* range over all possible values in the sets IdMPs and *ClAdm*. Therefore, the problem is decidable because it can be expressed as a combinatorial optimization problem with finitely many possibilities depending on the number of considered IdMPs, features supported by clients in *ClAdm* and on the cardinality of the considered atomic features in $\mathcal{AF}$: all the possibilities can be enumerated, the corresponding risk evaluated and the minimum value selected. Indeed, solving problem P_{risk} can be mechanized by using any available tools that are capable of solving optimization problems by specifying how the function to be minimized (the risk in our case) depends on the arguments of the operator $\mathcal{R}$.

4   OAuth/OIDC Solution DPIA-Compliant Risk Analysis

This section discusses our methodology to address the reported problem.

Running Example. *Graphy* is a client that gets the users' financial data from their bank accounts and makes a graphical representation of their financial status. Users need to make a profile with the *Graphy* and connect their bank accounts to it. *Graphy* utilizes OAuth/OIDC solutions to provide a smooth single sign-on login and access delegation experience for its users. This scenario is inspired by the example provided in [15].

Addressing P_{risk}. The problem assesses the risks to rights and freedoms of data subjects, for which controllers need to meet security and privacy goals (introduced in Sect. 2.3). This section details our three-step approach to address the problem, namely: *(I) assessing client, (II) evaluating risks of employed IdMPs*, and *(III) modeling and treating risks*.

I. Assessing Client. This sub-step identifies the roles; namely data subject, controller, and data processor; and data type category. Controller decides on data processing details and is responsible to demonstrate compliance with GDPR; data processor processes personal data on behalf of the controller. In this context, the client developer is the controller and, IdMPs are the data processor. We provide three ways for controllers to identify the employed IdMPs: finding the employed IdMP from list of popular IdMPs, indicating discovery endpoint URL, replying to a questionnaire. To identify data subjects and data type, we use the approach introduced in [2], that is, through specifying the economic sector. We call *Sensitive sector* the *sectors* that process special category of personal data and/or involve vulnerable data subject; and *non-sensitive sector* otherwise.

DPIA considers the impact of data processing *high*, when a large scale of data is involved, which impacts large scale of the data subject. For the sake of simplicity, we consider the impact level as a constant value 5 corresponding to a very high impact. *Application of the step on the scenario.* In this scenario, Graphy is the controller. The service it provides belongs to *Sensitive* sector as it processes financial data. As such, the impact is 5. Data subjects are natural persons. The controller employs *IBM* for single sign-on login and allows users of *Barclays* bank to link their bank accounts through access delegation; they are data processors.

II. Evaluating Risks of Employed IdMP(s). This sub-step assesses employed IdMPs and the implementation details w.r.t. their integration within the client, to identify any threat(s) they may pose to the right and freedoms of data subjects. For that, we introduce the following two components: *IdMP Processor* and *Client Processor*, detailed below. *IdMP Processor.* The processor assesses which are the features $af \in \mathcal{AF}_{common} \cup \mathcal{AF}_{idmp}$ supported by the employed IdMP(s). Thus, more formally, the IdMP processor aims to specify the $support_{idmp}$ Boolean mapping for the employed *idmp*.

Our methodology provides a database of known IdMPs. If the controller selects one of them, the known truth values are automatically retrieved. Otherwise, we identify two solutions to collect the missing information about the *af*s of an *idmp*. One option for the controller is to provide the IdMP's discovery endpoint URL (if available). Then, our tool sends an HTTP Get request to the endpoint and automatically extracts the information from the JSON response (e.g., subject_type_supported). (Please refer to the companion website (see footnote 1) for the list of the *af*s included in the JSON response). Finally, we ask the controller to reply to a dynamically generated Yes/No questionnaire to collect only the information that is still missing. To generate the questionnaire, we use the reference model (see Sect. 3.1), by filtering out the known information. For example, it asks "Does the IdMP support full redirect_uri?", "Does the IdMP support nonce?". Please find the details of the questionnaire in our website (see footnote 1).

The database of known IdMPs gets updated whenever a new IdMP is introduced either via the discovery endpoint or the questionnaire. To know the status of popular IdMPs and to start filling in the database, we have selected 19 popular IdMPs, according to their Alexa ranks and whether they have a free developer console to assess their features. The IdMPs are taken from the OIDF website [11]: for *Sensitive sector* they are taken from the list of *Certified FAPI OIDC*, and for *Non-sensitive sectors* from *certified*

OIDC Core. For each of them we studied the available documentation and/or the developer console to assess their features. Table 3 provides an excerpt of the IdMPs we have studied; the full table is available in our website (see footnote 1). Interesting enough, we discovered that the selected IdMPs do not provide the same level of security and privacy. For instance, *Yahoo* does not support the following *af*s: Plain, S256, claims, purpose, acr, verified_claims and vot, while the *IBM solution* supports Plain, S256 and acr.

Client Processor. The Client Processor aims to assess which features $af \in \mathcal{AF}_{common} \cup \mathcal{AF}_{cl}$ have been implemented by the client cl in the OAuth/OIDC solutions. Other words, it specifies the *implement*$_{cl}$ boolean mapping.

   To support the controller in this process, our methodology provides a Yes/No questionnaire. By leveraging the *support*$_{idmp}$ (obtained from the IdMP Processor) our approach automatically takes into account the choices that are constrained (as defined in Sect. 3.3) by the employed *idmp* and selects the relevant questions for the client. For example, it asks *"Have you implemented id token validation?"*, *"Have you set referrer parameter?"*.

Application of the Step on the Scenario. Given that *IBM* is in our database of known IdMPs then most of truth values of the features related to *IBM* are already available. For the missing information concerning *IBM* and the whole information concerning *Barclays*—that is not among the IdMP we analyzed—the controller has to reply to the questionnaire. The details about the collected information are on the website (see footnote 1).

III. Modeling and Treating Risks. The information collected from the previous two steps in Sect. 4.I and Sect. 4.II (namely the impact i, *support*$_{idmp}$, and *implement*$_{cl}$) as well as the information reported in Table 1 are used to model and evaluate the risk. Any risk modeling tool that can perform a what-if analysis can be employed. The what-if analysis allows controllers to observe how the risk level changes by considering different implementation choices. A good option is the RiskML tool [18], which provides a modeling language and a quantitative reasoning algorithm to analyze models, taking into account the risk propagation.

   Figure 1 illustrates an excerpt of the risk model, by considering, for simplicity, only two *af*s: referrer and mtls. We are assuming that, according to *implement*$_{cl}$, a controller did not implement referrer (dashed line in Fig. 1) and implemented mtls. As shown in Table 1, not implementing referrer leads to the threat *Obtain Code* (with the likelihood level 3), that consecutively allows an attacker to also *Obtain Access Token*. As such, it impacts the goals *confidentiality of the access token* and *confidentiality of the personal data*, respectively. At the same time, mtls is implemented, and thus contributes to protect the *access_token* (the protection level is 3), and, as a consequence, the *confidentiality of the personal data*.

Table 3. An excerpt list of known IdMPs and their supported features.

Atomic Feature	IBM	G	box	Y!
code	✓	✓	✓	✓
implicit	✓	✓	✗	✓
client credential	✓	✗	✗	✗
password	✓	✗	✗	✗
hybrid	✓	✗	✗	✗
mtls	✓	✗	✗	✗
client_secret_basic	✓	✓	✓	✓
client_secret_post	✓	✓	✓	✓
client_secret_jwt	✗	✗	✗	✗
private_key_jwt	✓	✗	✗	✗
plain	✓opt	✓opt	✗	✗
S256	✓opt	✓opt	✗	✗
request	✓	✗	✗	✗
request_uri	✓	✗	✗	✗
query	✓	✓	✓	✓
claim	✓	✗	✗	✗
scope	✓	✓	✓	✓
purpose	✗	✗	✗	✗
verfied_claims	✗	✗	✗	✗
acr	✓	✗	✗	✗
vot	✗	✗	✗	✗
public	✓	✓	NA	✓
pairwise	✗	✗	NA	✗
Sensitive Sector	*No*	*No*	*No*	*No*

LEGEND: ✓opt: Supported, but not enforced, NA: Not applicable

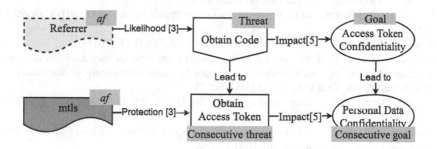

Fig. 1. An excerpt representation of risk model.

This simple excerpt shows that the effects of the implementation choices (expressed in terms of *af*s) propagate on threats and goals, and that the positive (protection) and negative (likelihood) effects contribute to the final risk level associated to each goal. Then, the specific operations to quantify the effects and calculate the risk levels are dependent on the employed risk assessment tool. Of course, when considering the whole set of *af*s the analysis is much more complex. By enabling controllers to perform a what-if analysis, our methodology allows them to take informed decisions on their IdMP and implementation choices.

5 Conclusion

Conducting a DPIA-compliant risk analysis for OAuth/OIDC solutions is complex. To assist controller with this task, we define a DPIA-compliant risk analysis as a security and privacy risk analysis (P_{risk}) problem, and proposed a methodology to solve it. The methodology is supported by a reference model that captures the OAuth/OIDC features that are required to solve the problem, respecting the security and privacy level of solution in question. Our analysis of the solution outputs a risk model that captures the system as-is, and provide the possibility to perform what-if analysis. Performing what-if analysis enables controllers to make an informed decision on their choice of IdMP and implementation to eliminate identified risks or minimize their impact. The model can be input into any risk model that can perform what-if analysis. As future work, we plan to extend the methodology to assess risks posed by (other components of) clients and introduce security controls and privacy-enhanced technology to address them.

References

1. Danezis, G., et al.: Privacy and data protection by design-from policy to engineering. arXiv preprint arXiv:1501.03726 (2015)
2. Dashti, S., Ranise, S.: A tool-assisted methodology for the data protection impact assessment. In: Proceedings of the International Conference on Security and Cryptography (2019)
3. Hansen, M., Jensen, M., Rost, M.: Protection goals for privacy engineering. In: IEEE SPW (2015)
4. Hardt, D.: The OAuth 2.0 authorization framework. IETF (2012)
5. Internet-Draft: International Government Assurance Profile (iGov) for OpenID Connect 1.0 (2018)
6. Jones, M., Bradley, J., Sakimura, N.: Json web token (JWT). IETF (2015)
7. Krebs, B.: Internet bank account takeover of +1m users without user interaction. https://mrbriankrebs.medium.com/internet-bank-account-takeover-of-1m-users-without-user-interaction-4fc9141740a3. Accessed 25 Mar 2021
8. Li, W., Mitchell, C.J.: User access privacy in OAuth 2.0 and OpenID connect. In: EuroS&PW. IEEE (2020)
9. Lodderstedt, T., Bradley, J., Labunets, A., Fett, D.: OAuth 2.0 Security Best Current Practice (draft-ietf-oauth-security-topics-16). IETF (2020)
10. Lodderstedt, T., McGloin, M., Hunt, P.: RFC 6819: OAuth 2.0 threat model and security considerations. IETF (2013)
11. OpenID Foundation: Financial-grade API - part 1: Baseline security profile. https://openid.net/certification/. Accessed 23 Nov 2020
12. Richer, J., Johansson, L.: Vector of trust (RFC 8485). IETF (2018)
13. Rost, M., Bock, K.: Privacy by design and the new protection goals. In: DuD, vol. 2009 (2011)
14. Rost, M., Pfitzmann, A.: Datenschutz-schutzziele–revisited. Datenschutz und Datensicherheit-DuD **33**(6), 353–358 (2009)
15. Sakimura, N.: Authorization delegation: a financial accounts aggregation use case. https://nat.sakimura.org/2016/01/29/authorization-delegation-a-financial-accounts-aggregation-use-case/. Accessed 25 Mar 2021
16. Sakimura, N., Bradley, J., Jay, E.: Financial-grade API - part 1: Baseline security profile. Accessed 23 Nov 2020

17. Sakimura, N., Bradley, J., Jones, M., De Medeiros, B., Mortimore, C.: OpenID connect core 1.0 incorporating errata set 1. The OpenID Foundation 335 (2014)
18. Siena, A., Morandini, M., Susi, A.: Modelling risks in open source software component selection. In: Yu, E., Dobbie, G., Jarke, M., Purao, S. (eds.) ER 2014. LNCS, vol. 8824, pp. 335–348. Springer, Cham (2014). https://doi.org/10.1007/978-3-319-12206-9_28
19. Similartech.com: Login providers. https://www.similartech.com/categories/login-provider. Accessed 29 Dec 2020
20. Sun, S.T., Beznosov, K.: The devil is in the (implementation) details: an empirical analysis of OAuth SSO systems. In: Proceedings of ACM ASIACCS (2012)
21. Torsten, L., Daniel, F.: OpenID connect for identity assurance 1.0. https://openid.net/specs/openid-connect-4-identity-assurance-1_0.html. Accessed 19 June 2019
22. Wuyts, K., Scandariato, R., Joosen, W., Deng, M., Preneel, B.: LINDDUN: a privacy threat analysis framework (2019)

Divide-and-Learn: A Random Indexing Approach to Attribute Inference Attacks in Online Social Networks

Sanaz Eidizadehakhcheloo[1], Bizhan Alipour Pijani[2(✉)], Abdessamad Imine[2], and Michaël Rusinowitch[2]

[1] Sapienza Università di Roma, 00185 Roma, Italy
eidizadehakhcheloo.1772528@studenti.uniroma1.it
[2] Lorraine University, CNRS, Inria, 54506 Vandœuvre-lès-Nancy, France
{bizhan.alipourpijani,abdessamad.imine,rusi}@loria.fr

Abstract. We present a Divide-and-Learn machine learning methodology to investigate a new class of attribute inference attacks against Online Social Networks (OSN) users. Our methodology analyzes commenters' preferences related to some user publications (e.g., posts or pictures) to infer sensitive attributes of that user. For classification performance, we tune Random Indexing (RI) to compute several embeddings for textual units (e.g., word, emoji), each one depending on a specific attribute value. RI guarantees the comparability of the generated vectors for the different values. To validate the approach, we consider three Facebook attributes: gender, age category and relationship status, which are highly relevant for targeted advertising or privacy threatening applications. By using an XGBoost classifier, we show that we can infer Facebook users' attributes from commenters' reactions to their publications with AUC from 94% to 98%, depending on the traits.

Keywords: Social networks · Privacy · Attribute inference attack · Random indexing

1 Introduction

Although OSN users are getting more cautious about their privacy, they remain vulnerable to attribute inference attacks where the principle is to illegitimately gain private attributes (such as age, gender or political orientation) from publicly available information. In general, attribute inference attacks are either based on data directly generated by the target user (such as pseudos) or data obtained by exploring the user vicinity network. The need to protect the sensitivity of some attributes compels users to conceal information that may disclose them. For instance, the authors of [6] investigate Facebook users' privacy awareness and show that age has the lowest exposure rate. Less than 3% of the users (about

This work is funded by DigiTrust (http://lue.univ-lorraine.fr/fr/article/digitrust/).

ⓒ IFIP International Federation for Information Processing 2021
Published by Springer Nature Switzerland AG 2021
K. Barker and K. Ghazinour (Eds.): DBSec 2021, LNCS 12840, pp. 338–354, 2021.
https://doi.org/10.1007/978-3-030-81242-3_20

495k users) reveal their age, which shows the sensitivity of this attribute. They also show that half of their members hide their gender and 37.3% conceal their friend list. Therefore collecting user-generated data is a difficult task in a real scenario. We note that most attribute inference attacks get inoperative in the case where no data are provided by target users in their profiles. For instance many attribute inference systems are based on network homophily [25] and do not apply without the availability of friend lists.

Even with more user awareness, the privacy risks can come from other sources. Indeed, we will show that an attribute can be inferred from non-user generated data such as the reactions of other social media users on the target user's shared contents. Users in OSN share contents, so-called *publications*, to give people a better sense of who they are and what they care about. For example, Twitter users share a tweet to express their opinion. As for Facebook, users publish pictures, posts or status updates to display their beliefs, favorite brands, grab attention and nourish relationships. While many users hide their sensitive attributes (e.g., gender, age or political view), these publications are still available to the public. However, many users do not realize that even though they are cautious about their writing style, other users' reactions to their publications can reveal their sensitive information. As a typical reaction, people spontaneously *comment* on publications. This option allows them to engage and impress their personal opinion, which might create a sense of connectedness between the publication author and commenters (especially when the commenters are friends). We consider these comments as part of the publication metadata. We also consider *alt text* as metadata which is freely available textual information describing the picture contents (faces, objects, and other themes) and which is generated by some OSN platforms (like Facebook) for blind people who use screen readers.

In this paper, we describe how to infer attributes from publications metadata. We suspect that OSN users' publications metadata convey sensitive information, even though the users did not contribute to generating them directly. We will show that a target user's attribute can indeed be leaked from other users' reactions (e.g., comments) to the target user's publications, even without analyzing the publication content. Note that this new type of attack is difficult to defeat as the user has no control over other users' reactions. Demonstrating the risks of attribute inference attacks raised by publication metadata gives us concrete grounds for alerting about their sensitivity.

A Central Problem and Our Solution. An important problem when learning attributes from metadata is that the same term often appears with different contexts in different attribute values. Such a term will be called an *overlapped* term and the value-specific contexts will be called *non-overlapped* contexts. Example 1 shows female and male-owned pictures with generated alt-text to describe the picture content and comments posted by commenters. The tag *1person* and the word *baby* are overlapped terms, while *you*, *miss*, and *cray* are non-overlapped contexts. The commenters employ different neighboring words for the term *baby* when commenting on female and male-owned pictures, which demonstrates the commenters' usage preferences. As a result, there is a variation in style and usage of the same term in the context of different attribute values.

Example 1. Metadata of two pictures.

Metadata of an image published by a female user
Generated alt-text: 1person
Comment: miss you baby

Metadata of an image published by a male user
Generated alt-text: 1person
Comment: cray cray baby 😆

To discover these variations, we apply a semantic space model, known as Distributional Semantic Model (DSM). Classical word embeddings (such as word2vec [20]) uncover the semantic relations among terms by scanning through the whole corpus and detecting co-occurrences in a fixed context window. They build a global view of terms co-occurrence in the entire dataset. In Example 1, *baby* is used as a romantic term of endearment in the female-owned picture, whereas in the male-owned picture it is about making fun of and teasing the picture owner. Hence generating a vector for each word using the entire dataset can mix and combine many possible word contexts.

We need to adapt the distributional semantic model so that an *overlapped* term with *non-overlapped* context will get different corresponding vector representations. These various representations should be comparable since attribute prediction often relies on computing similarities between vector representations of terms and users. However, due to different random initialization processes on the sub-datasets (corresponding to distinct attribute values), the generated vectors are not comparable by the standard similarity measures, such as cosine similarity [13]. To avoid this problem, we apply Random Indexing (RI) [26], which is an incremental and scalable method for constructing a vector space model. RI requires few computational resources for similarity computations and allows comparison of word spaces created over different attribute values [2]. Accordingly, from the same term, we can generate vectors in different attribute values and compare them.

As a case study, we conduct an intensive analysis of three Facebook attributes: age, gender, and relationship status, as they are recognized as key privacy concerns in the Internet era. Some Facebook users choose to hide gender to camouflage themselves against stalking, sexual harassment, or reducing discrimination [28]. Our attack receives promising results as we preserve the vector representation of each word for each attribute value. The result confirms that splitting the dataset can boost the attacker's performance.

Paper Organization. In Sect. 2 we review the related works. Section 3 presents our Divide-and-Learn methodology to incorporate attribute values in word vector generation. We outline our attribute inference attack steps in Sect. 4. We present a case study in Sect. 5 to evaluate our attacks. Finally, we conclude in Sect. 6 by discussing the capability of our attribute inference attacks on other OSN platforms and giving possible future work.

2   Related Works

User profiling based on their available data on social media has obtained remarkable attention in the past decade. It is a key ingredient of recommendation systems. Researchers have investigated popular social media platforms and have leveraged all possible available data such as content sharing [3], friendship [9], behavior [17] to perform their attribute inference attacks. [29] considers users' purchase data for predicting multiple demographic attributes simultaneously. The authors of [1] show how an attacker can leverage seemingly harmless interests to reveal sensitive information about users. In particular, they infer user private attributes based on music interest similarities. [30] shows that a movie rating recommender system can infer the user's gender without additional metadata. [10] combines network structure and node attribute information to perform link prediction and attribute inference attacks. These are motivated by the observed interaction and homophily between network structure and node attributes. An active attack on privacy-preserving publication of social graphs is presented in [19]. Demographic variable attacks on Twitter users based on whom they follow are presented in [4]. Research efforts have been also focused on user writing style (i.e., users' messages, posts, and status updates) [21] and word usage [27] to infer undisclosed attributes. They apply language analysis to the text generated by users and implement machine learning approaches to achieve the attack.

To sum up, the mentioned works mainly require exploration of user vicinity networks (e.g., friend lists) and digital records (e.g., profile attributes, joined groups and liked pages), which might be unavailable in a real scenario or computationally costly to collect. This personal data as well as writing styles can be modified by the target user to escape inference attacks. In contrast, we infer target user attributes from commenters and Facebook generated data (both called here *publication metadata*). This metadata is easily available. Since our approach does not need to explore the target user's vicinity network, groups and pages, inference attacks are efficient and can even be launched *online*.

Our work is related to [5, 23] where gender and age are inferred from Facebook picture metadata. However, here we do not limit ourselves to specific attributes and our attacks apply to many social media by leveraging commenters' preferences related to pictures, posts, and status updates. Our approach is inspired by recent techniques for analyzing word semantic changes over time [8].

3   Divide-and-Learn Methodology

In this section, we first explain how the training dataset is divided into sub-datasets according to the different values of the attribute to be inferred. Next, we introduce the distributional semantic space and our proposed value-based random indexing approach. The notations used in this paper are summarized in Table 1.

Table 1. Notations.

Notations	Descriptions
D	collected training dataset
U	set of users
$u \in U$	user in U
W	vocabulary of the training dataset
$w \in W$	word in W
$\boldsymbol{w}$	distributional vector of w
c	context
$C(w)$	set of contexts of w in D
$P(u)$	set of publications of $u \in U$
l	an attribute
l_m	mth value of l
U_m^l, U_m	set of users s.t. attribute l has mth value (U_m when l is implicit)
W_m	set of words in comments of publications from users in U_m
$C_m(w)$	set of contexts of w in D_m
$\boldsymbol{u}$	vector for u

3.1 Dividing Training Datasets

Here, we introduce our splitting conditions and their computation. In the following we use *term* as a shorthand for *word/emoji/tag*.

Criteria for Splitting. We can argue that splitting is not beneficial if the commenters use (i) a majority of different terms while commenting on users' publications in different attribute values, or (ii) more frequently the same terms co-occurring more often with similar contexts than dissimilar ones (see Subsect. 5.3). Examples 2 and 3 illustrate cases where splitting the dataset would not be beneficial.

Example 2. Metadata of two pictures.

Metadata of an image published by female user
Generated alt-text: 2people
Comment: Wooooooow, its NICE

Metadata of an image published by male user
Generated alt-text: outdoor, sunglasses
Comment: look at the long beard

Example 3. Metadata of two pictures.

Metadata of an image published by female user
Generated alt-text: selfie, closeup
Comment: great picture

Metadata of an image published by male user
Generated alt-text: selfie, closeup
Comment: great picture

However, in the complementary cases, our experiments have shown the neat benefit of splitting for accuracy (see Subsect. 5.3). For instance, if males and females are commented with the same terms and the contexts of those terms are mostly specific to an attribute value (see Example 1), we can take advantage of this variation and split the dataset to generate vectors that are biased towards that attribute value. We, therefore, propose two conditions to be jointly satisfied in order to split the training dataset. To express these conditions, we define the importance of a set L that contains terms or contexts w as

$$Q(L) = \sum_{w \in L} freq(w)$$

where *freq(w)* is the number of occurrences of w in the dataset.

Condition 1 is satisfied when the set of overlapped terms is more important than the set of non-overlapped terms.
Condition 2 is satisfied when the set of non-overlapped contexts is more important than the set of overlapped contexts.

In Example 1, the first condition is satisfied by the overlapped terms (*1person* and *baby*), and the second condition is met as the overlapped terms have different contexts (*miss you* and *cray cray*). None of the conditions are satisfied in Example 2 and only the first condition is satisfied in Example 3. Therefore, a unique term representation is sufficient in the two latter cases without missing the variations in the contextual meaning of the terms.

Dataset Dividing Algorithm. We label the original training dataset D in such a way that the ith sub-datasets D_i contains users labeled with the ith attribute value and words appearing in their publications comments. Algorithm 1 has for inputs D and D_is, and it returns a boolean that is true if D has to be split into sub-datasets D_i. We introduce the following notations:

1. $l_1, l_2, ..., l_k$ represents the attribute values. If l is gender attribute, then $l_1 = $ "male" and $l_2 = $ "female".
2. D_i is the ith sub-dataset containing the set of users with attribute value l_i (with $i \in \{1, .., k\}$). For gender attribute, we obtain D_1 and D_2 as sub-datasets annotated by male and female.
3. W_i is the set of words in comments of pictures published by users in U_i.
4. $UTop$, and $BTop$ are integer parameters.

5. $Unigram(D_i, UTop)$ computes Uni_i, the set of $UTop$ most frequent terms in D_i.
6. $Bigram(Uni_i, BTop)$ computes Big_i, the set of $BTop$ most frequent terms in D_i co-occurring with terms in Uni_i.
7. $Tcount(t) = |\{i \in \{1, \ldots k\} \mid t \in Uni_i\}|$, (resp. $Ccount(c) = |\{i \in \{1, \ldots k\} \mid c \in Big_i\}|$) is the number of sets Uni_i (resp. Big_i) where a term t (resp. a context c) appears.

Suitable values of $UTop$ and $BTop$ will be determined from experiments (see Subsect. 5.2).

Algorithm 1: Dataset dividing algorithm

---

    input : $D, D_1, \ldots, D_k$
    output: true iff the dataset D has to be split in $D_1, \ldots, D_k$

    Step1:
    for $i = 1, \ldots k$ **do**
        | $Uni_i \leftarrow Unigram(D_i, UTop)$
        | $Big_i \leftarrow Bigram(Uni_i, BTop)$
    Step 2:
    $OT \leftarrow \bigcap_{i=1}^{k} Uni_i$; $OC \leftarrow \bigcap_{i=1}^{k} Big_i$
    $T \leftarrow \bigcup_{i=1}^{k} Uni_i$; $C \leftarrow \bigcup_{i=1}^{k} Big_i$
    $NT \leftarrow \{t \in T \mid Tcount(t) = 1\}$; $NC \leftarrow \{c \in C \mid Ccount(c) = 1\}$
    if $Q(OT) > Q(NT)$ **and** $Q(OC) < Q(NC)$ **then** true;
    else false;

3.2 Random Indexing

Random Indexing (RI) is a fast dimensionality reduction method that transforms high-dimensional data into a lower-dimensional one by using a random matrix. It generates distributional representations that approximate similarities in sets of co-occurrence weights. RI assigns a randomly generated vector to each unique term in the text, the so-called index vector. These index vectors are sparse, n-dimensional, and ternary. They consist of a small number of randomly distributed non-zero elements, $\{-1, +1\}$. Each unique term is also represented by an n-dimensional initially empty vector, called distribution vector. RI incrementally updates the n-dimensional distribution vector of each word by summing the n-dimensional index vector(s) of all co-occurring words within a small window of text. As a result, terms appearing in a similar context tend to have a similar distributional vector. Let $c = [c_{-n}, \ldots c_{-1}, w, c_1, \ldots c_n]$ be the context of w with window from $-n$ to n (n chosen between 1 and 5) and let c be the vector obtained by accumulating word's index vector co-occurring with w in context c. We update the distributional vector w by using RI as follows:

$$w = \sum_{c \in C(w)} \sum_{\substack{-n \leqslant j \leqslant n \\ j \neq 0}} c_j \tag{1}$$

The problem with this approach is that the entire dataset potentially contributes to each term vector representation. Therefore, the vectors are affected by the different attribute values and lose their discriminating power for attribute inference attacks. To remedy this problem, we propose to generate several value-based vectors for each term [12].

Values-Based Random Indexing. Despite its simplicity, RI struggles to capture the relation between commenters' words/emojis usage preferences and the owners' publication profile. However, it can be adjusted for our task. Given a set of users U and an attribute l with k values $l_1, l_2, \ldots, l_k$, we introduce the subsets $U_1, U_2, \ldots, U_k$ of U, where U_m is the set of users whose attribute value is the mth value of l. Similarly, if W is the vocabulary of all comments for publications of users in U, we consider k sub-vocabularies $W_1, W_2, \ldots W_k$, such that each W_m records the commenters' preferences for a user in U_m.

In this way, we distinguish the different contexts of a term appearing within profiles with different attribute values. It is a key aspect of our inference attacks since the vector of a term occurring in W_m will be computed from its co-occurrences with other terms from W_m. Formally, instead of computing with standard RI, a single vector w from the entire W, we compute k vectors $w_1, w_2, \ldots w_k$, where w_m is derived from W_m, as follows:

$$w_m = \sum_{c \in C_m(w)} \sum_{\substack{-n \leqslant j \leqslant n \\ j \neq 0}} c_j \qquad (2)$$

From Eq. 2, we generate distinct vectors for the same term for different values. Previous word embedding approaches generate a single vector for each term appearing in Example 1 by combining different context terms. These approaches miss word semantical variations corresponding to different attribute values. In our approach, we can rely on Eq. 2 to compute several vector representations for the same term, each one corresponding to an attribute value.

Generating Index Vectors. RI relies on two important hypotheses. First, in high dimensional space, there exists a much larger number of almost orthogonal than orthogonal directions, according to Hecht-Nielsen [11]. Second, if we project points of a vector space into a randomly selected high dimensionality subspace, the distances between these points are approximately preserved (Johnson-Lindenstrauss-Schechtman lemma [16]). The choice of random matrix is an essential aspect of RI to satisfy these two hypotheses. In this work, we train a machine learning algorithm to find the best parameters of RI, namely the dimension and non-zero elements (see Subsect. 5.2). By following the steps mentioned above, our approach provides suitable vectors to perform attribute inference with higher accuracy than with alternative embeddings.

4   Attribute Inference Attacks

We consider an attacker who intends to infer OSN users' attributes from a set of publications P where each publication contains metadata. Thus, the OSN is exposed to potential privacy violations by an attacker (external or OSN user) collecting, storing and analyzing publications metadata from user profiles.

Once we learn the vector representations of terms for each attribute value (see Eq. 2), we compute a vector representation of u_m by aggregating all the terms that appear in his/her publications as follows:

$$u_m = \sum_{w \in P(u_m)} w_m \tag{3}$$

We introduce a set S of vectors computed by Eq. 3. For the target user t, we generate a set of user vectors $T = \{t_1, t_2, ...t_k\}$, where vector t_m is obtained as follows:

$$t_m = \sum_{w \in P(t)} w_m \tag{4}$$

We compute cosine similarity [20, 22] between T and S to check which user in S has the most similar vector to the vector of a user in T as follows:

$$(t_\mu, u_\mu) = \arg \max_{\substack{t \in T \\ u \in S}} (cosine(t, u))$$

The $arg\ max$ function outputs a tuple (t_μ, u_μ) and we infer that the target attribute value is l_μ. Cosine similarity is a common similarity measure for vectors when their magnitude is not relevant (e.g., they represent linguistic items in distributional semantics [14]).

As a concrete example, we consider a gender inference attack against a user named *Target*. Given two gender values l_1 = "male" and l_2 = "female", we generate two vectors w_1 and w_2 for *Target* where w_1 and w_2 correspond to vector of w in D_1 and D_2, respectively. Vectors *Target_F* and *Target_M* (red points in Fig. 1) represent female and male hypotheses vectors, respectively. Using cosine similarity, we compute in Fig. 1 the closest users of S to *Target_F* (blue dots) and the closest users of S to *Target_M* (blue dots). As *Raymond* is closer to *Target_M* than *Berkeley* to *Target_F*, we label therefore *Target* by *male*.

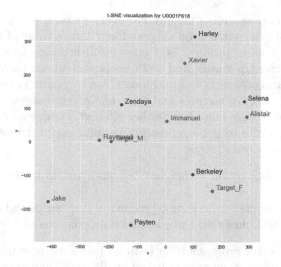

Fig. 1. Nearest users to female and male hypothesis vectors.

To sum up, the attacker can extract ground truth data and select the attribute to attack. Then, he checks if splitting the dataset is necessary by using Algorithm 1. Next, value-based random indexing from Subsect. 3.2 is applied to generate word vectors that are meaningful for each attribute value. Finally, the attribute inference is achieved by following the above steps.

5   Case Study: Facebook

In this section, we apply our methodology to implement attribute inference attacks on Facebook. We first describe the experimental setup containing our dataset, evaluation metrics and parameter settings related to the classifier and Algorithm 1. Next, we assess and compare our approach with word2vec, a state-of-the-art method for representing language semantics [20].

5.1   Dataset

As a case study, we concentrate on Facebook. More precisely, we consider photos published by their owners. Compared to other publications (such as post and status update), pictures on Facebook receive an extra comment from Facebook, namely alt-text. The generated alt-text has two advantages. First, it alleviates the image processing tasks. Second, it provides additional information for the attacker. We show the success of our attack on three Facebook sensitive attributes: age, gender and relationship status. The data collection was conducted from October 2019 to April 2020 by utilizing a python crawler for academic research purposes. For the ground truth, we focus on the profiles where these attributes are public, and we collect the required information from the

HTML files of the corresponding images. For every picture, we extract data such as comments, alt-texts, publication time and attribute's value of the owner. We have collected *9280* users' profiles: *7611* users published their gender, *4604* users shared their relationship status, and *3813* users announced their age. We randomly selected Facebook users to avoid usage bias by region or country. Overall we have collected *399,076* pictures and their *686,859* messages. We have leveraged the available picture metadata (either alt-text, comments or both) in our attack process. In order to get a representative and useful dataset, we perform the following pre-processing steps:

- Purifying the *conjunction*, *redundant tag*, and *text* from the generated alt-text.
- Cleaning the extracted comments by removing stop words and re-formulating flooded characters, misspelled words, and abbreviations.

Generating a vector representation from all user's pictures might be inaccurate when, for example, they are published at different time periods. Therefore, in addition to the above pre-processing steps, we use pictures' publication time range to prevent these impacts. For instance, if Alice shared *20* images in her profile from *January* to *June*, we create two users (*Alice_jan-mar* and *Alice_apr-jun*) by assigning pictures published from *January* to *March* to *Alice_jan-mar* and pictures of *April* to *June* to *Alice_apr-jun*. Moreover, Facebook users can share photos on each other profiles. Considering those pictures owned by someone else might hinder the inference attacks as the publisher owner and the user might have different attribute values. In this study, we only examine images published by the user and filter out photos owned by others.

5.2   Experimental Setup

Our attack relies on XGBoost classifier. We utilize different metrics to evaluate our approach. First, we use AUC (area under the ROC curve) as it is not sensitive to the label distribution [15]. Second, we use *macro* and *micro* averages to evaluate our inference attacks. A macro-average computes the metric independently for each class and then takes the average (hence treating all classes equally), whereas micro-average aggregates the contributions of all classes to compute the average metric.

Parameter Settings. We tune three different sets of parameters related to: classifier, RI and Algorithm 1.

For the classifier, we tune (i) the *learning_rate* to adjust weights on each step and make the model robust, (ii) *max_depth* and (ii) *min_child_weight* to control over-fitting. The *objective* specifies the learning task (e.g., binary classification) and the corresponding learning objective, *n_estimators* represents the number of trees to fit, and *subsample* indicates the fraction of observations to be randomly sampled for each tree. We set their default values and evaluate how different values affect our performance. Except for the objective depending on the number of classes, we create an array of different values for each parameter and use a python notebook tool called *GridSearchCV* to automatically find the best value of that array. For example, we assigned to *learning_rate* the array [0.01, 0.03, 0.05, 0.07], and use *GridSearchCV* to find the value giving the best result.

As for RI, [7] proposes a grid of sample values. We set *dimension= 500* and select two non-zero elements of the index vector to {-1, +1}, which maximize the result of our inference attacks.

Our dividing algorithm (see Algorithm 1) comprises parameters $UTop$ and $BTop$ to select the most frequent overlapped and non-overlapped *Unigram* and *Bigram*, respectively. The best parameter values are $UTop = 90$ and $BTop = 110$. They have been learned from our dataset by a grid search with $UTop, BTop \in \{10, 20, \ldots, 200\}$.

5.3   Inference Results

For the age inference attack, we consider the following classes: 20 to 25, 25 to 30, 30 to 35, 35 to 40, 40 to 45, and 45 to 50. We chose these age groups to have a compromise between the accuracy of age prediction and the balancing of datasets. The age categories in our dataset reflect, in general, the most active ones on Facebook. We do not consider ages under 20 or over 50 as it is time consuming to collect enough data and keep all age categories balanced. As for relationship status, we collect datasets for *single, married* and *engaged* users. Consequently, we consider three classes. Finally, we reduce gender inference attacks to a binary classification problem with *female* and *male* classes[1]. As for age and relationship status, we set the XGBoost classifier *objective* to *multi : softprob* that gives the probability of each class, while for gender we set it to *logistic*. We have used train-test splitting, as it performs faster, and split the entire dataset into the train, validation and test datasets. We have leveraged these datasets for training, parameter pruning and testing our XGBoost classifier, respectively. We have trained word2vec on the same dataset and compared its performance with our approach. Figure 2 shows the AUC result of word2vec and our approach over three different attributes.

Table 2. Comparison with word2vec when splitting conditions are not satisfied.

	C1				C2			
	AUC	Precision	Recall	Fscore	AUC	Precision	Recall	Fscore
word2vec/class Female	82	74.3	69.9	72.2	79	67.4	67.9	67.7
RI-split/class Female	61	53.8	49.1	51.3	60	53.4	49.5	51.4
word2vec/class Male	82	74.9	78.8	76.8	79	75.2	74.5	74.8
RI-split/class Male	61	62.6	67.1	64.6	60	61.6	65.3	63.4

---

[1] We do not ponder other genders and relationship status by lack of training samples.

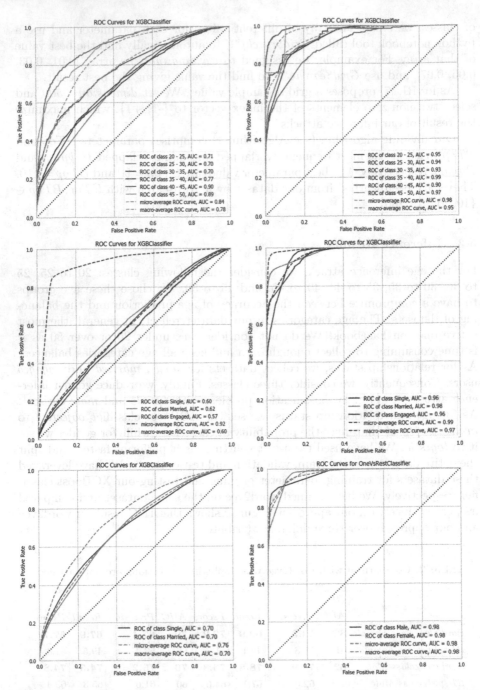

Fig. 2. Inference attacks performance measured by AUC for word2vec (left column) and our approach (right column): (line 1) age, (line 2) relationship status (line 3) gender

Figure 2 (a) shows the result of the age inference attack by using word2vec where the age classes are inferred with AUC from 70% to 90%. In contrast, Fig. 2 (b) represents the result of the same attribute inference attack by using our approach, which gets a tremendous boost with a substantial gain in performance. For example, our approach infers the class *35–40* with 99% AUC, where it was 77%. In addition to AUC, the *micro* and *macro* average increased to 98% and 95%, respectively. Figure 2 (c and d) display word2vec and our approach performance to relationship status inference attack. Our approach can accurately infer the relationship status attribute of the target user in comparison to word2vec. The class *Engaged* obtains 96% AUC in our approach, where word2vec infers this class inadequately (slightly better than random). Lastly, Fig. 2 (e and f) depict the gender inference attack. Similarly, our approach outstands word2vec model. The attacker can infer the user's gender by using our approach with 98% AUC, where it drops to 70% in word2vec.

As mentioned in Subsect. 3.1, we split the dataset only if two conditions are satisfied. To justify this, Table 2 represents some results of RI with splitting when the conditions are not satisfied. From a crawled dataset labeled by gender, we have synthetized two datasets $C1$ and $C2$. In $C1$, the first condition is satisfied and the second one is not satisfied. In $C2$, the first condition does not hold (and we ignore the status of second condition). We note that in both cases it is better to use word2vec than RI with splitting. Moreover, word2vec generates only one vector for each word which is space economical compared to RI with splitting. To sum up, this result confirms that by relying on Algorithm 1, splitting is applied in our approach only when it is beneficial. When Conditions 1 and 2 (see Subsect. 3.1) are satisfied, RI with splitting captures the commenters' words/emojis usage preferences adequately and this boosts the accuracy result. Otherwise, word2vec is more performant.

6   Conclusion

In this paper, we have presented a new perspective on attribute inference attacks based on reactions to target user publications. We have shown that if a term appears in diverse contexts, it can be represented by different vectors. To capture these diverse contexts, we have divided the dataset based on attribute values. We have also defined some conditions to prevent useless splittings. We have relied on the Random Indexing method to compute the term vectors in each attribute value, as the generated vectors need to be comparable. Based on the intensive analysis of *399,076* pictures and their *686,859* comments on Facebook, we have demonstrated that picture metadata conveys sensitive information and some private attributes such as gender, age category and relationship status are leaked by the variations in commenter's words/emojis usage preferences and picture owner sharing style. Our attacks are suitable for online execution as they do not require exploring user behavioral data and vicinity networks. They generalize easily to other social media platforms such as Twitter.

Even though some Facebook users limit themselves to sharing pictures with friends only, it is often not hard for an attacker to be added to a large list of friends and have access to the picture metadata.

As future work, we plan to extend our tool by explainable machine learning techniques [18,24] to offer users several means to reduce attribute inference risks (e.g., deleting or reducing the influence of leaking terms in comments). We also intend to enlarge the user age boundary and consider users over 50. Furthermore, using recent state-of-the-art tools such as BERT or ELMo for word embeddings would a promising direction to improve our work.

Ethical Statement. Our experiments have been performed on publicly available OSN data collected from Facebook. Although this data is public, it may lead to infer private information and we are therefore committed to keeping it in secure storage and only for the time necessary to carry out this work.

References

1. Abdelberi, C., Ács, G., Kâafar, M.A.: You are what you like! information leakage through users' interests. In: 19th Annual Network and Distributed System Security Symposium, NDSS. The Internet Society, San Diego, California, USA (2012)
2. Basile, P., Caputo, A., Semeraro, G.: Temporal random indexing: a system for analysing word meaning over time. Italian J. Comput. Linguist. 1(1), 55–68 (2015)
3. Choudhury, M.D., Sharma, S.S., Logar, T., Eekhout, W., Nielsen, R.C.: Gender and cross-cultural differences in social media disclosures of mental illness. In: Proceedings of the Conference on Computer Supported Cooperative Work and Social Computing. CSCW, pp. 353–369. ACM, Portland, OR, USA (2017)
4. Culotta, A., Kumar, N.R., Cutler, J.: Predicting the demographics of twitter users from website traffic data. In: Proceedings of the Twenty-Ninth AAAI Conference on Artificial Intelligence, Austin, Texas, USA, pp. 72–78 (2015)
5. Eidizadehakhcheloo, S., Pijani, B.A., Imine, A., Rusinowitch, M.: Your age revealed by Facebook picture metadata. In: Bellatreche, L., et al. (eds.) TPDL/ADBIS - 2020. CCIS, vol. 1260, pp. 259–270. Springer, Cham (2020). https://doi.org/10.1007/978-3-030-55814-7_22
6. Farahbakhsh, R., Han, X., Cuevas, Á., Crespi, N.: Analysis of Publicly Disclosed Information in Facebook Profiles. CoRR abs/1705.00515 (2017)
7. Fernández, A.M., Esuli, A., Sebastiani, F.: Lightweight random indexing for polylingual text classification. J. Artif. Intell. Res. 57, 151–185 (2016)
8. Giulianelli, M., Tredici, M.D., Fernández, R.: Analysing lexical semantic change with contextualised word representations. In: Proceedings of the 58th Annual Meeting of the Association for Computational Linguistics. ACL 2020, pp. 3960–3973. Association for Computational Linguistics (2020)
9. Gong, N.Z., Liu, B.: You are who you know and how you behave: attribute inference attacks via users' social friends and behaviors. In: 25th Security Symposium, pp. 979–995. USENIX, Austin, TX, USA (2016)
10. Gong, N.Z., et al.: Joint link prediction and attribute inference using a social-attribute network. ACM Trans. Intell. Syst. Technol. 5(2), 27:1–27:20 (2014)
11. Hecht-Nielsen, R., et al.: Context vectors: general purpose approximate meaning representations self-organized from raw data. Comput. Intell. Imitating Life 3(11), 43–56 (1994)

12. Jurgens, D., Stevens, K.: Event detection in blogs using temporal random indexing. In: Proceedings of the Workshop on Events in Emerging Text Types, pp. 9–16 (2009)

13. Kutuzov, A., Øvrelid, L., Szymanski, T., Velldal, E.: Diachronic word embeddings and semantic shifts: a survey. In: Proceedings of the 27th International Conference on Computational Linguistics. COLING 2018, Santa Fe, New Mexico, USA, pp. 1384–1397 (2018)

14. Levy, O., Goldberg, Y.: Linguistic regularities in sparse and explicit word representations. In: Proceedings of the Eighteenth Conference on Computational Natural Language Learning, CoNLL 2014, Baltimore, Maryland, USA, pp. 171–180. ACL (2014)

15. Lichtenwalter, R., Lussier, J.T., Chawla, N.V.: New perspectives and methods in link prediction. In: Proceedings of the 16th ACM SIGKDD International Conference on Knowledge Discovery and Data Mining, Washington, DC, USA, pp. 243–252 (2010)

16. Lindenstrauss, W.J.J.: Extensions of Lipschitz maps into a Hilbert space. Contemp. Math. **26**, 189–206 (1984)

17. Ludu, P.S.: Inferring gender of a Twitter user using celebrities it follows. CoRR abs/1405.6667 (2014)

18. Lundberg, S.M., Lee, S.: A unified approach to interpreting model predictions. In: Advances in Neural Information Processing Systems 30: Annual Conference on Neural Information Processing Systems 2017, Long Beach, CA, USA, pp. 4765–4774 (2017)

19. Mauw, S., Ramírez-Cruz, Y., Trujillo-Rasua, R.: Robust active attacks on social graphs. Data Min. Knowl. Discov. **33**(5), 1357–1392 (2019)

20. Mikolov, T., Chen, K., Corrado, G., Dean, J.: Efficient estimation of word representations in vector space. arXiv preprint arXiv:1301.3781 (2013)

21. Nguyen, D., Gravel, R., Trieschnigg, D., Meder, T.: How old do you think I am? A study of language and age in twitter. In: Proceedings of the Seventh International Conference on Weblogs and Social Media. ICWSM 2013, Cambridge, Massachusetts, USA. The AAAI Press (2013)

22. Pennington, J., Socher, R., Manning, C.D.: GloVe: global vectors for word representation. In: Proceedings of the 2014 Conference on Empirical Methods in Natural Language Processing. EMNLP 2014, Doha, Qatar, A meeting of SIGDAT, a Special Interest Group of the ACL, pp. 1532–1543 (2014)

23. Pijani, B.A., Imine, A., Rusinowitch, M.: You are what emojis say about your pictures: language-independent gender inference attack on Facebook. In: SAC 2020: The 35th ACM/SIGAPP Symposium on Applied Computing, Online Event, Brno, Czech Republic, pp. 1826–1834. ACM (2020)

24. Ribeiro, M.T., Singh, S., Guestrin, C.: "Why should I trust you?": Explaining the predictions of any classifier. In: Proceedings of the 22nd ACM SIGKDD International Conference on Knowledge Discovery and Data Mining, San Francisco, CA, USA, pp. 1135–1144 (2016)

25. Ryu, E., Rong, Y., Li, J., Machanavajjhala, A.: CURSO: protect yourself from curse of attribute inference: a social network privacy-analyzer. In: Proceedings of the 3rd ACM SIGMOD Workshop on Databases and Social Networks. DBSocial 2013, New York, NY, USA, pp. 13–18. ACM (2013)

26. Sahlgren, M.: An introduction to random indexing. In: Methods and Applications of Semantic Indexing Workshop at the 7th International Conference on Terminology and Knowledge Engineering (2005)

27. Sap, M., et al.: Developing age and gender predictive lexica over social media. In: Proceedings of the 2014 Conference on Empirical Methods in Natural Language Processing. EMNLP, pp. 1146–1151. ACL, Doha, Qatar (2014)
28. Sherwin, G., Bhandari, E.: Facebook settles civil rights cases by making sweeping changes to its online ad platform (2019). https://www.aclu.org/blog/womens-rights/womens-rights-workplace/facebook-settles-civil-rights-cases-making-sweeping
29. Wang, P., Guo, J., Lan, Y., Xu, J., Cheng, X.: Your cart tells you: inferring demographic attributes from purchase data. In: Proceedings of the Ninth ACM International Conference on Web Search and Data Mining, San Francisco, CA, USA, pp. 173–182 (2016)
30. Weinsberg, U., Bhagat, S., Ioannidis, S., Taft, N.: BlurMe: inferring and obfuscating user gender based on ratings. In: Sixth ACM Conference on Recommender Systems, RecSys 2012, Dublin, Ireland, pp. 195–202 (2012)

Access Control

Verifiable Hierarchical Key Assignment Schemes

Anna Lisa Ferrara[1]([✉]), Federica Paci[2], and Chiara Ricciardi[1]

[1] Università degli Studi del Molise, Campobasso, Italy
annalisa.ferrara@unimol.it, c.ricciardi1@studenti.unimol.it
[2] Università degli Studi di Verona, Verona, Italy
federicamariafrancesca.paci@univr.it

Abstract. A Hierarchical Key Assignment Scheme (HKAS) is a method to assign some private information and secret keys to a set of classes in a partially ordered hierarchy, so that the private information of a higher class together with some public information can be used to derive the keys of all classes lower down in the hierarchy. Historically, HKAS has been introduced to enforce multi-level access control, where it can be safely assumed that the public information is made available in some authenticated form. Subsequently, HKAS has found application in several other contexts where, instead, it would be convenient to certify the trustworthiness of public information. Such application contexts include key management for IoT and for emerging distributed data acquisition systems such as wireless sensor networks. In this paper, motivated by the need of accommodating this additional security requirement, we first introduce a new cryptographic primitive: *Verifiable Hierarchical Key Assignment Scheme* (VHKAS). A VHKAS is a key assignment scheme with a verification procedure that allows honest users to verify whether public information has been maliciously modified to induce an honest user to obtain an incorrect key. Then, we design and analyse VHKASs which are provably secure. Our solutions support key update for compromised secret keys by making a limited number of changes to public and private information.

Keywords: Hierarchical key assignment · Access control · Applied cryptography

1 Introduction

Users of a computer system could be organized into a hierarchy consisting of a number of separate classes. These classes, called security classes, are positioned and ordered within the hierarchy according to the fact that some users have more access rights than others. For instance, in a hospital, doctors can access their patients' medical records, while researchers can only consult anonymous clinical information for studies.

A hierarchical key assignment (HKAS) scheme is a method to assign a secret key and some private information to each class in the hierarchy so that keys for

© IFIP International Federation for Information Processing 2021
Published by Springer Nature Switzerland AG 2021
K. Barker and K. Ghazinour (Eds.): DBSec 2021, LNCS 12840, pp. 357–376, 2021.
https://doi.org/10.1007/978-3-030-81242-3_21

descendant classes can be obtained via a key derivation procedure. This assignment is carried out by a central authority, the Trusted Authority (TA). Following the seminal work by Akl and Taylor [2], many researchers have proposed different HKASs that either have better performances or allow dynamic updates to the hierarchy (e.g., [3,5,10,12,14,16,19]). Crampton et al. in [15] provided a detailed classification for HKASs, according to several parameters, including memory requirements for public and private information and the complexity of handling dynamic updates. In particular, they identified families of schemes where the public information is used to store secret keys in order to reduce the amount of private information users need to manage[1]. The use of HKASs belonging to such families is desirable to prevent or limit the change of private information and its redistribution when handling encryption key updates. Indeed, for these schemes, the key update procedure which is necessary to replace compromised keys, often requires changing only the public information without the need of interacting with the involved users. In the remainder of the paper, we refer to HKAS schemes belonging to such families.

Historically, HKASs have been introduced to enforce multi-level access control in scenarios where it can be safely assumed that the public information is made available to everyone via a publicly accessible repository for which only the TA has write permissions. However, key assignment schemes have recently been employed in different application context where the public information may be exposed to changes by malicious users. These application contexts include key management for IoT and distributed data acquisition systems such as wireless sensor networks [4,10,24] as well as sensitive data outsourcing to the cloud [9,11,17,18,23]. For instance, consider sensitive data outsourcing in cloud; the data owner encrypts the data before outsourcing them at the server and distributes the private information (i.e., secret keys used to encrypt the data and derivation material) to the users by means of an HKAS according to the access policy. Only the data owner and users who know the appropriate encryption and derivation keys will be able to decrypt the data. However, metadata which includes the public information will also be stored at the server and thus may be modified voluntarily or involuntarily by those who have access to it, including the cloud service provider which is not necessarily trusted. Unfortunately, a change in the public information will prevent an honest user to derive a correct decryption key. Similarly, in wireless sensor networks, the cluster head nodes are responsible for forwarding any public information that has been changed as a result of key updates. If a cluster head node is corrupted, this information may be maliciously modified before reaching its destination.

The scenarios outlined above introduce the need for an honest user to verify the trustworthiness of the public information. In order to accommodate this additional security requirement, we introduce a new cryptographic primitive: *Verifiable Hierarchical Key Assignment Scheme* (VHKAS). A VHKAS is a key assignment scheme with a verification procedure that allows honest users to verify whether public information has been maliciously modified. In order to

---

[1] Such families include IKEKAS, DKEKAS, and TKEKAS [15].

capture a notion of security against an adversary who has the ability to replace or modify the public information, we introduce the notion of *strong key-consistency*. This notion models the fact that even the TA is unable to maliciously modify the public information once distributed the private information.

More in detail, our contributions are as follows:

- We first give formal definitions for VHKAS; and the notion of security *strong key-consistency*;
- subsequently, we present a construction of VHKAS that uses as building block a Message Locked Encryption (MLE) scheme. We show that the construction is provably-secure with respect to strong key-consistency and *key indistinguishability*, which corresponds to the requirement that an adversary is not able to learn any information about a key that it should not have access to;
- afterwards, we show how to handle key replacement for compromised secret keys by making a limited number of changes to public and private information;
- finally, we instantiate our MLE-based construction with the deterministic MLE scheme proposed by Abadi et al. [1].

The paper is organized as follows: in Sect. 2 we review the definitions of HKAS, and MLE as well as their notions of security. In Sect. 3 we define VHKAS and introduce the security notion of strong key-consistency. In Sect. 4 we show our MLE-based construction, prove it to be provably-secure with respect to key-consistency and key indistinguishability and instantiate it with the deterministic MLE scheme proposed by Abadi et al. [1]. In Sect. 5 we show how to handle key replacements. In Sect. 6 we evaluate the performance of our construction by comparing it with that of popular HKAS from the literature, while Sect. 7 concludes the paper.

2 Preliminaries

Notation. We use the standard notation to describe probabilistic algorithm and experiments. If $A(\cdot, \cdot, \dots)$ is any probabilistic algorithm then $a \leftarrow A(x, y, \dots)$ denotes the experiment of running A on inputs $x, y, \dots$ and letting a be the outcome, the probability being over the coins of A. Similarly, if X is a set then $x \leftarrow X$ denotes the experiment of selecting an element uniformly from X and assigning x this value. If w is neither an algorithm nor a set then $x \leftarrow w$ is a simple assignment statement. For two bit-strings x and y we denote by $x \| y$ their concatenation. A function $\epsilon : \mathbb{N} \to \mathbb{R}$ is *negligible* if for every constant $c > 0$ there exists an integer n_c such that $\epsilon(n) < n^{-c}$ for all $n \geq n_c$.

2.1 Hierarchical Key Assignment Schemes

Consider a set of users divided into a number of disjoint classes, called *security classes*. A binary relation $\preceq$ that partially orders the set of classes V is defined in accordance with authority, position, or power of each class in V. The poset $(V, \preceq)$ is called a *partially ordered hierarchy*. For any two classes u and v, the notation

$u \preceq v$ is used to indicate that the users in v can access u's data. The partially ordered hierarchy $(V, \preceq)$ can be represented by the directed graph $G = (V, E)$, where each class corresponds to a vertex and there is a path from class v to class u if and only if $u \preceq v$. A hierarchical key assignment scheme is a method to assign a secret key and some private information to each class in the hierarchy. The private information will be used by each class to compute the keys assigned to all classes lower down in the hierarchy. This assignment is carried out by the TA. Formally:

Definition 1 [20]. *Let Γ be a family of graphs corresponding to partially ordered hierarchies. A HKAS for Γ is a pair (Gen, Der) of algorithms satisfying the following conditions:*

1. *The* information generation algorithm *Gen is probabilistic polynomial-time. It takes as input the security parameter 1^τ and a graph $G = (V, E)$ in Γ, and produces as outputs*
   - *a private information s_u, for any class $u \in V$;*
   - *a key k_u, for any class $u \in V$;*
   - *a public information pub.*
   We denote by (s, k, pub) the output of the algorithm Gen, where s and k denote the sequences of private information and of keys, respectively.
2. *The* key derivation algorithm *Der is deterministic polynomial-time. It takes as input the security parameter 1^τ, a graph $G = (V, E)$ in Γ, two classes u, v in V, the private information s_u assigned to class u and the public information pub, and produces as output the key k_v assigned to class v if $v \preceq u$, or a special rejection symbol $\perp$ otherwise.*
   We require that for each class $u \in V$, each class $v \preceq u$, each private information s_u, each key k_v, each public information pub which can be computed by Gen on inputs 1^τ and G, it holds that $Der(1^\tau, G, u, v, s_u, pub) = k_v$.

Security Notions. Atallah et al. [5] first introduced two different security goals for HKASs: security with respect to key indistinguishability and security against key recovery. In this paper, we only consider the stronger notion of key indistinguishability [5,6].

$STAT_u$ is a static adversary who wants to attack a class $u \in V$ and who is able to corrupt *all* users not entitled to compute the key of class u. Algorithm $Corrupt_u$ which, on input the private information s generated by the algorithm Gen, extracts the secret values s_v associated to each class that the adversary is able to corrupt. In the indistinguishability game, the adversary must distinguish the key of class u from a random value.

Definition 2 [IND-ST]. *Let Γ be a family of graphs corresponding to partially ordered hierarchies, let $G = (V, E)$ be a graph in Γ, let (Gen, Der) be a HKAS for Γ and let $STAT_u$ be a static adversary who attacks a class u. Consider the following two experiments:*

$$\begin{array}{l|l}
\text{Experiment } \mathbf{Exp}_{\mathtt{STAT}_u}^{\mathtt{IND}-1}(1^\tau, G) & \text{Experiment } \mathbf{Exp}_{\mathtt{STAT}_u}^{\mathtt{IND}-0}(1^\tau, G) \\
\quad (s, k, pub) \leftarrow Gen(1^\tau, G) & \quad (s, k, pub) \leftarrow Gen(1^\tau, G) \\
\quad corr \leftarrow Corrupt_u(s) & \quad corr \leftarrow Corrupt_u(s) \\
\quad d \leftarrow \mathtt{STAT}_u(1^\tau, G, pub, corr, k_u) & \quad \rho \leftarrow \{0,1\}^{length(k_u)} \\
\quad \textbf{return } d & \quad d \leftarrow \mathtt{STAT}_u(1^\tau, G, pub, corr, \rho) \\
& \quad \textbf{return } d
\end{array}$$

The advantage of $\mathtt{STAT}_u$ is defined as $\mathbf{Adv}_{\mathtt{STAT}_u}^{\mathtt{IND}}(1^\tau, G) = |Pr[\mathbf{Exp}_{\mathtt{STAT}_u}^{\mathtt{IND}-1}(1^\tau, G) = 1] - Pr[\mathbf{Exp}_{\mathtt{STAT}_u}^{\mathtt{IND}-0}(1^\tau, G) = 1]|$. The scheme is secure in the sense of IND-ST if, for each graph $G = (V, E)$ in Γ and each $u \in V$, the function $\mathbf{Adv}_{\mathtt{STAT}_u}^{\mathtt{IND}}(1^\tau, G)$ is negligible, for each adversary $\mathtt{STAT}_u$ with time complexity polynomial in τ^2.

2.2 Message-Locked Encryption

An Message-Locked Encryption (MLE) is a symmetric encryption scheme in which the key is itself derived from the message. Provably-secure MLE schemes have been first proposed by Bellare et al. in [8].

Definition 3 [1]. *A MLE scheme is a tuple* $(PPGen, KD, Enc, Dec, Valid)^3$ *of algorithms satisfying the following conditions:*

1. *The* parameter generation algorithm *PPGen on input* 1^τ *returns a public parameter pp.*
2. *The* key derivation function *KD takes as input the message m in the message space $\mathcal{M}$ and pp and produces as output message-derived key k_m.*
3. *The* encryption algorithm *Enc takes as input pp, the key k_m, and the message m in $\mathcal{M}$, and produces as output the ciphertext c.*
4. *The* decryption algorithm *Dec takes as input pp, the key k_m and the ciphertext c and produces as output the message m or $\perp$.*
5. *The* validity-test *Valid takes as input public parameters pp and a ciphertext c and outputs 1 if the ciphertext c is a valid ciphertext, and 0 otherwise.*

Security Notions. The definitions below make use of some parameters that are functions of the security parameter. Specifically, $k = k(\tau)$ denoting min-entropy requirements over message sources, and $T = T(\tau)$ representing the number of blocks in the message.

Entropy. The min-entropy of a random variable X is defined as $H_\infty(X) = -log$ $(max_x Pr[X = x])$. In other words, $H_\infty(X) = k$, if $max_x Pr[X = x] = 2^{-k}$. A k-source is a random variable X with $H_\infty(X) \geq k$. A $(k_1, \ldots, k_T)$-source is a random variable $\mathbf{X} = (X_1, \ldots, X_T)$ where each X_i is a k_i-source. A (T, k)-source is a random variable $\mathbf{X} = (X_1, \ldots, X_T)$ where, for each $i = 1, \ldots, T$, it holds that X_i is a k-source. Next, we recall the definitions of real-or-random encryption oracle, polynomial-size X-source adversary, and, for schemes that rely on random oracles, q-query X-source adversary.

---

[2] In [6] it has been proven that security against adaptive adversaries is (polynomially) equivalent to security against static adversaries.

[3] MLE definition also includes an equality algorithm. We omit it since it is not necessary for our goals.

362   A. L. Ferrara et al.

Definition 4 [1]. **Real or Random encryption oracle.** *The real-or-random encryption oracle,* RoR, *takes as input triplets of the form* $(\text{mode}, pp, \mathbf{M})$, *where* $\text{mode} \in \{\text{real}, \text{rand}\}$, pp *denotes public parameters, and* $\mathbf{M}$ *is a polynomial size circuit representing a joint distribution over T messages. If* $\text{mode} = \text{real}$ *then the oracle samples* $(m_1, \ldots, m_T) \leftarrow \mathbf{M}$, *and if* $\text{mode} = \text{rand}$ *then the oracle samples uniform and independent messages* $(m_1, \ldots, m_T) \leftarrow \mathcal{M}$. *Next, for each* $i = 1, \ldots, T$, *it samples* $k_i \leftarrow KD(pp, m_i)$, *computes* $c_i \leftarrow Enc(pp, k_i, m_i)$ *and outputs the ciphertext vector* $(c_1, \ldots, c_T)$.

Definition 5 [1]. **Poly-sampling complexity adversary.** *Let $\mathcal{A}$ be a probabilistic polynomial-time algorithm that is given as input a pair* $(1^\tau, pp)$ *and oracle access to* $(1^\tau, pp)$ *for some* $\text{mode} \in \{\text{real}, \text{rand}\}$. *Then, $\mathcal{A}$ is a polynomial-size* (T, k)-source adversary if for each of $\mathcal{A}$'s RoR-queries $\mathbf{M}$ it holds that $\mathbf{M}$ is an (T, k)-source that is samplable by a circuit of (an arbitrary) polynomial size in the security parameter.*

Definition 6 [1]. **PRV-CDA2 security**[4]. *An MLE scheme* $\Pi = (PPGen, KD, Enc, Dec, EQ, Valid)$ *is* (T, k)-source PRV–CDA2 *secure, if for any probabilistic polynomial-time polynomial-size* (T, k)-source adversary $\mathcal{A}$, *there exists a negligible function* $\epsilon(\tau)$ *such that the advantage of $\mathcal{A}$ is defined as*

$$\mathbf{Adv}_{\mathcal{A}}^{\text{PRV-CDA2}}(1^\tau) = |Pr[\mathbf{Exp}_{\mathcal{A}}^{\text{real}}(1^\tau) = 1] - Pr[\mathbf{Exp}_{\mathcal{A}}^{\text{rand}}(1^\tau) = 1]| \leq \epsilon(\tau)$$

where for each $\text{mode} \in \{\text{real}, \text{rand}\}$ *the experiment* $\mathbf{Exp}_{\mathcal{A}}^{\text{mode}}(\tau)$ *is defined in the following game:*

$$\text{Experiment } \mathbf{Exp}_{\mathcal{A}}^{\text{PRV-CDA2}}$$
$$pp \leftarrow PPGen(1^\tau)$$
$$\text{return } \mathcal{A}^{\text{RoR(mode}, pp, \cdot)}(1^\tau, pp)$$

3   Verifiable Hierarchical Key Assignment Schemes

In this section, we introduce a novel cryptographic primitive that we call *Verifiable Hierarchical Key Assignment Scheme* (VHKAS). A VHKAS is a hierarchical key assignment scheme equipped with a verification procedure that allows honest users to check whether the public information has been maliciously changed.

A VHKAS for a family Γ of graphs, corresponding to partially ordered hierarchies, is defined as follows:

Definition 7. *A VHKAS is a triple* (Gen, Der, Ver) *of algorithms satisfying the following conditions:*

1. *The* information generation algorithm *Gen is probabilistic polynomial-time defined as in Definition 1.*
2. *The* key derivation algorithm *Der is deterministic polynomial-time defined as in Definition 1.*

---

[4] Notice that PRV-CDA2 notion enables adversaries to query the oracle with message distributions that depend on the public parameters pp.

3. *The* verification algorithm Ver *is deterministic polynomial-time. It takes as input the security parameter* 1^τ, *a graph* $G = (V, E)$ *in* Γ, *a class u in V, the private information* s_u, *a public information pub and it outputs 1 if for each class* $v \in V$ *such that* $v \preceq u$, $Der(1^\tau, G, u, v, s_u, pub)$ *return a valid key for the class v, 0 otherwise.*

Security Notions. In order to capture a notion of security against an adversary who has the ability to replace or modify the public information so as to mislead an honest user into deriving an incorrect key, we introduce the notion of *Strong Key-Consistency* (Strong-KC). Specifically, an adversary SSTAT is able to generate the secret information s, the set of keys k, and two different public values pub and pub' in such a way that, given a class u, the key of some class $v \preceq u$ derived by a user in class u according to pub differs from that derived according to pub' while the verification procedure succeeds in both cases. This notion models the fact that even the TA is not able to maliciously modify public information once private information has been distributed.

Now, we formally define the notion of Strong-KC:

Definition 8 [Strong-KC]. *Let* Γ *be a family of graphs corresponding to partially ordered hierarchies, let* $G = (V, E)$ *be a graph in* Γ, *let* (Gen, Der, Ver) *be a VHKAS for* Γ *and let* SSTAT *be an adversary. Consider the following experiment:*

Experiment $\mathbf{Exp}_{\mathrm{SSTAT}}^{\mathrm{Strong-KC}}(1^\tau, G)$
  $(s, k, pub, pub', u) \leftarrow \mathrm{SSTAT}(1^\tau, G)$
  if $(Ver(1^\tau, G, u, s_u, pub) = 0 \vee Ver(1^\tau, G, u, s_u, pub') = 0)$ **return** 0
  for each $v \preceq u$
    if $(Der(1^\tau, G, u, v, s_u, pub) \neq Der(1^\tau, G, u, v, s_u, pub'))$
      return 1
  return 0

The advantage of SSTAT *is defined as* $\mathbf{Adv}_{\mathrm{SSTAT}}^{\mathrm{Strong-KC}}(1^\tau) = |Pr[\mathbf{Exp}_{\mathrm{SSTAT}}^{\mathrm{Strong-KC}}(1^\tau) = 1]|$. *The scheme is* Strong-KC*secure if, the function* $\mathbf{Adv}_{\mathrm{SSTAT}}^{\mathrm{Strong-KC}}(1^\tau)$ *is negligible, for each adversary* SSTAT *whose time complexity is polynomial in* τ.

Remark 1. To help understand the extra feature provided by a VHKAS over an HKAS, in the following we consider the Encryption-Based Construction (EBC) proposed in [20] and show an example of how an attacker who manages to modify the public values is able to induce honest users to derive an incorrect key. In the EBC every class u is assigned a private information s_u, a secret key k_u, and a public information $\pi(u, u)$, which is the encryption of the key k_u with the private information s_u; furthermore, for each edge (u, v), there is a public value $p(u, v)$, which allows class u to compute the private information s_v held by class v. Indeed, $p(u, v)$ consists of encrypting the private information s_v with the private information s_u. This allows any user of a class u to compute the key k_v held by any class v lower down in the hierarchy.

The public value $\pi(u, u)$ can be considered as being associated with an additional edge connecting the class u to a dummy class u'. The Fig. 1 illustrates

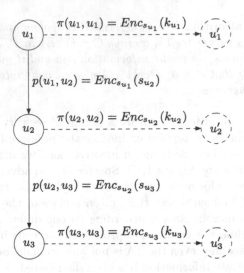

Fig. 1. A chain of length 3 along with the public information generated by the EBC.

a chain of length $t = 3$ together with the public information associated by the EBC with the edges of the chain, as well as that associated with the additional edges, which are represented by dashed lines.

A malicious user belonging to the class u_2 who manages to modify the public value $\pi(u_2, u_2)$ is able to induce honest users of u_2 and u_1 to derive a secret key k of his choice instead of the real key k_{u_2} associated with the class u_2. Indeed, such a malicious user, as part of the class u_2, holds the private information s_{u_2} and can use it to compute the value $Enc_{s_{u_2}}(k)$ which can then be substituted for the public value $\pi(u_2, u_2)$ associated with the edge (u_2, u_2').

Note that even if the TA digitally signs the public values, the EBC construction does not achieve the notion of Strong-KC. Indeed, the adversary could reuse old public values and induce honest users to derive old keys. Furthermore, digitally signing public values does not prevent scenarios where the TA is involved in maliciously modifying public information.

A VHKAS will be able to withstand such attacks because public and private information will be crafted in such a way that any malicious changes to public values will be identified through the verification procedure.

4   An MLE-Based Construction

In this section, we present a VHKAS which uses as a building block an MLE scheme. The scheme assumes that the partially ordered hierarchy has been partitioned into chains (i.e. totally ordered sets) [13,22]. This gives a method of constructing a HKAS, represented by a directed acyclic graph $G = (V, E)$, from a HKAS for a simple chain by partitioning the poset into chains. This approach has the nice property that the amount of private storage needed per class, is

Let Γ be a family of graphs corresponding to chains. Let $G = (V, E) \in \Gamma$ and let Π $= (PPGen, KD, Enc, Dec, Valid)$ be an MLE scheme whose key derivation function KD is collision-resistant and let $\mathcal{F} : \{0,1\}^\tau \times \{0,1\}^\tau \rightarrow \{0,1\}^\tau$ be a PRF.

Algorithm $Gen(1^\tau, G)$

1. Let $u_1, \ldots, u_t$ be the classes in the chain.
2. Let $pp \leftarrow PPGen(1^\tau)$;
3. Let $\pi_{u_{t+1}} \leftarrow \{0,1\}^\tau$;
4. For each class u_i, for $i = t, \ldots, 1$, let
   (a) $r_{u_i} \leftarrow \{0,1\}^\tau$;
   (b) $\pi_{u_i} \leftarrow KD(pp, \pi_{u_{i+1}} || r_{u_i})$;
   (c) $k_{u_i} \leftarrow \mathcal{F}(\pi_{u_i}, r_{u_i})$;
   (d) $s_{u_i} = (pp, r_{u_i}, \pi_{u_i}, k_{u_i})$;
5. Let s and k be the sequences of private information $s_{u_1}, s_{u_2}, \ldots, s_{u_t}$ and keys $k_{u_1}, k_{u_2}, \ldots, k_{u_t}$, respectively, computed in the previous steps;
6. For each $i = 1, \ldots, t$, compute the public information

$$p_{(u_i, u_{i+1})} \leftarrow Enc(pp, \pi_{u_i}, \pi_{u_{i+1}} || r_{u_i});$$

7. Let pub be the sequence of public information $p_{(u_1, u_2)}, p_{(u_2, u_3)}, \ldots, p_{(u_t, u_{t+1})}$ computed in the previous step;
8. Output(s, k, pub).

Algorithm $Der(1^\tau, G, u, v, s_u, pub)$

1. Let $u = u_i$ and $v = u_j$, for some $j \geq i$, $j = 1 \ldots, t$.
2. Parse s_u as $(pp, r_{u_i}, \pi_{u_i}, k_{u_i})$;
3. For any $z = i, \ldots, j$, extract the public value $p_{(u_z, u_{z+1})}$ from pub
   (a) if $Valid(pp, p_{(u_z, u_{z+1})}) = 0$, return $\perp$;
   (b) compute
$$(\pi_{u_{z+1}} || r_{u_z}) \leftarrow Dec(pp, \pi_{u_z}, p_{(u_z, u_{z+1})});$$
4. Output $k_v \leftarrow \mathcal{F}(\pi_{u_j}, r_{u_j})$.

Algorithm $Ver(1^\tau, G, u, s_u, pub)$

1. Let $u = u_i$, for some $i = 1, \ldots, t$.
2. Parse s_u as $(pp, r_{u_i}, \pi_{u_i}, k_{u_i})$;
3. For each $j = i, \ldots, t$,
   (a) extract the public value $p_{(u_j, u_{j+1})}$ from pub;
   (b) if $Valid(pp, p_{(u_j, u_{j+1})}) = 0$, return 0;
   (c) let $\alpha_i = \pi_{u_i}$, compute

$$(\alpha_{j+1} || \beta_j) \leftarrow Dec(pp, \alpha_j, p_{(u_j, u_{j+1})});$$

   (d) if $KD(pp, \alpha_{j+1} || \beta_j)$ is different from α_j, return 0;
4. return 1.

Fig. 2. The MLE-based Construction.

bounded by the width of the poset[5]. Thus, in the following we will only consider a family Γ of graphs corresponding to chains. Figure 2 shows the MLE-based construction for a chain of t classes $u_1, \ldots, u_t$. In order to simplify the presentation, we consider a dummy class u_{t+1}. This will enable us to consider all public information as values associated to the edges of a chain.

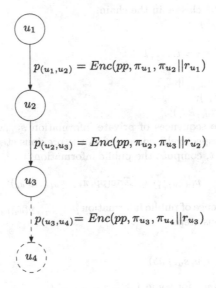

Fig. 3. Public information generated by the MLE-based construction for a chain of length $t = 3$.

The public information, associated to the edges of the chain, is used to store the secret keys in an encrypted form. Indeed, for each $i = 1, \ldots, t$, $k_{u_i} \leftarrow \mathcal{F}(\pi_{u_i}, r_{u_i})$ can be obtained by retrieving π_{u_i} and r_{u_i}, respectively from $p_{(u_{i-1}, u_i)}$ and $p_{(u_i, u_{i+1})}$. Figure 3 illustrates the public information computed by the scheme for a chain of length $t = 3$. The verification procedure allows to check whether a honest user is able to derive valid keys. Specifically, in the MLE-based construction a honest user in some class u_i will derive valid keys $k_{u_j} = Der(1^\tau, G, u_i, u_j, s_{u_i}, pub)$, if for each $j = i, \ldots, t$, the public information $p_{(u_j, u_{j+1})}$ stores a value x such that $\pi_{u_j} = KD(pp, x)$. Intuitively, any changes to public values will be detected since KD is collision resistant. For example, if a malicious user belonging to class u_1 (see Fig. 3) changes $p(u_1, u_2)$ in such a way that honest users in u_1 derive a value $\pi'_{u_2} \neq \pi_{u_2}$, the verification procedure fails since $KD(pp, \pi'_{u_2} || r_{u_1})$ is different from π_{u_1}.

---

[5] The width is the cardinality of the largest antichain in V. $A \subseteq V$ is an antichain in V if for all $u, v \in A$, where $u \neq v$, we have $v \preceq u$ and $u \preceq v$.

4.1   Analysis of the Scheme

In this section we show that the security of the MLE-based construction depends upon the security properties of the underlying MLE scheme.

We first show that the MLE-based construction is secure in the sense of IND-ST.

Theorem 1. *Let* $\Pi = (PPGen, KD, Enc, Dec, Valid)$ *be a* (T, μ)*-source* PRV-CDA2 *secure MLE scheme where* $\mu = \omega(\log \tau)$ *and let* $\mathcal{F} : \{0,1\}^\tau \times \{0,1\}^\tau \to \{0,1\}^\tau$ *be a PRF. The MLE-based VHKAS of Fig. 2 is secure in the sense of* IND-ST.

Proof. Let STAT_u be a static adversary attacking class u. Let $V = \{u_1, \ldots, u_t\}$ and $(u_i, u_{i+1}) \in E$, for $i = 1, \ldots, t-1$, and, w.l.o.g., let $u = u_j$ for some $1 \leq j < t$. In order to prove the theorem, we need to show that the adversary's views in experiments $\mathbf{Exp}_{\text{STAT}_u}^{\text{IND}-1}$ and $\mathbf{Exp}_{\text{STAT}_u}^{\text{IND}-0}$ are indistinguishable. Notice that the only difference between $\mathbf{Exp}_{\text{STAT}_u}^{\text{IND}-1}$ and $\mathbf{Exp}_{\text{STAT}_u}^{\text{IND}-0}$ is the last input of STAT_u, which corresponds to the key k_u in the former experiment and to a random value in the latter. Thus, while in $\mathbf{Exp}_{\text{STAT}_u}^{\text{IND}-1}$ the public information is related to the last input of STAT_u, in $\mathbf{Exp}_{\text{STAT}_u}^{\text{IND}-0}$ it is completely independent on such a value.

We construct a sequence of 4 experiments $\mathbf{Exp}_u^1, \ldots, \mathbf{Exp}_u^4$, all defined over the same probability space, where the first and the last experiments of the sequence correspond to $\mathbf{Exp}_{\text{STAT}_u}^{\text{IND}-1}$ and $\mathbf{Exp}_{\text{STAT}_u}^{\text{IND}-0}$, respectively. In each experiment we modify the way the view of STAT_u is computed, while maintaining the view's distributions indistinguishable among any two consecutive experiments. For any $q \in \{2, 3\}$, experiment $\mathbf{Exp}_u^q$ is defined as follows:

$$\text{Experiment } \mathbf{Exp}_u^q(1^\tau, G)$$
$$(s, k, pub^q) \leftarrow Gen^q(1^\tau, G)$$
$$corr \leftarrow Corrupt_u(s)$$
$$\textbf{return } \text{STAT}_u(1^\tau, G, pub^q, corr, \alpha_u)$$

The algorithm Gen^q used in $\mathbf{Exp}_u^q(1^\tau, G)$ differs from Gen for the way part of the public information pub^q is computed. Indeed, for any $i = 1, \ldots, j$, the public values associated to the edge (u_i, u_{i+1}) is computed as the encryption $Enc(pp, \pi_i, \gamma_{u_{i+1}} || r_{u_i})$ where $\gamma_{u_{i+1}}$ and r_{u_i} are random values in $\{0,1\}^\tau$ and $\pi_i \leftarrow KD(pp, \gamma_{u_{i+1}} || r_{u_i})$. Moreover, $\mathbf{Exp}_u^2(1^\tau, G)$ differs from $\mathbf{Exp}_u^3(1^\tau, G)$ for the way α_{u_j} is constructed. Specifically, $\alpha_{u_j} \leftarrow \mathcal{F}(\gamma_{u_j}, r_{u_j})$ in $\mathbf{Exp}_u^2$ while $\alpha_{u_j} \leftarrow \{0,1\}^\tau$ in $\mathbf{Exp}_u^3$.

Now we show that, the adversary's view in $\mathbf{Exp}_u^1$ is indistinguishable from the adversary's view in $\mathbf{Exp}_u^2$. Assume by contradiction that there exists a polynomial-time adversary A which is able to distinguish between the adversary STAT_u's views in experiments $\mathbf{Exp}_u^1$ and $\mathbf{Exp}_u^2$ with non-negligible advantage. We show how to construct a polynomial-time distinguisher D which uses A to break the security of the (T, μ)-source PRV-CDA2 MLE scheme.

In particular, the distinguisher D constructs the public values associated to the edges (u_i, u_{i+1}), for $i = 1, \ldots, j$, calling the RoR oracle where $\mathbf{M}$ implements

---

Let $\Pi = (PPGen, KD, Enc, Dec, Valid)$ be an MLE scheme.

Algorithm for M

On input the public parameter pp output by $PPGen$, and a τ-length value π, produces T messages $m_1, \ldots, m_T$ as follows:

- choose $r_T \in \{0,1\}^\tau$;
- compute $m_T \leftarrow \pi \| r_T$;
- compute $\pi_T \leftarrow KD(pp, \pi \| r_T)$;
- for $i = 1, \ldots, T-1$, choose $r_i \in \{0,1\}^\tau$;
- for $i = 1, \ldots, T-1$, compute $\pi_i \leftarrow KD(pp, \pi_{i+1} \| r_i)$;
- for $i = 1, \ldots, T-1$, compute $m_i \leftarrow \pi_{i+1} \| r_i$.

Fig. 4. Polynomial size circuit M representing a joint distribution over T messages.

the circuit of Fig. 4 with $\pi = \pi_{u_{j+1}}$ and $T = j$. Notice that M is (T, μ)-source. Indeed, since r_{u_i} is chosen at random by M, it is easy to see that the min-entropy of m_i is at least τ, for each $i = 1, \ldots, T$. Distinguisher D is defined in Fig. 5.

Notice that if $\mathsf{mode} = \mathsf{real}$, then STAT_u's view is that of $\mathbf{Exp}_u^1(1^\tau, G)$ while when $\mathsf{mode} = \mathsf{rand}$, STAT_u's view is that of $\mathbf{Exp}_u^2(1^\tau, G)$. Therefore, if the algorithm A is able to distinguish between such views with non negligible advantage, it follows that D is able to break the $\mathsf{PRV\text{-}CDA2}$ security of the MLE scheme.

Algorithm $D^{\mathcal{ROR}(mode,pp,\cdot)}(1^\tau)$

$\pi_{u_{t+1}} \leftarrow \{0,1\}^\tau$
$corr \leftarrow \emptyset$
for each $i = t, \ldots, j+1$
$\quad r_{u_i} \leftarrow \{0,1\}^\tau$
$\quad \pi_{u_i} \leftarrow KD(pp, \pi_{u_{i+1}} \| r_{u_i})$
$\quad k_{u_i} \leftarrow \mathcal{F}(\pi_{u_i}, r_{u_i})$
$\quad p_{(u_i, u_{i+1})} \leftarrow Enc(pp, \pi_{u_i}, \pi_{u_{i+1}} \| r_{u_i})$
$\quad corr \leftarrow corr \cup (\pi_{u_i}, k_{u_i}, r_{u_i})$
Let $c_i = p_{(u_i, u_{i+1})}$, for $i = 1, \ldots, j$
$(c_1, \ldots, c_j) \leftarrow \mathsf{RoR}(mode, pp, M(\pi_{u_{j+1}}, j))$
$b \leftarrow A(1^\tau, G, pub, corr, k_{u_j})$

Algorithm $D'^{f(\cdot)}(1^\tau)$

$\pi_{u_{t+1}} \leftarrow \{0,1\}^\tau$
for each $i = t, \ldots, j+1$
$\quad r_{u_i} \leftarrow \{0,1\}^\tau$
$\quad \pi_{u_i} \leftarrow KD(pp, \pi_{u_{i+1}} \| r_{u_i})$
$\quad k_{u_i} \leftarrow \mathcal{F}(\pi_{u_i}, r_{u_i})$
$\quad p_{(u_i, u_{i+1})} \leftarrow Enc(pp, \pi_{u_i}, \pi_{u_{i+1}} \| r_{u_i})$
$corr \leftarrow (\pi_{u_{j+1}}, k_{u_{j+1}}, r_{u_{j+1}})$
for each $i = j, \ldots, 1$
$\quad r_{u_i} \leftarrow \{0,1\}^\tau$
$\quad \pi_{u_i} \leftarrow \{0,1\}^\tau$
$\quad p_{(u_i, u_{i+1})} \leftarrow Enc(pp, \pi_{u_i}, \pi_{u_{i+1}} \| r_{u_i})$
$\alpha_{u_j} \leftarrow f(r_{u_j})$
$b \leftarrow B(1^\tau, G, pub, corr, \alpha_{u_j})$

Fig. 5. Distinguishers D and D'.

Now we show that, the adversary's view in $\mathbf{Exp}_u^2$ is indistinguishable from the adversary's view in $\mathbf{Exp}_u^3$. Assume by contradiction that there exists a polynomial-time algorithm B which is able to distinguish between the adversary STAT_u's views in experiments $\mathbf{Exp}_u^2$ and $\mathbf{Exp}_u^3$ with non-negligible advantage.

We show how to construct a polynomial-time distinguisher D' which uses B to distinguish whether its oracle $f(\cdot)$ corresponds to the pseudorandom function $\mathcal{F}(k, \cdot)$ or to a random function $\mathcal{F}(\cdot)$.

Notice that if α_{u_j} corresponds to the evaluation of the pseudorandom function $\mathcal{F}(k, \cdot)$ on r_{u_j} then $\mathtt{STAT}_u$'s view is that of $\mathbf{Exp}_u^2(1^\tau, G)$ while when it is the output of a random value, $\mathtt{STAT}_u$'s view is that of $\mathbf{Exp}_u^3(1^\tau, G)$. Therefore, if the algorithm B is able to distinguish between such views with non negligible advantage, it follows that the distinguisher D' is able to break the pseudorandomness of $\mathcal{F}$. Figure 5 defines distinguisher D'.

Algorithm $D''^{\mathcal{ROR}(mode, pp, \cdot)}(1^\tau)$

$\quad \pi_{u_{t+1}} \leftarrow \{0, 1\}^\tau$
$\quad corr \leftarrow \emptyset$
$\quad \textbf{for each } i = t, \dots, j+1$
$\quad\quad r_{u_i} \leftarrow \{0, 1\}^\tau$
$\quad\quad \pi_{u_i} \leftarrow KD(pp, \pi_{u_{i+1}} \| r_{u_i})$
$\quad\quad k_{u_i} \leftarrow \mathcal{F}(\pi_{u_i}, r_{u_i})$
$\quad\quad p_{(u_i, u_{i+1})} \leftarrow Enc(pp, \pi_{u_i}, \pi_{u_{i+1}} \| r_{u_i})$
$\quad\quad corr \leftarrow corr \cup (\pi_{u_i}, k_{u_i}, r_{u_i})$
$\quad \text{Let } c_i = p_{(u_i, u_{i+1})}, \text{ for } i = 1, \dots, j$
$\quad (c_1, \dots, c_j) \leftarrow \mathtt{RoR}(mode, pp, \mathtt{M}(\pi_{u_{j+1}}, j))$
$\quad \rho \leftarrow \{0, 1\}^\tau$
$\quad b \leftarrow C(1^\tau, G, pub, corr, \rho)$

Fig. 6. Distinguisher D''.

We finally show that, the adversary's view in $\mathbf{Exp}_u^4$ is indistinguishable from the adversary's view in $\mathbf{Exp}_u^3$.

Assume by contradiction that there exists a polynomial-time algorithm C which is able to distinguish between the adversary $\mathtt{STAT}_u$'s views in experiments $\mathbf{Exp}_u^4$ and $\mathbf{Exp}_u^3$ with non-negligible advantage.

Notice that such views differ only for the public values associated to the edges (u_i, u_{i+1}) for $i = 1, \dots, j$. We show how to construct a polynomial-time distinguisher D'' which uses C to break the PRV-CDA2 security of the MLE scheme. In particular, the algorithm D'', on input 1^τ, constructs the public values associated to the edges (u_i, u_{i+1}), for $i = 1, \dots, j$, calling the RoR oracle where $\mathbf{M}$ implements the circuit of Fig. 4 with $\pi = \pi_{u_{j+1}}$ and $T = j$.

Formally, distinguisher D'' is defined in Fig. 6. Notice that if $\mathtt{mode} = \mathtt{real}$, then $\mathtt{STAT}_u$'s view is that of $\mathbf{Exp}_u^4$ while when $\mathtt{mode} = \mathtt{rand}$, $\mathtt{STAT}_u$'s view is that of $\mathbf{Exp}_u^3$.

Thus, if C distinguishes such views with non negligible advantage, it follows that algorithm D'' breaks the PRV-CDA2 security of the MLE scheme. $\qquad \square$

We now show that the MLE-based construction is secure in the sense of $\mathtt{Strong\text{-}KC}$.

Theorem 2. *Let $\Pi = (PPGen, KD, Enc, Dec, Valid)$ be an MLE scheme whose key derivation function KD is collision-resistant. The MLE-based VHKAS of Fig. 2 is secure in the sense of* Strong–KC.

Sketch of the Proof. Let $V = \{u_1, \ldots, u_t\}$ and $(u_i, u_{i+1}) \in E$, for $i = 1, \ldots, t-1$. We show by contradiction that if there exists a static adversary SSTAT whose advantage $\mathbf{Adv}_{\text{SSTAT}}^{\text{Strong-KC}}$ is non-negligible, then there exists a PPT adversary $\mathcal{A}$ such that $Pr[(x_0, x_1) \leftarrow \mathcal{A}(1^\tau, KD(pp, \cdot)) : x_0 \neq x_1 \wedge KD(pp, x_0) = KD(pp, x_1)]$ is non-negligible.

The adversary SSTAT first produces two public values pub and pub' along with the private information. Then, it chooses a class $u \in V$ such that $Ver(1^\tau, G, u, s_u, pub) = Ver(1^\tau, G, u, s_u, pub') = 1$ while there exists a class $v \prec u$ where $Der(1^\tau, G, u, v, s_u, pub) = k_v$, $Der(1^\tau, G, u, v, s_u, pub') = k'_v$ and $k_v \neq k'_v$.

W.l.o.g., let $u = u_i$ and $v = u_j$, for some $1 \leq i < j \leq t$ where j is the smallest index such that $k_{u_j} = \mathcal{F}(\pi_{u_j}, r_{u_j})$ is different from $k'_{u_j} = \mathcal{F}(\pi'_{u_j}, r'_{u_j})$.

We distinguish the following two cases:

1. $\pi_{u_j} = \pi'_{u_j}$ and $r_{u_j} \neq r'_{u_j}$;
2. $\pi_{u_j} \neq \pi'_{u_j}$.

In the following, for each class u_s we will denote by $s_{u_z} = (pp, r_{u_s}, \pi_{u_s}, k_{u_s})$ and $s_{u'_z} = (pp, r'_{u_s}, \pi'_{u_s}, k'_{u_s})$, the private information computed by using the derivation procedure with respect to pub and pub'. Consider the first case. In order for the verification procedure to succeed on both pub and pub' it must be $\pi_{u_j} = KD(pp, \pi_{u_{j+1}} || r_{u_j}) = KD(pp, \pi'_{u_{j+1}} || r'_{u_j})$ where $\pi_{u_{j+1}}$ may or may not be equal to $\pi'_{u_{j+1}}$. Thus, the adversary $\mathcal{A}$ wins his game by exhibiting $x_0 = \pi_{u_{j+1}} || r_{u_j}$ and $x_1 = \pi'_{u_{j+1}} || r'_{u_j}$.

Now, consider the second case. In order for the verification procedure to succeed on both pub and pub' it must be

$$
\begin{aligned}
\pi_{u_i} &= KD(pp, \pi_{u_{i+1}} || r_{u_i}) \\
&= KD(pp, KD(pp, \pi_{u_{i+2}} || r_{u_{i+1}}) || r_{u_i}) \\
&= KD(pp, KD(\ldots (KD(pp, \pi_{u_j} || r_{u_{j-1}}) || r_{u_{j-2}}) \ldots) || r_{u_i})
\end{aligned}
$$

and

$$
\begin{aligned}
\pi_{u_i} &= KD(pp, \pi'_{u_{i+1}} || r'_{u_i}) \\
&= KD(pp, KD(pp, \pi'_{u_{i+2}} || r'_{u_{i+1}}) || r'_{u_i}) \\
&= KD(pp, KD(\ldots (KD(pp, \pi'_{u_j} || r'_{u_{j-1}}) || r'_{u_{j-2}}) \ldots) || r'_{u_i}).
\end{aligned}
$$

Since π_{u_j} is different from π'_{u_j} and $\pi_{u_i} = KD(pp, \pi_{u_{i+1}} || r_{u_i}) = KD(pp, \pi'_{u_{i+1}} || r'_{u_i})$ it holds that KD is not collision resistant, thus if SSTAT wins his game with non-negligible probability also $\mathcal{A}$ succeeds with overwhelming probability. $\square$

4.2 A Concrete Instance

In Fig. 7 we instantiate the scheme of Fig. 8 with the deterministic MLE scheme $\Pi_{det}^{(q)}$ proposed in [1].

Let $G = (V, E) \in \Gamma$ (family of graphs corresponding to chains.) Let $(PPGen, KD, Enc, Dec, Valid)$ be the $\Pi_{det}^{(q)}$ scheme and let $\mathcal{F} : \{0,1\}^\tau \times \{0,1\}^\tau \to \{0,1\}^\tau$ be a PRF.

Algorithm $Gen(1^\tau, G)$

1. Let $u_1, \ldots, u_t$ be the classes in the chain.
2. Let $pp = (H_1, H_2, q) \leftarrow PPGen(1^\tau)$. Let $\pi_{u_{t+1}} \leftarrow \{0,1\}^\tau$;
3. For each class u_i, for $i = t, \ldots, 1$, let
   (a) $r_{u_i} \leftarrow \{0,1\}^\tau$;
   (b) $\pi_{u_i} \leftarrow KD(pp, \pi_{u_{i+1}} \| r_{u_i}) = H_1(\pi_{u_{i+1}} \| r_{u_i} \| 1) \oplus \ldots \oplus H_1(\pi_{u_{i+1}} \| r_{u_i} \| t+1)$
   (c) $k_{u_i} \leftarrow \mathcal{F}(\pi_{u_i}, r_{u_i})$; $s_{u_i} = (pp, r_{u_i}, \pi_{u_i}, k_{u_i})$;
4. Let s and k be the sequences of private information $s_{u_1}, s_{u_1}, \ldots, s_{u_t}$ and keys $k_{u_1}, k_{u_1}, \ldots, k_{u_t}$, respectively, computed in the previous step;
5. For each $i = 1, \ldots, t$, let $w_{u_{i+1}} = H_2(\pi_{u_{i+1}} \| r_{u_i} \| 1) \oplus \ldots \oplus H_2(\pi_{u_{i+1}} \| r_{u_i} \| t+1)$, compute the public information $p_{(u_i, u_{i+1})} \leftarrow Enc(pp, \pi_{u_i}, \pi_{u_{i+1}} \| r_{u_i} \| w_{u_{i+1}})$.

Algorithm $Der(1^\tau, G, u, v, s_u, pub)$

1. Let $u = u_i$ and $v = u_j$, for some $j \geq i$, $j = 1, \ldots, t$;
2. For any $z = i, \ldots, j$, extract the public value $p_{(u_z, u_{z+1})}$ from pub
   (a) if $Valid(pp, p_{(u_z, u_{z+1})}) = 0$, return $\perp$;
   (b) otherwise, compute $(\pi_{u_{z+1}} \| r_{u_z} \| w_{u_{z+1}}) \leftarrow Dec(pp, \pi_{u_z}, p_{(u_z, u_{z+1})})$
3. Output $k_v \leftarrow \mathcal{F}(\pi_{u_j}, r_{u_j})$.

Algorithm $Ver(1^\tau, G, u, v, s_u, pub)$

1. Let $u = u_i$, for $i = 1, \ldots, t$;
2. For any $j = i, \ldots, t$,
   (a) extract the public value $p_{(u_j, u_{j+1})}$ from pub
   (b) if $Valid(pp, p_{(u_j, u_{j+1})}) = 0$, return 0;
   (c) let $\alpha_i = \pi_{u_i}$, compute $(\alpha_{j+1} \| \beta_j \| \gamma_{j+1}) \leftarrow Dec(pp, \alpha_j, p_{(u_j, u_{j+1})})$;
   (d) if $KD(pp, \alpha_{j+1} \| \beta_j)$ is different from α_j, return 0;
3. return 1.

Fig. 7. An instance of the MLE-based Construction.

$\Pi_{det}^{(q)}$ uses as a building block a symmetric-key encryption scheme $\mathcal{SE} = (\mathcal{K}, \mathcal{E}, \mathcal{D})$ and two hash functions $H_1 : \{0,1\}^* \to \{0,1\}^\tau$ and $H_2 : \{0,1\}^* \to \{0,1\}^\rho$ with randomness length ρ. If $\mathcal{SE}$ is an IND-CPA secure scheme and H_1 and H_2 are modeled as random oracles, then, for any $T = poly(\tau)$ and any $k = \omega(log\tau)$, $\Pi_{det}^{(q)}$ is q-query (T, k)-source PRV-CDA2-secure.

The scheme $\Pi_{det}^{(q)} = (PPGen, KD, Enc, Dec, EQ, Valid)$ is defined as follows:

- **Parameter-generation algorithm:** On input 1^λ, the algorithm $PPGen$ chooses two hash functions $H_1 : \{0,1\}^* \to \mathcal{K}$ and $H_2 : \{0,1\}^* \to \{0,1\}^\rho$. It outputs the public parameters $pp = (H_1, H_2, q)$.

- **Key-derivation function:** The algorithm KD takes as input public parameters pp, the message m and outputs the message-derived key $k_m = H_1(m||1) \oplus H_1(m||2) \oplus \ldots \oplus H_1(m||q+1) \in \mathcal{K}$.

- **Encryption algorithm:** The algorithm Enc takes as input public parameters pp, a message m, and a message-derived key k_m. It computes $w_m = H_2(m||1) \oplus H_2(m||2) \oplus \ldots \oplus H_2(m||q+1)$ and outputs $\mathrm{E}_{k_m}(m||w_m) \in C$.

- **Validity test:** The algorithm $Valid$ outputs 1 on any input $c \in C$.

- **Decryption algorithm:** Dec takes as input public parameters pp, a ciphertext c, and a message-derived key k_m and outputs $m \leftarrow D_{k_m}(c)$.

- **Equality algorithm:** Algorithm EQ on input public parameters pp and ciphertexts c_1 and c_2 outputs 1 if and only if $c_1 = c_2$.

From the PRV-CDA2 security of $\Pi_{det}^{(q)}$ the instance of Fig. 7 is secure in the random oracle model.

5   Handling Key Replacement

Cryptographic keys need to be periodically changed. Thus, a key assignment scheme should feature an efficient procedure for the TA to handle key replacements.

A VHKAS which handles key replacement is a tuple $(Gen, Der, Ver, KReplace)$, where (Gen, Der, Ver) is defined in Definition 7 and algorithm $KReplace$ satisfies the following conditions:

- The *key replacement algorithm* $KReplace$ is probabilistic polynomial-time. It takes as input 1^τ and a graph $G = (V, E)$ in Γ, a class u, the secret information s, the public information pub and produces as output (s, k, pub).

A key replacement procedure may require both public information and private values to be changed. Ideally, such a procedure will only change public information so that private values are not redistributed. In general, it is desirable to design the key replacement algorithm to modify as few private values as possible.

Let (Gen, Der, Ver) be the scheme of Figure 2. The MLE-based scheme with key replacement is the tuple $(Gen, Der, Ver, KReplace)$ where $KReplace$ is as follows:

Algorithm $KReplace(1^\tau, G, u, s, pub)$

1. Let $u_1, \ldots, u_t$ be the classes in the chain.
2. Let $u = u_j$, for some $1 \leq j \leq t$;
3. For $i = j, \ldots, 1$, let
   (a) choose a new $r_{u_i} \leftarrow \{0,1\}^\tau$;
   (b) $\pi_{u_i} \leftarrow KD(pp, \pi_{u_{i+1}} \| r_{u_i})$;
   (c) $p_{(u_i, u_{i+1})} \leftarrow Enc(pp, \pi_{u_i}, \pi_{u_{i+1}} \| r_{u_i})$;
   (d) $k_{u_i} \leftarrow \mathcal{F}(\pi_{u_i}, r_{u_i})$.
4. Let s, k, and pub be the new sequences of private information, keys and public values, respectively;
5. Output(s, k, pub).

Fig. 8. MLE-based construction with key replacement.

Figure 8 shows how the TA can handle the replacement of a key k_u for a class u in the MLE-based construction of Fig. 2. In such a procedure only the classes higher than u in the chain are affected by the change.

In Fig. 9 we describe the key replacement procedure when the MLE-based construction is instantiated with the deterministic MLE scheme $\Pi_{det}^{(q)}$ proposed in [1].

Algorithm $KReplace(1^\tau, G, u, s, pub)$

1. Let $u_1, \ldots, u_t$ be the classes in the chain.
2. Let $u = u_j$, for some $1 \leq j \leq t$;
3. Choose a new $r_{u_j} \leftarrow \{0,1\}^\tau$;
4. For $i = j, \ldots, 1$, let $w_{u_{i+1}} = H_2(\pi_{u_{i+1}} \| r_{u_i} \| 1) \oplus \ldots \oplus H_2(\pi_{u_{i+1}} \| r_{u_i} \| t + 1)$
   (a) $\pi_{u_i} \leftarrow KD(pp, \pi_{u_{i+1}} \| r_{u_i})$;
   (b) $p_{(u_i, u_{i+1})} \leftarrow Enc(pp, \pi_{u_i}, \pi_{u_{i+1}} \| r_{u_i} \| w_{u_{i+1}})$;
   (c) $k_{u_i} \leftarrow \mathcal{F}(\pi_{u_i}, r_{u_i})$.
5. Let s, k, and pub be the new sequences of private information, keys and public values, respectively;
6. Output(s, k, pub).

Fig. 9. The key replacement procedure for the instance of Fig. 7.

6 Comparisons with Hierarchical Key Assignment Schemes

We evaluate the performance of our construction by comparing it with that of popular IND-ST HKAS from the literature having the feature that the public information is used to store encryption keys. The comparison shows that the performance of our construction is similar to that of such schemes despite the fact that it also achieves the notion of Strong-KC.

Figure 10 shows the comparison. The summary takes into account several parameters, such as the size of the public and private information, the number and the type of operations required by a class $u \in V$ to compute the key of a class v lower down in the hierarchy. Moreover, it specifies the notions of security achieved. In Fig. 10, τ and τ_1 correspond respectively to the size of the secret key in symmetric encryption based constructions and in schemes obtained from factoring[6]. The value c is a constant depending on the underlying encryption scheme. For instance c is equal to 2 for the so called XOR construction in [7]. Notice that, to describe the parameters of the MLE-based construction we refer to the concrete instance described in Sect. 4.2 where the MLE considered uses as a building block a symmetric encryption scheme. Finally, w represents the width of the poset which corresponds to the number of chains in the partition.

Scheme	Public info.	Private info.	Key derivation	Security Notions
MLE-based §4	$|E|c\tau$	$\tau O(w)$	$Path(u,v)$ MLE decryptions +1 PRF eval.	IND-ST Strong-KC.
EBC [20]	$(|E|+|V|)c\tau$	τ	$Path(u,v)+1$ decryptions	IND-ST.
Freire et al. [21]	τ_1	$\tau_1 O(w)$	$Path(u,v)+1$ Modular squaring operations	IND-ST.
Atallah et al. [5]	$2|E|c\tau+|V|\tau$	τ	$2 \cdot Path(u,v)+1$ operations: -$Path(u,v)$ decryptions -$Path(u,v)+1$ PRF eval	IND-ST.

Fig. 10. Comparisons with Hierarchical Key Assignment Schemes.

In [18], De Capitani di Vimercati et al. proposed a data-outsourcing architecture which employs the HKAS in [5] for representing an authorization policy trough an equivalent encryption policy. The positive results provided by the experimental analysis performed in [18], are encouraging in order to evaluate the feasibility of our approach. Indeed, as shown above the performance of our MLE-based construction in terms of space required to store public information and key derivation operations are similar to that exhibited by the HKAS in [5]. Only, the users need to manage a bigger number of secrets. Also, the verification procedure should not add significant overhead, indeed, it consists of computing the value of a collision resistant function (step 3(c) of the Ver Algorithm of Fig. 2) at each step of derivation beside the decryption operation (step 3(b) of Der Algorithm of Fig. 2).

---

[6] 1024-bit, 2048-bit, 3072-bit RSA keys are equivalent in strength to respectively, 80-bit, 112-bit, 128-bit symmetric keys.

7   Conclusions

In this paper we have introduced Verifiable HKAS and have designed it using an MLE scheme as a building block. The security properties of our construction depends on those of the underlying MLE scheme. The concrete instance described in Sect. 4.2 achieves the notions IND-ST and Strong-KC in the random oracle model. Our proposal also manages with the replacement of compromised encryption keys by making a limited number of changes to public and private information. We leave as an interesting open problem that of building VHKAS secure in the standard model. Our solution produces a scheme which belongs to the family of IKEKAS, it would also be interesting to consider other HKAS families identified in [15] to construct VHKAS and study their performance.

Acknowledgements. The work of Ferrara and Ricciardi is partially supported by the project PON-ARS01 00860 titled *Ambient-intelligent Tele-monitoring and Telemetry for Incepting and Catering over hUman Sustainability - ATTICUS* funded by the Italian Ministry of Education and Research - RNA/COR 576347.

References

1. Abadi, M., Boneh, D., Mironov, I., Raghunathan, A., Segev, G.: Message-locked encryption for lock-dependent messages. In: Canetti, R., Garay, J.A. (eds.) CRYPTO 2013. LNCS, vol. 8042, pp. 374–391. Springer, Heidelberg (2013). https://doi.org/10.1007/978-3-642-40041-4_21
2. Akl, S.G., Taylor, P.D.: Cryptographic solution to a multilevel security problem. In: Advances in Cryptology: Proceedings of CRYPTO 1982, pp. 237–249. Plenum Press, New York (1982)
3. Alderman, J., Farley, N., Crampton, J.: Tree-based cryptographic access control. In: Foley, S.N., Gollmann, D., Snekkenes, E. (eds.) ESORICS 2017. LNCS, vol. 10492, pp. 47–64. Springer, Cham (2017). https://doi.org/10.1007/978-3-319-66402-6_5
4. Altaha, M., Muhajjar, R.: Lightweight key management scheme for hierarchical wireless sensor networks, pp. 139–147 (2017)
5. Atallah, M.J., Blanton, M., Fazio, N., Frikken, K.B.: Dynamic and efficient key management for access hierarchies. ACM Trans. Inf. Syst. Secur. **12**(3), 18:1–18:43 (2009)
6. Ateniese, G., De Santis, A., Ferrara, A.L., Masucci, B.: Provably-secure time-bound hierarchical key assignment schemes. J. Cryptol. **25**(2), 243–270 (2012)
7. Bellare, M., Desai, A., Jokipii, E., Rogaway, P.: A concrete security treatment of symmetric encryption. FOCS **1997**, 394–403 (1997)
8. Bellare, M., Keelveedhi, S., Ristenpart, T.: Message-locked encryption and secure deduplication. In: Johansson, T., Nguyen, P.Q. (eds.) EUROCRYPT 2013. LNCS, vol. 7881, pp. 296–312. Springer, Heidelberg (2013). https://doi.org/10.1007/978-3-642-38348-9_18
9. Stefano, B., Carbone Roberto, L.A.J., Silvio, R.: Exploring architectures for cryptographic access control enforcement in the cloud for fun and optimization (ASIA CCS 2020) (2020)

10. Castiglione, A., De Santis, A., Masucci, B., Palmieri, F., Castiglione, A., Huang, X.: Cryptographic hierarchical access control for dynamic structures. IEEE Trans. Inf. Forensics Secur. **11**(10), 2349–2364 (2016)
11. Castiglione, A., De Santis, A., Masucci, B., Palmieri, F., Huang, X., Castiglione, A.: Supporting dynamic updates in storage clouds with the Akl-Taylor scheme. Inf. Sci. **387**, 56–74 (2017)
12. Chang, C.C., Hwang, R.J., Wu, T.C.: Cryptographic key assignment scheme for access control in a hierarchy. Inf. Syst. **17**(3), 243–247 (1992)
13. Crampton, J., Daud, R., Martin, K.M.: Constructing key assignment schemes from chain partitions. In: Foresti, S., Jajodia, S. (eds.) DBSec 2010. LNCS, vol. 6166, pp. 130–145. Springer, Heidelberg (2010). https://doi.org/10.1007/978-3-642-13739-6_9
14. Crampton, J., Farley, N., Gutin, G., Jones, M., Poettering, B.: Cryptographic enforcement of information flow policies without public information. In: Malkin, T., Kolesnikov, V., Lewko, A.B., Polychronakis, M. (eds.) ACNS 2015. LNCS, vol. 9092, pp. 389–408. Springer, Cham (2015). https://doi.org/10.1007/978-3-319-28166-7_19
15. Crampton, J., Martin, K.M., Wild, P.R.: On key assignment for hierarchical access control. In: 19th IEEE Computer Security Foundations Workshop, (CSFW-19 2006), pp. 98–111. IEEE Computer Society (2006)
16. D'Arco, P., De Santis, A., Ferrara, A.L., Masucci, B.: Variations on a theme by akl and Taylor: Security and tradeoffs. Theor. Comput. Sci. **411**(1), 213–227 (2010)
17. De Capitani di Vimercati, S., Foresti, S., Jajodia, S., Paraboschi, S., Samarati, P.: A data outsourcing architecture combining cryptography and access control. In: CSAW 2007, pp. 63–69. ACM (2007)
18. De Capitani di Vimercati, S., Foresti, S., Jajodia, S., Paraboschi, S., Samarati, P.: Encryption policies for regulating access to outsourced data. ACM Trans. Database Syst. **35**(2), 12:1–12:46 (2010)
19. De Santis, A., Ferrara, A.L., Masucci, B.: New constructions for provably-secure time-bound hierarchical key assignment schemes. Theor. Comput. Sci. **407**(1–3), 213–230 (2008)
20. De Santis, A., Ferrara, A.L., Masucci, B.: Efficient provably-secure hierarchical key assignment schemes. Theor. Comput. Sci. **412**(41), 5684–5699 (2011)
21. Freire, E.S.V., Paterson, K.G.: Provably secure key assignment schemes from factoring. In: Parampalli, U., Hawkes, P. (eds.) ACISP 2011. LNCS, vol. 6812, pp. 292–309. Springer, Heidelberg (2011). https://doi.org/10.1007/978-3-642-22497-3_19
22. Freire, E.S.V., Paterson, K.G., Poettering, B.: Simple, efficient and strongly KI-secure hierarchical key assignment schemes. In: Dawson, E. (ed.) CT-RSA 2013. LNCS, vol. 7779, pp. 101–114. Springer, Heidelberg (2013). https://doi.org/10.1007/978-3-642-36095-4_7
23. Zarandioon, S., Yao, D.D., Ganapathy, V.: K2C: cryptographic cloud storage with lazy revocation and anonymous access. In: 7th International ICST Conference. SecureComm 2011, vol. 96, pp. 59–76 (2011)
24. Zhu, W.T., Deng, R.H., Zhou, J., Bao, F.: Applying time-bound hierarchical key assignment in wireless sensor networks. In: 13th International Conference. ICICS 2011, vol. 7043, pp. 306–318 (2011)

An ABAC Model with Trust and Gossiping (ABAC–TG) for Online Social Networks

Adi Swissa[1]($\boxtimes$) and Ehud Gudes[1,2]

[1] Department of Mathematics and Computer Science, The Open University, Raanana, Israel
ehudgu@openu.ac.il
[2] Department of Computer Science, Ben-Gurion University of the Negev, Beer-Sheva, Israel

Abstract. In this paper, we propose an attribute-based access control model called ABAC–TG for online social networks (OSNs). This model comprehensively considers user and object attributes and two main social attributes: trust and gossip, which are calculated based on the Ego-node (the user sharing the information) point of view. Each user is evaluated trust and gossip wise by several criteria, such as total number of friends, number of interactions between two users, and more. A new algorithm for calculating user gossiping value by graph clustering is defined, and this gossiping value can also be used for trust calculation. The ABAC model is formally presented, including rules and attribute definitions, and is demonstrated by several use case scenarios. The gossip and trust assessments provide more accurate and viable information-sharing decisions that serve the purpose of more precise and flexible authorizations.

This work is novel in two respects. First, we are using trust and gossip as dynamic attribute calculations. And second, we present a new algorithm for calculating the user's gossip value from the ego user point of view and use it either as part of the trust attribute calculation or as a separate attribute in the ABAC model.

Keywords: Attribute based access control (ABAC) · Gossip · Trust · Online social networks

1 Introduction

As online social networks (OSNs) increase in size and more people use them as their primary Internet website, the volume of information shared in OSNs keeps on growing.

The public accessibility of such networks with the ability to share opinions, thoughts, information, and experience offers great promise to people and communities. In addition to individuals using such networks to connect to their friends and families, governments and enterprises have started exploiting these platforms for delivering their services to citizens and customers [4]. Because of the sensitive and private information that is commonly stored in these networks, controlling access to this information is becoming very important and that depends largely on the level of trust that members have with each other. Several access control models for OSN based on trust have recently appeared [4, 5, 11, 14], but none of them uses the attribute of gossiping as a significant factor in the

© IFIP International Federation for Information Processing 2021
Published by Springer Nature Switzerland AG 2021
K. Barker and K. Ghazinour (Eds.): DBSec 2021, LNCS 12840, pp. 377–392, 2021.
https://doi.org/10.1007/978-3-030-81242-3_22

378   A. Swissa and E. Gudes

access control model. Zhang et al. [1] present an attribute-based access control (ABAC) model for OSN, but does not use either Trust nor Gossiping attributes. Since gossip is one of the oldest and most common means of information sharing among people, we consider it also very important for influencing access control.

This paper aims to demonstrate a new ABAC model called ABAC–TG, for an online social network, which combines privacy, trust, and a gossip model.

The general idea is to use the ABAC model with additional complex attributes such as user trust and gossip, calculated by clustering. The gossip attribute may be used as part of the trust calculation or as a separate attribute in an ABAC rule. The user selects attributes and defines rules for defining the access for a specific object. This model is extensible by adding additional dynamic attributes. The examples demonstrating the model use actions provided by a Facebook like network but are not limited to it.

This new model has three significant advantages: First, the model calculation is dynamic - the trust is calculated based on user selection and network parameters, and the gossip is dependent on specific network interactions.

The second is flexibility and scalability – we can add or remove attributes and decide on threshold values for trust and gossip calculations. The third is simplicity – the user will choose simple attributes and define the access to his objects in terms of these attributes. Thus, our model provides a solution to one of the most significant social network problems, the control and prevention of the spread of sensitive private information in the network.

The rest of this paper is structured as follows: Sect. 2 provides background information and an overview of related work. Sections 3 and 4 describe the new ABAC model, where Sect. 3 describes the model, trust, and gossip attributes calculations, and Sect. 4 discusses the rules and attributes used in detail and presents several use-cases of using the model. The last section presents the relevant conclusions and discusses future work.

2 Background and Related Work

As mentioned earlier, this paper's main goal is to propose a new ABAC model called ABAC–TG, which combines privacy, trust, and gossip attributes. In this section, the relevant background is provided.

An ABAC model relies upon evaluating attributes of the subject, attributes of the object, environment conditions, and the formal relationship or access control rule or policy defining the allowable operations for subject-object attribute combinations. All ABAC solutions contain these basic core capabilities to evaluate attributes and enforce rules or relationships between those attributes [1]. Examples of such rules for a social network are presented in this paper in Sect. 4.3.

The access control model in Facebook is based on roles. It does a reasonable job of access control while handling millions of operations/seconds from its billion users. The mechanism of Facebook is a function of communication history among users (for instance, the existence of friendship is necessary for certain policies), however even though it's a quite simple model, users often do not use it properly. An analysis of Facebook access control model and its privacy problems has recently appeared in [2, 8,

11]. However, Facebook access control does not use trust or gossip and lacks reliance on users' specific attributes and objects for Access control.

Recently, a trust-based model for a social network called RTBAC was presented in [5]. The RTBAC model is a combination of User-Trust attributes, based on real OSN characteristics, within an RBAC model that usually grants permissions solely to roles.

The trust value is defined on a scale of 0 to 1 since the decision of sharing information with a certain user is defined as a probability variable, 0 being no sharing willingness at all, 1 being definite sharing willingness. The trust model is based on several criteria such as quality of Friendship, connection strength, and users' similarity. We will use this model in our new ABAC-TG model to calculate the trust value described in Sect. 3.3. Another relevant work [14] describes the way trust can be used to identify adversaries and limit information flow to them.

The fast spread of information is a common and essential feature of social networks. A simple model of diffusion shows how bounded rational individuals can, just by tracking gossip about people, identify those who are most central in a network according to "diffusion centrality." [7, 9]. Gossip can essentially be defined as information passed from one individual (originator) to another (gossiper) about an absent third individual [13].

In [6], an algorithm is given for provably finding the clusters, provided there is a sufficiently large gap between internal density and external sparsity. This clustering is used to build knots of trust between users in [12]. Knots of trust are groups of community members having overall "strong" trust relations between them. In order to provide a member with reputation information relative to her viewpoint, the system must identify the knot to which that member belongs and interpret its reputation data correctly. Such clustering can be used to identify and measure the amount of "gossipness" of users and groups of users since users tend to gossip with other users they trust on. We'll use this clustering to define a "gossipness" measure in our proposed ABAC model.

The most relevant article to this work is by Zhang et al. [10]. They present an ABAC model for social networks with many examples for rules involving various attributes. Our model is similar but more general than [10] since it explicitly includes two important new attributes: Trust and Gossiping attributes. Our model is dynamic and extensible and can be used for any online social network.

3   A New ABAC Model with Trust and Gossip for Online Social Network

This section presents the main theme of this paper. The aim is to define a new ABAC model for online social networks (for instance, Facebook, Twitter, etc.) and to incorporate user trust calculation and gossip calculation in the ABAC model.

3.1   Establishment of ABAC-TG Model for Online Social Network

Today, Facebook, the world's most popular social network, uses the RBAC model to manage user roles on a specific object or action [2]. Our new ABAC-TG model will replace the RBAC model. The key difference with ABAC is the concept of policies that express a complex Boolean rule set that can evaluate many different attributes [1].

In the ABAC-TG model, the user will choose attributes from a predefined list and define rules. For instance, the user doesn't want the specific objectA to be visible to users who are mostly gossiping; or the user doesn't want the particular objectA to be visible to teenagers or to users who have low trust value.

Relevant attributes may be: a number of friends, age, education, job title, work, family status, friends type (e.g., in comparing to ego user's age, education, work, city, etc.), action types (e.g., comment to friends with the same age, work, city), and attributes of the objects themselves. More attributes will be described in Chapter 4.

Figure 1 demonstrates the access decision-making of the ABAC-TG model. For each object, the model checks if the object has roles restrictions, if yes, the model checks if the user has fitting attributes and grants access accordingly.

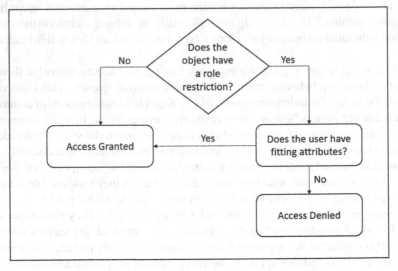

Fig. 1. Access decision in the ABAC-TG model

The new ABAC model includes user trust and user gossiping. Therefore we define two new attributes: GossipingAtt and TrustAtt.

GossipingAtt – this attribute describes the gossiping level of the user. The value will be calculated using a network-directed graph containing the interactions (messages, likes, comments, shares, etc.) between the user to other users, applying the clustering algorithm, and deciding if the user is gossiping. The gossiping value is on a scale of 0 to 1, where 0 means a total gossiping user, and 1 means not gossiping at all. The gossiping algorithm is described in Sect. 3.4.

TrustAtt – this attribute describes the social trust level of the user. The value is calculated by the model presented in [5]. The trust calculation is described in Sect. 3.3. Note that in [5] the main goal is to compute the user's trust, which is the base for access control, while in our model, trust is just one attribute among several attributes of the

ABAC model. Besides, we can adjust the trust calculation by using the GossipingAtt attribute. In case that the user chooses TrustAtt and GossipAtt in the rules, we do not use them twice in the trust model calculation.

Therefore, our model is dynamic due to the dependency on the user selection in calculating trust and ABAC attributes that may change in time. This capability defines a new perspective on the trust model from [5] with the new GossipingAtt and user selection, which affects the trust calculation.

3.2 Dynamic ABAC-TG Model

The decision algorithm of ABAC-TG is the same as the original ABAC model, by evaluating different attribute conditions and getting a Boolean value as a result for "allow access" or "deny access" (see Fig. 1).

If the user chooses the TrustAtt and an attribute1 that is already part of the trust formula, we have two options: first is to adjust the trust formula and remove attribute1 in order not to use it twice. The second is not to change the trust formula and let the user influence the weight of the attributes. In this paper, we choose the first option by removing it from the trust formula and use it in the ABAC rules.

For example, if the user defines a rule of – "I want that only my trusted friends (0.8) and my non-gossiping friends (0.8) will be able to see my image1". In this case, we'll adjust the trust formula and remove the gossiping parameter and use it only once in the rule. As a result, the user affects the trust value as a dynamic trust value and changes the gossiping value's weight as a more effective attribute (due to the 50% of the rule and not just a small part in the trust value). This use case is demonstrated in Sect. 4.3 on use case 3.

3.3 Trust Calculation

In our ABAC-TG model, we're using the trust attribute (denotes as TrustAtt) as yet another attribute. In order to calculate the TrustAtt value, we'll use the formula described in [5], with an extension to include the new friendship characteristic of gossiping value. The original formula from [5] to compute the user trust value (denotes as UTV) is shown in Fig. 2.

The different factors and their corresponding weights are explained in detail in [5]: taken into consideration user credibility factors (knowledge factors such as the total number of friends) denotes as u, and connection-based factors (friendship characteristics such as mutual friends) denotes as c. In order to add the gossiping value into the formula we just add another factor and its corresponding weight, such that the total sum of weights (denotes as w) is still 1.

The gossiping value will affect the trust value in the same manner of every friendship characteristic. The gossiping value calculation is described in Sect. 3.4, and the value is the probability from a scale of 0 to 1, 0 mainly being gossiping, 1 being not gossiping at all.

The trust value is also expressed on a scale of 0 to 1, 0 being no sharing willingness at all, 1 being definite sharing willingness.

$$u = \langle WiUi \rangle = \frac{\sum_{i=1}^{|u|} WiUi}{\langle W \rangle |u|} \qquad (8)$$

$$c = \langle WiCi \rangle = \frac{\sum_{i=1}^{|c|} WiCi}{\langle W \rangle |c|} \qquad (9)$$

$$UTV = \frac{c \cdot |c| + u \cdot |u|}{|c+u|} \qquad (10)$$

Fig. 2. The formula from [5] to calculate the trustAtt value. User Trust Value (UTV) is calculated as the weighted average of user credibility and connection strength. The weight is set according to the relative number of attributes in each category.

For example, in Table 1, we have two users: UserA and UserB, which are friends of Alice. Alice gave them different scores regarding what she thinks, based on her friendship experience. In this case, UserA and UserB have different gossiping values that affect the calculated trust value.

Table 1. An example of trust value including gossiping attribute

Username	Knowledge value (u)	GossipingAtt	Friendship value (c)	TrustAtt (UTV)	Result
UserA	0.51	0.23	0.35	0.43	The trust value is lower with a gossiping user
UserB	0.77	0.98	0.92	0.845	The trust value is higher with non-gossiping user

3.4 Gossiping Calculation

This section defines the algorithm for calculating the gossiping value for users in a social network. The gossiping attribute's goal is to identify a set of friends who would leak the shared information to an adversary. The gossiping value is the probability from a scale of 0 to 1, 0 being mostly gossiping, 1 being not gossiping at all.

The GossipAtt is calculated by taking into consideration connection-based factors: the number of human interactions between two users. The gossiping calculation is in the context of an Ego user perspective (the user sharing the information) to his friends and not general to the whole network. For example, from Alice's perspective, Bob's gossiping value is 0.8, but from David's perspective, it's 0.2, which leads to that Alice will share more information with Bob than David will share with Bob.

We assume, for this paper, that gossip serves to strengthen the relationship between gossipers and weakens the relationship between the victim and each gossiper [13]. Therefore, the friends with which the Ego user has fewer interactions will have the potential to gossip about him. Thus, in our algorithm, we focus on gossiping friends and exclude the "Ego's best friends" from the graph. We define "best friends" as users who have more than R interactions with each other. For example, a relationship of 100 interactions and above is with a high probability of being a best friend.

In our model, we consider only two levels of ego user's friends and friends of friends, for three reasons: the first reason is to have a comprehensive perspective on user's interactions and an extensive network graph. The second reason is to enable the ego user to share his post object with a wide enough forum (but not to a huge subnetwork) and limit it to non-gossiping users. The third reason is to restrict the network size to a reasonable amount of nodes to achieve better running time.

To compute the gossiping value, we use graph clustering based on the logic described in [6, 12]. A knot [12] is a subset of community members identified as having overall strong interaction relations. Two members i and j should belong to the same knot if i has high direct interaction in j denoted $I\_M (i, j)$. Knots are groups of members with strong interactions, sharing the same gossiping value from the Ego user perspective. Ego is an individual focal node, which is the specific user from which we consider the gossiping flow. It is, therefore, plausible that the gossiping of the same user may differ significantly between different Ego users.

A community is modeled as a directed graph $G = (V, E)$ that describes a social network, where V is the set of network's users, and E is the set of directed and weighted edges representing the users' interactions. The weight on a directed edge from vertex i to vertex j is the level of direct interactions i has in j at time t and is denoted by $I\_Mt(i, j)$. Since we deal with the state of the graph at time period t, we omit the time indicator for simplicity. An edge $(ui, uj) \in E$ exists only if ui has interactions with uj.

We refer to the task of identifying knots as graph clustering. In a social networking graph, these clusters could represent users with similar interactions. More specifically, we aim to find a partition of the community graph based on the direct interactions between pairs of members. For this purpose, we replace the interaction relations between any two members $I\_M(i, j)$ and $I\_M(j,i)$ with a weaker relation named Mutual Interactions in Member (MIM) which is the minimum of the above two values, that is, the two directed edges (i, j) and (j,i), are replaced by a single, undirected edge whose weight is $MIM(i, j) = MIM(j,i) = \min\{IM(i, j), I M(j,i)\}$.

This way, we can use the edge relation as the input for the clustering algorithm, which must decide if its two end-vertices should reside in the same cluster or not. Intuitively, the new relation is more stringent because it considers the minimum level of mutual interactions between any two members as the representing value of gossiping between them.

Gossip Algorithms:
Algorithm 1. Calculates the gossiping value for Ego user's friends. The algorithm returns a map of clusters and their gossip value.

Algorithm 1. CalcGossipGraph(G, u$_e$)

Input: G = (V, E) an undirected graph that describes the social network of Ego's user, which vertices represent users and edges represent the interactions relations between the users at their end-point vertices

      u$_e$ the ego user.

Output: M: a map of clusters and their gossip value.

1: R = 100
2: **for** each ego's direct friends u$_i$ **do**
3:      **if** the MIM (u$_e$, u$_i$) >= R **then**
4:          Remove u$_i$ from graph G.
5:          Add u$_i$ to cluster "best friends" and set in the map M with gossipValue equal
            to 1.
6:      **end if**
7: **end for**
8: Remove u$_e$ from graph G
9: Create clusters based on the graph G.
10: **for** each cluster C **do**
11:     set gossipValue = min {Sum of MIM in C / (number of vertices in C * r), 1}.
12: **end for**
13: Return M as map of clusters and gossipValues.

In Algorithm 1, we defined Ego's best friends as friends which MIM bigger than 100 interactions, denotes as R, (a parameter which obviously can be changed), as gossip serves to strengthen the relationship between gossipers and weakens the relationship between the victim and each gossiper [13].

Lines 1–8 remove the ego user and his best friends from the graph. Line 9 creates the clusters based on the algorithm described in [12]. Finally, line 11 sets the gossiping value of each user to the average MIM in the cluster, normalized by the factor R. An example for this calculation is shown below.

Algorithm 2. Returns the gossiping value for a specific user from the Ego user perspective. This algorithm is calls algorithm 1 to calculate the gossiping clusters.

Algorithm 2. GetUserGossipValue(G, u$_e$, u$_i$)

Input: G = (V,E) an undirected graph that describes the social network of Ego's user, which vertices represent users and edges represent the interactions relations between the users at their end-point vertices

      u$_e$ the ego user.

      u$_i$ – the user to evaluate the gossip value.

Output: V: U u$_i$'s gossiping value

1: M = CalcGossipGraph (G, u$_e$).
2: Find u$_i$ in M.
3: V = u$_i$'s clustering gossiping value.
4: Return V.

Table 2 and Fig. 3 that appears below present an example of Algorithm 1, with ten users: friends and friends of friends of Alice. Based on their interactions, we calculate the gossiping value of each user from Alice's perspective. The edges represent the interactions with the value of MIM.

Users 1,2,3 have more than R interactions with u_e and defined as "Ego's user best friends" with a gossiping value of 1 (see cluster "best friends" in figure).

Users 4,5 has less than R MIM value, and they are sharing the same cluster1, with gossiping value of C1 = min {60/(3 * 100), 1} = 0.2.

Users 7,8,10 sharing the same cluster2 with gossiping value of C2 = min {245/(6 * 100), 1} = 0.4.

Users 9,6 sharing the same cluster3 with gossiping value of C3 = min {250/(4 * 100), 1} = 0.62.

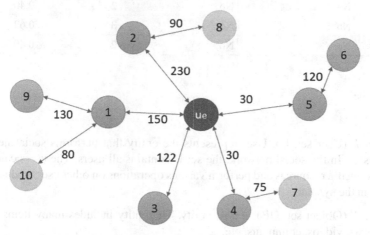

Fig. 3. An example of CalcGossipingGraph is described in table2. Nodes 1,2,3 are in cluster "best friends", nodes 4,5 are in cluster1, nodes 7,8,10 are in cluster2, and nodes 9 and 6 are in cluster3.

Note that the gossiping value calculation may run once for each ego user and stored in cache memory. The recalculation of it, which reflects the model dynamics, may be a parameter that depends on the number of new interactions in the ego user subnetwork, which the network administrator can set.

4 Formalization and Applications of the ABAC-TG Model

This section describes ABAC-TG model's formalization by defining user's attributes, object's attributes, and rules. We also represent five main use cases for using this model in an online social network.

4.1 Definition and Formal Descriptions of ABAC-TG Model Components

We define a rule definition for our ABAC-TG model. Our syntax is based on article [10], but it's simpler and focuses on trust value, gossiping value, and the dynamic model, which the user defines.

Table 2. An example of gossiping calculation for Alice's friends

Username	Is friend of ego user?	Is best friends of ego user?	Part of cluster #number	GossipingAtt
User1	Yes	Yes	---	1
User2	Yes	Yes	---	1
User3	Yes	Yes	---	1
User4	Yes	No	1	0.2
User5	Yes	No	1	0.2
User6	No	No	3	0.62
User7	No	No	2	0.40
User8	No	No	2	0.40
User9	No	No	3	0.62
User10	No	No	2	0.40

Definitions:

Definition 1. (User set, U): User represents the entity that performs social network's user accesses. In the social network, the set U contains all users. The users can upload and access media resources and perform various operations on other users and resources available in the system.

Definition 2. (Object set, OB): a post entity. The entity includes many items such as images, texts, videos, comments, etc.

Definition 3. (Actions set, AC): a post action. The actions include different activities such as display, share, post, like, comment, etc.

Definition 4. (User attribute set, AU): The set AU includes user basic information attributes, user social relationships attributes, and user community attributes. The user basic information attributes include name, age, identity, hobbies, and user-level attributes (as described below in Sect. 4.2).

Definition 5. (Object attribute set, OU): The set OU includes object basic information attributes. The object's basic information attributes include attributes such as publish date time, location, object type, related objects (e.g., comment on an object has a related object of the original object – post). Image object has corresponding objects of the users who appear in the image, check-in object has a location type attribute such as restaurant, work office, etc.).

Definition 6. (Attribute expression set, AE): The set AE includes AU and OU expressions, separated by the *and* ($\wedge$) and *or* ($\vee$) operations. The not sign is allowed by adding '!' before an expression.

Definition 7. (Basic Rule set, BR): a rule definition in ABAC-TG model as <U, OB, AC, AE>. If the rule condition is true, the user can access the object. If the rule is false, the user cannot access the object.

Definition 8. An ABAC-TG instance is a tuple of <U, OB, AC, AE>, which is a combination of user, object, actions, and attributes. Figure 4 describes the ABAC-TG model and the connections between the various components.

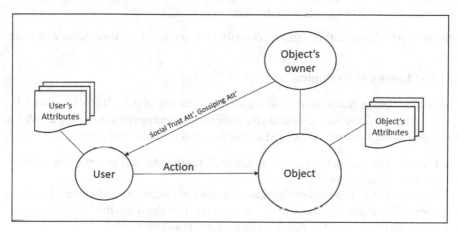

Fig. 4. ABAC-TG model components

4.2 Attributes Definitions

In this section, we define the main attributes that will be used in our ABAC-TG model. These attributes are examples, and we can add any attribute that is relevant to online social networks.

Attributes Definitions:

Definition 1. (TrustAtt): this attribute describes the user's social trust level, calculated as described in Sect. 3.3.

Definition 2. (GossipingAtt): this attribute describes the user's gossiping level, calculated as described in Sect. 3.4.

Definition 3. (AgeLevel): this attribute describes the user's age level (e.g. level1 is age 10–20, level 2 is age 20–40, level3 is age 40–60, level 4 is 60+).

Definition 4. (Education): this attribute describes the education of the user, for example: "Bachelor of Science in Computer Science," "Bachelor of Laws in Law", etc.

Definition 5. (Job title): this attribute describes the job title of the user, for example: "development manager", "software developer", "product owner", etc.

Definition 6. (Work): this attribute describes the workplace of the user, for example: "Microsoft", "Google", "Amazon", etc.

Definition 7. (Family status): this attribute describes the family status of the user, for example, single, married, divorced, etc.

Definition 8. (Friends type): this attribute describes the "friends' type", for example: in comparison to a user's age, education, work, city etc.

Definition 9. (Action types): this attribute describes the "Action types" of the user, for example, comments to friends with the same age, work, city, etc.

Definition 10. (Gender): this attribute describes the gender of the user: female or male

4.3   Use Cases and Examples

This section demonstrates the expressiveness and usability of the ABAC-TG model. We design several real-life scenarios and give their corresponding rules in our logic. We can define additional scenarios as needed as this model is dynamic and extensible.

Use Case 1. The user defines a rule which includes trustAtt and few attributes that are not part of the trust formula.

Scenario 1. Alice, a student 26 years old, post her locations and events, but she is suspicious. Therefore, she wants to share with users that she trusts them and at the same age and same university. The rule for this scenario is as follows:

S1: $<R = \{Alice\}, \{obj1\}, \{display\}, \{(trustAtt > 0.7) \wedge (ageLevel = myAgeLevel) \wedge (education = myEducation)\}>$.

Table 3 shows an example of two users with different social trust values, which leads to different access results in scenario1.

Table 3. An example of using scenario 1

Username	TrustAtt (including gossiping value)	GossipingAtt	AgeLevelAtt	Education	Result
UserA	0.6	0.5	27	Harvard U	Deny access
UserB	0.8	0.8	28	Harvard U	Allow access

Use Case 2. The user defines a rule which includes gossipingAtt, and more attributes.
Scenario 2. Bob, a political man in the United States that doesn't care who will see his posts in his country. He wants to get the most likes and comments; therefore, he shares his post with gossiping users. The rule for this scenario is as follows:

S2: $<R = \{Bob\}, \{obj2\}, \{display, comment, like\}, \{(gossipingAtt < 0.7) \wedge (friendTypeCountry = mycountry)\}>$

Table 4 shows an example of two users with different gossiping values, which leads to different access results in scenario2.

Table 4. An example of using scenario 2

Username	gossipingAtt	friendTypeCountry	Result
UserA	0.5	USA	Allow access
UserB	0.8	USA	Deny access

Use Case 3. The user defines a rule which includes trustAtt, gossipingAtt, and more attributes.

Scenario 3. Carlos, a COO of "FX" high-tech company, doesn't want to share his locations (check-in posts and event posts) with untrusted nor gossiping users but to share with employees who work with him. The rule for this scenario is as follows:

S3: <R = {Carlos}, {obj3}, {display}, {(gossipingLevel > 0.7) ∧ (trustLevel > 0.7) ^ (friendTypeWork = mywork)}>

Table 5 shows an example of two users with different gossiping values and social trust value, which leads to different access results in scenario3.

Table 5. An example of using scenario 3

Username	TrustAtt (without gossiping value)	gossipingAtt	friendTypeWork	Result
UserA	0.7	0.5	FX	Deny access
UserB	0.9	0.8	FX	Allow access

Use Case 4. The user defines a rule which includes trustAtt and few attributes that are part of the trust formula.

Scenario 4. David, a seller in the Facebook marketplace, would like to share his new items for selling with trusted users and popular users that will share his items. The rule for this scenario is as follows:

S4: <R = {David}, {obj4}, {display, like, comment, share}, {(TrustAtt > 0.7) ∧ (numOfFriends > 300)}>

Table 6 shows an example of two users with different social trust values and a number of friend's values, which leads to different access results in scenario4.

Table 6. An example of using scenario 4

Username	TrustAtt (including gossiping value, without numOfFriends value)	numOfFriends	Result
UserA	0.55	85	Deny access
UserB	0.75	350	Allow access

Use Case 5. The user defines a rule which includes simple attributes.

Scenario 5. Erin, a teenager 22 old, looks for a girlfriend and wants all the girls in his city to see and like his post. The rule for this scenario is as follows:

S5: <R = {Erin}, {obj5}, {display, like}, {(friendTypeCity = myCity) ∧ (FamilyStatus = Single) ∧ (ageLevel = myAgeLevel)}>

Table 7 shows an example of two users with different city attribute values, which leads to different access results in scenario5.

Table 7. An example of using scenario 5

Username	City	Family status	AgeLevelAtt	Result
UserA	Palo Alto	Single	28	Deny access
UserB	San Francisco	Single	22	Allow access

5 Conclusions and Future Work

5.1 Conclusion

In this paper, we have presented a new Access-Control model for an online social network. Our ABAC-TG model's novelty is its combination of user-attributes that includes trust and gossiping values based on real online social network characteristics.

The algorithm for computing the trust attribute is based on [5] but enables the addition and removal of attributes in its formulation based on what appears in the ABAC rules. We described a new algorithm for calculating the gossiping value uses graph clustering, and this value may be either included in the trust calculation or treated separately in the ABAC rule. This makes the model very flexible and adaptive. The attributes of this model were carefully picked, but there could be flexibility in these choices and their values that are debatable. This model can help to make important permission decisions and prevent unwanted information leakage from users, making online social network privacy better in many ways.

Our ABAC-TG model is dynamic and extensible and can be used for any social network that would like to enable users to choose who will see their data. We decided to demonstrate the new model by five different use cases on Facebook, as it's the world's most popular social network.

5.2 Future Work

In future work, we intend to continue exploring in several directions. First, we plan to conduct an extensive evaluation experiment. We like to evaluate the new ABAC model with Trust and Gossiping on a Facebook DB, as it's the world's most popular social network. We plan to build a database that includes users, objects, and user actions and attributes. These items will be extracted from a real Facebook network of at least 100 users. We plan to do three experiments, in one we let the users define their own rules using our model and get their feedback on its usability. In the second, we'll set ourselves rules and threshold values and compare our model results with the user's perceptions and expectations. Third, we like to evaluate the new gossiping algorithms on the same Facebook DB.

Second, we like to define an anonymity mechanism for Facebook objects for protecting shared objects by using summarization, filtering, blurring, and other techniques. Recently, initial work on this was published in [15]. We plan to extend it in several ways and integrate it with the ABAC-TG model. For example – define a filtering model – for sharing entities. The filter model will anonymize the data. For example, hide the username, hide the age and display range of ages, hide the gender, location, etc.

This mechanism will be part of the ABAC attribute definition by the user. For instance, this action is relevant for the "sharing" action on Facebook. Today the user adds a post, and his friends can share it, and the original user doesn't know who will see his post. Therefore, we would like to protect the objects, so that the other users will not be able to see the full object, but only part of it (the anonymization result).

References

1. Hu, V.C., et al.: Guide to attribute based access control (ABAC) definition and considerations. NIST Spec. Publ. **800**, 162 (2014)
2. Patil, V.T., Shyamasundar, R.K.: Undoing of privacy policies on facebook. In: Livraga, G., Zhu, S. (eds.) DBSec 2017. LNCS, vol. 10359, pp. 239–255. Springer, Cham (2017). https://doi.org/10.1007/978-3-319-61176-1_13
3. Lavi, T., Gudes, E.: Trust-based dynamic RBAC. ICISSP 317–324 (2016)
4. Sherchan, W., Nepal, S., Paris, C.: A survey of trust in social networks. ACM Comput. Surv. **45**(4), 47, 1–47, 33 (2013)
5. Voloch, N., Levy, P., Elmakies, M., Gudes, E.: An access control model for data security in online social networks based on role and user credibility. CSCML 156–168 (2019)
6. Mishra, N., Schreiber, R., Stanton, I., Tarjan, R.E.: Clustering social networks. WAW 56–67 (2007)
7. Banerjee, A., Chandrasekhar, A., Duflo, E., Jackson, M.O.: Gossip: identifying central individuals in a social network. CoRR abs/1406.2293 (2014)
8. Patil, V.T., Jatain, N., Shyamasundar, R.K.: Role of apps in undoing of privacy policies on facebook. In: Kerschbaum, F., Paraboschi, S. (eds.) DBSec 2018. LNCS, vol. 10980, pp. 85–98. Springer, Cham (2018). https://doi.org/10.1007/978-3-319-95729-6_6
9. banerjee, A., Chandrasekhar, A.G., Duflos, E., Jackoson, M.O.: Using gossips to spread information: theory and evidence from two randomized controlled trial, Rev. Econ. Stud. (2016).
10. Zhang, Z., Han, L., Li, C., Wang, J.: A novel attribute-based access control model or multimedia social netwroks. Neural Netw. World **26**(6), 543–557 (2016)

11. Pang, J., Zhang, Y.: A new access control scheme for Facebook-style social networks. Comput. Secur. **54**, 44–59 (2015)
12. Gal-Oz, N., Yahalom, R., Gudes, E.: Identifying knots of trust in virtual communities. In: Wakeman, I., Gudes, E., Jensen, C.D., Crampton, J. (eds.) IFIPTM 2011. IAICT, vol. 358, pp. 67–81. Springer, Heidelberg (2011). https://doi.org/10.1007/978-3-642-22200-9_8
13. Shaw, A.K., Tsvetkova, K., Daneshvar, R.: The effect of gossip on social networks. CompLex **16**(4), 39–47 (2011)
14. Voloch, N., Gudes, E.: An MST-based information flow model for security in online social networks. In: Proceedings of ICUFN, pp. 460–465 (2019)
15. Voloch, N., Nissim, P., Elmakies, M., Gudes, E.: A role and trust access control model for preserving privacy and image anonymization in social networks. In: Meng, W., Cofta, P., Jensen, C.D., Grandison, T. (eds.) IFIPTM 2019. IAICT, vol. 563, pp. 19–27. Springer, Cham (2019). https://doi.org/10.1007/978-3-030-33716-2_2

On Feasibility of Attribute-Aware Relationship-Based Access Control Policy Mining

Shuvra Chakraborty[✉] and Ravi Sandhu

Institute for Cyber Security (ICS) and NSF Center for Security and Privacy
Enhanced Cloud Computing (C-SPECC), Department of Computer Science,
University of Texas at San Antonio, San Antonio, TX, USA
{shuvra.chakraborty,ravi.sandhu}@utsa.edu

Abstract. This paper studies whether exact conversion to an AReBAC (Attribute-aware Relationship-Based Access Control) system is possible from an Enumerated Authorization System (EAS), given supporting attribute and relationship data. The Attribute-aware ReBAC Ruleset Existence Problem (ARREP) is defined formally and solved algorithmically, along with complexity analysis. Approaches to resolve infeasibility using exact solutions are discussed.

Keywords: Access control policy mining · Attribute-aware relationship-based access control · Attribute-based access control · Relationship-based access control

1 Introduction

Relationship-Based Access Control (ReBAC) [7] emerged from the access control requirements of Online Social Networks. ReBAC expresses authorization policy in terms of various relationship parameters such as type and depth, whereas Attribute-Based Access Control (ABAC) [8] has been motivated by its generalized structure and versatility in access control policy specification through attributes of users, resources and environment. Although ReBAC expresses authorization through direct and indirect relationships, there are cases where using relationships only is insufficient. Consider a social network policy where only adults (18 or higher age) can send a friend request to anyone who lives in the same location as themselves. Here, both age and location of each user in the network must be known. Formally, these two required characteristics/attribute can be incorporated with users to make policy generation more expressive and flexible. Integrating attributes with ReBAC components certainly add more expressiveness [6], formally named as Attribute-aware ReBAC (AReBAC).

Generally, access control policy mining facilitates automation of migrating from one access control system to another with certain set of assumptions such

K. Barker and K. Ghazinour (Eds.): DBSec 2021, LNCS 12840, pp. 393–405, 2021.
https://doi.org/10.1007/978-3-030-81242-3_23

as allowing direct use of entity ID in rule generation, strict or approximate equivalency between the source and generated policy, and availability of appropriate supporting data. Deployment of manual effort to convert from one access control system to another could be tedious, labor-intensive and error-prone.

This paper analyzes the feasibility of AReBAC policy mining from a given Enumerated Authorization System (EAS) under certain assumption, for example, no user ID will be allowed in the generated AReBAC rule. Note that, rule generation is always possible with use of entity IDs. The major contributions made in this paper are as follows.

- The first formal notion of Attribute-aware ReBAC RuleSet Existence Problem (ARREP) is developed. A novel algorithm for AReBAC policy mining feasibility detection is presented along with complexity analysis.
- Infeasibility problem in ARREP is formulated. Furthermore, exact solutions are proposed.
- Rule structure generality and unrepresented path label problem are noted.

2   Related Works

Although both ABAC and ReBAC have their own advantages to express authorization policies (see [1] for a rigorous comparison of their expressive power), integrating ABAC with ReBAC can provide finer-grained controls and improve the expressiveness of standalone ABAC or ReBAC. For example, [6] presents an attribute-aware ReBAC access control model.

Although the policy specification language in this paper is very different from [2,9], these two works are relevant related work. In [2], an approach to mine ABAC and ReBAC policies has been proposed where access control lists and incomplete information about entities are given. A few significant points about [2] are i) the proposed algorithm prefers the context of ReBAC mining because ReBAC is more general than ABAC, ii) entity ids are allowed to be used (which makes the generated policy less general), and iii) there is a policy quality metric available. Compared to [2], entity ids are strictly prohibited in the attribute-aware context of this paper. On the other hand, [9] presents an attribute-supporting ReBAC model for Neo4j (a popular graph database) that provides finer-grained access control by operating over resources.

While this paper introduces feasibility in the field of AReBAC policy mining for the first time, there are a few similar prior feasibility studies as follows.

- The work in [4] introduces ABAC RuleSet Existence Problem for the first time. Besides, the notion of infeasibility correction has been discussed.
- The work in [5] adapts and extends ABAC RuleSet Existence Problem for RBAC input. Additionally, it proposes infeasibility solution, with and without presence of supporting attribute data.
- In [3], feasibility of ReBAC policy mining has been investigated for the first time, assuming user to user relations are given by a static relationship graph.

3   Attribute-Aware ReBAC RuleSet Existence Problem

This section defines the ARREP along with a feasibility detection algorithm, complexity analysis and related issues.

3.1   Preliminaries

A user/subject is an entity who performs operation on a resource/object. The set of users is represented by U. A user requests to perform an operation on another user. An operation is an action performed by a user on another user. The set of operations in the system is represented by OP. Without loss of generality it is assumed that OP is a singleton given by $\{op\}$, since each operation has its specific policy or rules. An access request is a tuple $\langle u, v \rangle$ where user u is asking permission to perform operation op on user v where $u, v \in U, op \in OP, u \neq v$. An access request is either granted or denied, based on the access control policy. In any access control system, a logical construct is required to decide the outcome of an access request. The logical construct is formally defined as, $checkAccess: U \times U \rightarrow \{True, False\}$, where the result $True$ grants access while $False$ denies it.

We define a simple authorization system, EAS as follows:

Definition 1. *Enumerated Authorization System (EAS)*
An EAS is a tuple $\langle U, AUTH, checkAccess_{EAS} \rangle$ where, U is the finite sets of users and $AUTH \subseteq U \times U$, is a specified authorization relation where

$$checkAccess_{EAS}(u, v) \equiv (u, v) \in AUTH$$

For example, given U = $\{Alice, Bob\}$ and OP = $\{readData\}$, Bob can read Alice's data iff (Bob, Alice) belongs to AUTH.

In order to define an Attribute-aware ReBAC system, the key component is Attribute-aware Relationship Graph (ARG), which is defined as follows.

Definition 2. *Attribute-aware Relationship Graph (ARG)*
The Attribute-aware Relationship Graph
   $ARG = (V, VA, VA - RangeSet, UATTValue, EA, EA - RangeSet, E)$
is a directed labeled graph where,

a. *V is the set of vertices in ARG, representing the set of users in the system.*
b. *VA is the finite set of atomic user attribute function names $\{va_1, va_2, ..., va_m\}$.*
c. *For each $va_i \in VA$, $Range(va_i)$ specifies a finite set of atomic values for user attribute va_i. $VA\text{-}RangeSet = \{(va_i, value)|va_i \in VA \wedge value \in Range(va_i)\}$.*
d. *UATTValue denotes the user attribute value assignments. $UATTValue = \{UATTValue_{va_i}|va_i \in VA\}$ where $UATTValue_{va_i}: V \rightarrow Range(va_i)$. For convenience, we understand $va_i(a)$ to denote $UATTValue_{va_i}(a)$, that is the attribute value assignment of an actual user a for attribute va_i.*

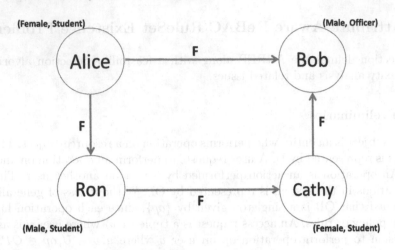

Fig. 1. Example ARG

e. *EA is the finite set of edge attribute function names, $\{ea_1, ea_2, ..., ea_n\}$.*
f. *For each $ea_i \in EA$, Range(ea_i) specifies a finite set of atomic values for edge attribute ea_i. EA-RangeSet $= \{(ea_i, value)|ea_i \in EA \wedge value \in Range(ea_i)\}$.*
g. *$E \subseteq V \times V \times Range(ea_1) \times Range(ea_2) \times ... \times Range(ea_n)$ is a finite set of directed edges where, an edge $(u, v, \sigma_1, \sigma_2, ..., \sigma_n) \in E$, $u \neq v$, represents the relations $\sigma_1, \sigma_2, ..., \sigma_n$ from user $u \in V$ to $v \in V$ in ARG where $\sigma_1 \in Range(ea_1), \sigma_2 \in Range(ea_2), ..., \sigma_n \in Range(ea_n)$.*
   Note: For a directed edge e from vertex a to vertex b in ARG, $ea_i(e)$ specifies the associated edge attribute value assignment for $ea_i \in EA$.

Figure 1 presents an ARG where the set of users V = {Alice, Bob, Cathy, Ron}, the set of user attribute function names, VA = {*Gender, Profession*}, the set of edge attribute function names, EA = {*Relation − type*}, and the set of edges E = {(Alice, Ron, F), (Alice, Bob, F), (Ron, Cathy, F), (Cathy, Bob, F)}. The user and edge attribute value assignments are shown in Fig. 1. The notion of a path in an ARG is defined as follows:

Definition 3. *Path in ARG*
Given ARG as in Definition 2 and a vertex pair $(u, v) \in V \times V$ where $u \neq v$, a path from u to v is a sequence of edges where the terminating (i.e., second) vertex of each edge is same as the starting (i.e., first) vertex of the next edge given by $\langle (u, v_i, \sigma_{w1}, \sigma_{w2}, ..., \sigma_{wn}), (v_i, v_j, \sigma_{x1}, \sigma_{x2}, ..., \sigma_{xn}), ..., (v_k, v_l, \sigma_{y1}, \sigma_{y2}, ..., \sigma_{yn}), (v_l, v, \sigma_{z1}, \sigma_{z2}, ..., \sigma_{zn}) \rangle$, where
a. *$u, v_i, v_j, ..., v_k, v_l, v \in V$*
b. *$\sigma_{w1}, \sigma_{x1}, ..., \sigma_{y1}, \sigma_{z1} \in Range(ea_1)$, $\sigma_{w2}, \sigma_{x2}, ..., \sigma_{y2}, \sigma_{z2} \in Range(ea_2)$, ..., $\sigma_{wn}, \sigma_{xn}, ..., \sigma_{yn}, \sigma_{zn} \in Range(ea_n)$.*

A path p from u to v is said to be simple iff $u, v_i, v_j, ..., v_k, v_l, v \in V$ are distinct. The length of p, denoted by $|p|$, is the number of edges in the path. The attribute aware path label of the path p from u to v, denoted by $pathLabel_{att}(p)$, is $(va_1(u), va_2(u), ..., va_m(u)).(\sigma_{w1}, \sigma_{w2}, ..., \sigma_{wn}).(va_1(v_i), va_2(v_i), ..., va_m(v_i)).(\sigma_{x1}, \sigma_{x2}, ..., \sigma_{xn}).(va_1(v_j), va_2(v_j), ..., va_m(v_j)).....(va_1(v_k), va_2(v_k), ..., va_m(v_k)).(\sigma_{y1}, \sigma_{y2}, ..., \sigma_{yn}).(va_1(v_l), va_2(v_l), ..., va_m(v_l)).(\sigma_{z1}, \sigma_{z2}, ..., \sigma_{zn}).(va_1(v), va_2(v), ..., va_m(v))$.

Clearly, $pathLabel_{att}(p)$ is a string, consisting of concatenated tuples of vertex and edge attribute value assignments, traversed in the same order as the vertices and edges appear in path p. Note that, the vertex and edge attribute values follow specific orders, given by $\langle va_1, va_2, ..., va_m \rangle$ and $\langle ea_1, ea_2, ..., ea_n \rangle$, respectively. For sth edge in path p where $1 \leq s \leq |p|$, starting vertex, edge, and terminating vertex attribute value assignments are represented by $(2 \times s - 1)$th, $(2 \times s)$th, and $(2 \times s + 1)$th tuples in $pathLabel_{att}(p)$, respectively.

Given ARG in Fig. 1, the only path p from Cathy to Bob is $\langle (Cathy, Bob, F) \rangle$ with $pathLabel_{att}(p) = (Female, Student).(F).(Male, Officer)$. Henceforth, we understand path to mean simple path.

Definition 4. *Attribute aware ReBAC Policy*
An Attribute aware ReBAC policy, POL_{AAR} is a tuple, given by $\langle OP, VA, EA, RuleSet \rangle$ where,

a. *OP, VA, and EA are as defined in Definition 2.*
b. *RuleSet is a set of rules where, for each operation $op \in OP$, RuleSet contains a rule $Rule_{op}$. Each $Rule_{op}$ is specified using the grammar below.*
   $Rule_{op} ::= Rule_{op} \vee Rule_{op} \mid pathRuleExpr \mid Attexp$
   $pathRuleExpr ::= pathRuleExpr \wedge pathRuleExpr \mid (pathLabelExpr)$
   $pathLabelExpr ::= pathLabelExpr.pathLabelExpr \mid edgeExp$
   $Attexp ::= Attexp \wedge Attexp \mid uexp = value \mid vexp = value$
   $edgeExp ::= edgeExp \wedge edgeExp \mid edgeuexp = value \mid edgevexp = value \mid edgeattexp = value$
   where, value is a atomic constant.
   $uexp \in \{va(u) \mid va \in VA\}$, *u is a formal parameter.*
   $vexp \in \{va(v) \mid va \in VA\}$, *v is a formal parameter.*
   $edgeuexp \in \{va(e.u) \mid va \in VA\}$, *e.u is a formal parameter.*
   $edgevexp \in \{va(e.v) \mid va \in VA\}$, *e.v is a formal parameter.*
   $edgeattexp \in \{ea(e) \mid ea \in EA\}$, *e is a formal parameter.*

Here "." is the concatenation operator. The length of a pathLabelExpr is given by the number of concatenation operators plus 1. A pathLabelExpr can be split at the point of each .operator into edgeExp, and numbered sequentially, starting from 1 to the length of the pathLabelExpr.

Based on the stated POL_{AAR}, the following defines an access control system:

Definition 5. *Attribute aware ReBAC System*

An Attribute aware ReBAC system is a tuple, $\langle ARG, POL_{AAR}, check$ $Access_{AAR} \rangle$ where ARG and POL_{AAR} are as in Definition 2 and 4, respectively. For an access request (a, b), $checkAccess_{AAR}(a{:}V, b{:}V) \equiv Rule_{op}(a{:}V,$ $b{:}V)$ where $Rule_{op}$ is evaluated as follows:

Step 1:

a. *for each Attexp in $Rule_{op}$, substitute the values va(a) for va(u) and va(b) for va(v), where $va \in VA$.*

b. *For a pathLabelExpr in $Rule_{op}$, substitute True iff i) there exists a simple path p from a to b in ARG such that $|p| = $ length of pathLabelExpr, and ii) each sth edgeExpr of the pathLabelExpr where $1 \leq s \leq$ length of pathLabelExpr, evaluates to True. To evaluate sth edgeExpr, substitute va(e.u), ea(e), and va(e.v) by the corresponding $va \in VA$, $ea \in EA$, and $va \in VA$ attribute value assignments from $(2 \times s - 1)th$, $(2 \times s)th$, and $(2 \times s + 1)th$ tuples in $pathLabel_{att}(p)$, respectively.*

Step 2:

Evaluate the resulting boolean expression.

User a is permitted to do operation op on object b if and only if $Rule_{op}(a, b)$ evaluates to True.

For example, given ARG in Fig. 1, and $Rule_{op} = $ (Gender(e.u)=Female ∧ Profession(e.u)=Student ∧ Relation-type(e) = F ∧ Gender(e.v)=Male ∧ Profession(e.v)=Student), $Rule_{op}(Alice, Ron)$ evaluates to True.

Although both ReBAC and ABAC are powerful, flexible and comparable [1] in expressing authorization policies, relying solely on one is often insufficient. An example case will be used in order to compare ABAC policy presented in [4] and ReBAC policy in [3] with the proposed AReBAC policy in Definition 4. Consider the ARG in Fig. 1 and Table 1. Each row of Table 1 represents a case and an associated authorization state example.

1. Row 1 indicates that both AReBAC and ReBAC policies can express the authorization state (Alice, Bob) whereas only ABAC rules cannot. ABAC rule fails because Alice and Cathy have the same attribute value combination. The generated ReBAC and AReBAC rules are "F.F.F" and "(Relation-type(e) = F.Relation-type(e) = F.Relation-type(e) = F)", respectively.

2. Row 2 indicates that both ABAC and AReBAC policies can express the authorization state (Ron, Bob) whereas only ReBAC rules cannot. ReBAC rule fails because there is only one path labeled "F.F" from Ron to Bob which is satisfied by unauthorized pair, such as, (Alice, Cathy). The generated ABAC and AReBAC rule is the same, "Gender(u)=Male ∧ Profession(u)=Student ∧ Gender(v)=Male ∧ Profession(v)=Officer".

3. The 3rd row, authorization state (Alice, Ron), cannot be expressed by both ABAC and ReBAC. ABAC rule fails because Alice and Cathy have the same attribute value combination. ReBAC rule fails because the only path label

Table 1. Example data

ReBAC	ABAC	AReBAC	AUTH
Yes	No	Yes	{(Alice, Bob)}
No	Yes	Yes	{(Ron, Bob)}
No	No	Yes	{(Alice, Ron)}
No	No	No	{(Bob, Alice)}

"F" is satisfied by other unauthorized pairs, such as (Alice, Bob). The AReBAC rule is "(Gender(e.u) = Female ∧ Profession(e.u) = Student ∧ Relationtype(e) = F ∧ Gender(e.v)=Male ∧ Profession(e.v)=Student)".

4. The 4th row, Auth = {(Bob, Alice)} is not expressible by only ABAC (Since (Bob, Cathy) will be allowed), only ReBAC (since no path exists from Bob to Alice), and AReBAC ((Bob, Cathy) will be allowed and no path exists).

According to the used policy specification language, AReBAC is more expressive than ABAC [4] and ReBAC [3]. Additionally, it can be clearly observed that, if entity ids are allowed, AReBAC policy will never fail (such as [2]). However, imposing this condition conflicts with core principles of ABAC and ReBAC. Therefore, the AReBAC policy specification in this paper checks whether the target access control system could be generated avoiding explicit use of unique entity id. Based on this motivation, ARREP problem is defined as follows:

Definition 6. *Attribute aware ReBAC RuleSet Existence Problem (ARREP) Given an EAS and an ARG as in Definition 1 and 2, respectively, where V=U, does there exist a RuleSet as in Definition 4 so that the resulting Attribute aware ReBAC system satisfies:*

$$(\forall u, v \in U)[checkAccess_{AAR}(u,v) \Leftrightarrow checkAccess_{EAS}(u,v)]$$

Such a RuleSet, if it exists, is said to be a suitable RuleSet, otherwise the problem is said to be infeasible.

The following subsection develops a ARREP solution algorithm.

3.2   ARREP Solution Algorithm

Algorithm 1 resolves the ARREP problem. Given an ARREP instance, it returns either feasible status and $Rule_{op}$, or infeasible status, incomplete $Rule_{op}$ and failed authorizations. Given any graph, the task finding all possible simple paths from a source vertex to a target vertex is well known, hence, details of function FindAllSimplePath() in Algorithm 1 are not provided (it can be adapted from [3]). The overall complexity of computing all possible paths from a vertex to another in ARG is $O(|E|!)$ as it considers only simple paths.

Theorem 1. *The overall complexity of ARREP feasibility detection Algorithm 1 is $O(|V|^4 \times (|E|!))$.*

Algorithm 1. ARREP Solution Algorithm

Input: An EAS and an ARG where V=U.
Output: Feasible/infeasible status and $Rule_{op}$. If infeasible, failedAuthPairs.

1:   $Rule_{op}$:= NULL
2:   failedAuthPairs := $\emptyset$
3:   tempAUTH := AUTH
4:   **for** each $(a, b) \in tempAUTH$ **do**
5:     **if** ABAC-Expr(EAS, VA, UATTValue, a, b) == SUCCESS **then**
6:      **if** $Rule_{op}$ is NULL **then** $Rule_{op} := \bigwedge\limits_{va \in VA} va(u) = va(a) \wedge \bigwedge\limits_{va \in VA} va(v) = va(b)$

      else $Rule_{op} := Rule_{op} \vee \bigwedge\limits_{va \in VA} va(u) = va(a) \wedge \bigwedge\limits_{va \in VA} va(v) = va(b)$

7:      $tempAUTH\backslash := \{(a, b)\}$
8:   **while** $\exists (a, b) \in tempAUTH$ **do**
9:     $SP(a, b)$:= FindAllSimplePath(a,b, ARG)
10:   **if** $SP(a, b) = \emptyset$ **then**
11:     $failedAuthPairs := failedAuthPairs \cup \{(a, b)\}$ //Not Feasible for (a,b) tuple

12:     $tempAUTH\backslash := \{(a, b)\}$ and **Continue**
13:     $PATHLABEL_{att}(a.b) := \{pathLabel_{att}(p) | p \in SP(a, b)\}$
14:     **for** each $pl \in PATHLABEL_{att}(a.b)$ **do**
15:      $SAT_{ab}(pl) = \{(c, d) \in V \times V |$ there exists a simple path s from c to d in ARG,
      c$\neq$d, (c,d)$\notin$AUTH, pl=$pathLabel_{att}(s)\}$
16:      $Q_{ab} := \bigcap\limits_{pl \in PATHLABEL_{att}(a.b)} SAT_{ab}(pl)$
17:     **if** $Q_{ab} \neq \emptyset$ **then**
18:      $failedAuthPairs := failedAuthPairs \cup \{(a, b)\}$ //Not Feasible for (a,b) tuple

19:     $tempAUTH\backslash := \{(a, b)\}$ and **Continue**
20:     **if** $Rule_{op}$ is NULL **then** $Rule_{op} := \bigwedge\limits_{pl \in PATHLABEL_{att}(a.b)} (generateRule(pl))$

      else $Rule_{op} := Rule_{op} \vee \bigwedge\limits_{pl \in PATHLABEL_{att}(a.b)} (generateRule(pl))$

21:     $tempAUTH\backslash := \{(a, b)\}$
22:   **if** failedAuthPairs is $\emptyset$ **then**
23:     **return** "feasible" and $Rule_{op}$
24:   **else**
25:     **return** "infeasible" and failedAuthPairs and $Rule_{op}$

Algorithm 2. ABAC-Expr

Input: EAS, VA, UATTValue, vertex a, vertex b.
Output: SUCCESS or FAILURE
1: $R1 = \{u1|\forall va \in VA.va(a) = va(u1)\}$
2: **if** $\exists u1, u2 \in R1.(u1, u3) \in Auth \wedge (u2, u3) \in \overline{Auth}$ where $u3 \in V$ **then**
3:   **return** FAILURE
4: $R2 = \{u4|(\forall va \in VA.va(b) = va(u4)\}$
5: **if** $\exists u4, u5 \in R2.(u4, u6) \in Auth \wedge (u5, u6) \in \overline{Auth}$ where $u6 \in V$ **then**
6:   **return** FAILURE
7: **return** SUCCESS

Algorithm 3. generateRule

Input: String pathlabel
Output: String rule
1: rule := NULL
2: SubStr := splitStr(pathlabel,".") // The splitStr function splits pathlabel using .
   into an ordered list of substrings, and return the saved substrings into an array.
3: numEdges := (number of elements in SubStr-1)÷ 2
4: //rm function returns the given string after removal of leading "(" and trailing ")"

5: **for** i = 1 to numEdges **do**
6:   tempu := splitStr(rm(SubStr[2*i-1]), ",")
7:   tempv := splitStr(rm(SubStr[2*i+1]), ",")
8:   tempe := splitStr(rm(SubStr[2*i]), ",")
9:   **if** rule is NULL **then** rule := $\bigwedge_{1\leq j\leq m} va_j(e.u) = tempu[j] \wedge va_j(e.v) = tempv[j] \wedge$
     $\bigwedge_{1\leq k\leq n} ea_k(e) = tempe[k]$ **else** rule := rule . $\bigwedge_{1\leq j\leq m} va_j(e.u) = tempu[j] \wedge$
     $va_j(e.v) = tempv[j] \wedge \bigwedge_{1\leq k\leq n} ea_k(e) = tempe[k]$ //. means the concatenation
10: **return** rule

Proof. In Algorithm 2, overall complexity of Lines 1, 4, 2–3 and 5–6 are $O(|U|)$, $O(|U|)$, $O(|AUTH|)$, and $O(|AUTH|)$, respectively. Therefore, overall complexity of Algorithm 2 is $O(|AUTH|)$. The overall complexity of Algorithm 3 is $O(|V|)$ since the maximum number of edges allowed in a simple path of ARG is $|V|$-1. Combining all these, the computational complexity of Algorithm 1 as follows: Lines 4–7 of Algorithm 1 give $O(|AUTH|^2)$ complexity. According to the complexity of FindAllSimplePath() noted before, Lines 9 and 13, both give $O(|E|!)$ complexity. The overall complexity of Lines 14–15 is $O(|V|^2 \times (|E|!))$, and the set intersection in Line 16 takes $O(|E|!)$. Lines 17–21 can be ignored compared to others, therefore, the loop from Lines 8–21 takes overall $O(|V|^4 \times (|E|!))$ complexity as the loop may iterate $|AUTH| \leq |V|^2$ times. Hence, the worst case complexity of Algorithm 1 is $O(|V|^4 \times (|E|!))$.

The correctness proof of Algorithm 1 is similar to the feasibility detection algorithm in [3], and is therefore omitted. Although overall complexity of feasibility detection algorithm in [3] and Algorithm 1 are same, however, the latter may have more or less computation time. If Algorithm 2 succeeds $\forall (a, b) \in AUTH$, only $O(|AUTH|^2)$ will be the real computational complexity, which is linear compared to the computed worst case complexity. The computational complexity significantly reduces even if Algorithm 2 succeeds for some $(a, b) \in AUTH$ since avoiding all possible path generation from a source vertex to target vertex in ARG (FindAllSimplePath() in Line 9) to any extent helps. Otherwise, taking both user and edge attribute value combination into consideration certainly adds overhead to the computation time of Algorithm 1, compared to feasibility detection algorithm in [3].

Let us consider the ARG in Fig. 1 where Range(Relation-type) is changed from $\{Friendship\}$ to $\{Friendship, Parent\}$. Since the "Parent" relation is not present anywhere as edge attribute in the ARG, the effect of introducing a new user with "Parent" relation in ARG remains undetermined. This might happen to any ARG with a particular rule structure as change in relationships or adding a new user may effect the validity of the current rule set. We call this "unrepresented path labels" problem in ARG. The rule structure in this paper compares direct values, the $Rule_{op}$ generated by Algorithm 1 does consider all user and edge attributes, and ARG is static by nature. Thereby, unrepresented path labels does not impact the $Rule_{op}$.

In order to show a comparison with our AReBAC policy language in user to user relationship context, the model presented in [6] is compared as follows:

- By construction, the policy language in this paper does not support inverse relationship and count attribute as in [6].
- The policy language in Definition 4 is unable to count the number of existing paths between access initiator and target users. Another example is, the policy language in Definition 4 is unable to compare attribute value assignments of any two particular users along the path from initiator to target in ARG.
- The policy language in [6] supports the common regular expression feature, wildcard (* means to 0 to any number), optional (? means 0 or 1) notation, and negative path expression, while this paper completely ignores them.

Clearly the AReBAC rule structure presented in this paper is not the most general one. More expressiveness can be added such as in [6] and current feasibility problem statement could be correspondingly reformulated.

4 ARREP Infeasibility Solution

Given an infeasible ARREP instance as in Definition 6, an infeasibility solution basically generates a RuleSet which completes the AReBAC system. Formally, given an infeasible ARREP instance as in Definition 6, an infeasibility solution is said to be exact iff: $(\forall u, v \in U)[checkAccess_{AAR}(u, v) \Leftrightarrow checkAccess_{EAS}(u, v)]$.

In this section, an exact solution to infeasibility in ARREP will be discussed with computational complexity as well as shortcomings. It is accomplished by adding edges to the given ARG as follows:

Definition 7. *Add Relationship Edge*
Given an ARREP infeasible instance, Algorithm 1 returns a set of failed autho-rization pairs, failedAuthPairs. Subsequently, the following steps are used:

1. *It is assumed that, $\forall ea \in EA.op \notin Range(ea)$.*
2. *$\forall ea \in EA, Range(ea)\cup := op$, where $op \in OP$.*
3. *For each $(a, b) \in failedAuthPairs$, $E := E \cup \{(a,b,op,op,...,op)\}$.*
   Note*: for each newly added edge, say e, $\forall ea \in EA.ea(e) = op$.
4. *$Rule_{op} := Rule_{op} \vee (\bigwedge_{ea \in EA} ea(e) = op)$, $Rule_{op}$ is returned by Algorithm 1.*

For example, given the previous infeasible example where Auth $= \{(Bob, Alice)\}$ and ARG as in Fig. 1, an additional relationship edge from Bob to Alice, labeled by the operation op$\in OP$ where op is added to Range(Relation-type), solves the problem. The following theorem proves the correctness of the stated infeasibility correction approach in Definition 7.

Theorem 2. *Definition 7 provides an exact solution to infeasibility in ARREP.*

Proof. As stated, for all $(a, b) \in failedAuthPairs$, adding an edge from vertex a to b in ARG creates a path of length 1. By the checkAccess evaluation presented in Definition 5, all $(a, b) \in failedAuthPairs$ satisfy ($\bigwedge_{ea \in EA} ea(e) = op$), and therefore, adding a term is sufficient for a operation $op \in OP$. Since it is assumed that, $\forall ea \in EA.op \notin Range(ea)$, therefore, no other $U \times U \setminus failedAuthPairs$ satisfies ($\bigwedge_{ea \in EA} ea(e) = op$). Hence, the claim is correct.

As stated in Definition 7, the solution adds $|AUTH|$ edges to the ARG at most. Hence, the worst case complexity is linear to $|AUTH|$. However, this solution approach has limitations. For example, less number of additional edges could be used to resolve the infeasibility [3]. Furthermore, there might be cases where it is undesirable to alter the given ARG and Range of attributes at all. We leave considering such cases as future work.

5   Conclusion

This paper provides an insightful discussion regarding attribute-aware ReBAC policy mining. It introduces the ARREP problem and formalizes infeasibil-ity issues in ARREP. A few simple rule optimization technique may reduce the generated rule size. For instance, rule minimization is limited to find-ing minimal number of path labels in conjunctive terms only. As per Algo-rithm 1, for a tuple (a, b) in AUTH, the conjunctive term is formed by AND'ing all possible path labels from a to b iff i) Algorithm 2 fails, and ii) the conjunctive term evaluates false for all unauthorized tuples. Instead of using all possible path labels in the conjunctive term of such (a, b), the smallest possible subset (except empty set) of those is used to form the conjunctive term, ensuring that the minimal size conjunctive evaluates false for all unauthorized tuples. For instance, given ARG in Fig. 1 and AUTH = $\{(Alice, Bob)\}$, i) Algorithm 2 returns FAILURE for (Alice, Bob), ii) there exist two paths, say p1 and p2, from Alice to Bob in ARG where $pathLabel_{att}(p1)$ and $pathLabel_{att}(p2)$ are (Female, Student).(F).(Male, Officer) and (Female, Student).(F).(Male, Student).(F).(Female, Student).(F).(Male, Officer). Without any rule minimization, $Rule_{op}$ generated by Algorithm 1 is given by the conjunction of generateRule($pathLabel_{att}(p1)$) and generateRule($pathLabel_{att}(p2)$): (Gender(e.u) = Female $\wedge$ Profession(e.u) = Student $\wedge$ Relation-type(e) = F $\wedge$ Gender(e.v) = Male $\wedge$ Profession(e.v) = Officer) $\wedge$ (Gender(e.u) = Female $\wedge$ Profession(e.u) = Student $\wedge$ Relation-type(e) = F $\wedge$ Gender(e.v) = Male $\wedge$ Profession(e.v) = Student . Gender(e.u) = Male $\wedge$ Profession(e.u) = Student $\wedge$ Relation-type(e) = F $\wedge$ Gender(e.v) = Female $\wedge$ Profession(e.v) = Student . Gender(e.u) = Female $\wedge$ Profession(e.u) = Student $\wedge$ Relation-type(e) = F $\wedge$ Gender(e.v) = Male $\wedge$ Profession(e.v) = Officer). The possible subset of path labels in this case is: either one or both. It is evident that, i) only $pathLabel_{att}(p1)$ is not possible because it is satisfied by unauthorized pair (Cathy, Bob) ii) only $pathLabel_{att}(p2)$ is possible since it is not satisfied by unauthorized pairs. Thereby, $Rule_{op}$ reduces to (Gender(e.u) = Female $\wedge$ Profession(e.u) = Student $\wedge$ Relation-type(e) = F $\wedge$ Gender(e.v) = Male $\wedge$ Profession(e.v) = Student . Gender(e.u) = Male $\wedge$ Profession(e.u) = Student $\wedge$ Relation-type(e) = F $\wedge$ Gender(e.v) = Female $\wedge$ Profession(e.v) = Student. Gender(e.u) = Female $\wedge$ Profession(e.u) = Student $\wedge$ Relation-type(e) = F $\wedge$ Gender(e.v) = Male $\wedge$ Profession(e.v) = Officer).

Acknowledgement. This work is partially supported by NSF CREST Grant HRD-1736209.

References

1. Ahmed, T., Sandhu, R., Park, J.: Classifying and comparing attribute-based and relationship-based access control. In: 7th ACM CODASPY 2017, pp. 59–70 (2017)
2. Bui, T., Stoller, S.D.: Learning attribute-based and relationship-based access control policies with unknown values. In: Kanhere, S., Patil, V.T., Sural, S., Gaur, M.S. (eds.) ICISS 2020. LNCS, vol. 12553, pp. 23–44. Springer, Cham (2020). https://doi.org/10.1007/978-3-030-65610-2_2
3. Chakraborty, S., Sandhu, R.: Formal analysis of rebac policy mining feasibility. In: Proceedings of the 11th ACM CODASPY, pp. 197–207 (2021)
4. Chakraborty, S., Sandhu, R., Krishnan, R.: On the feasibility of attribute-based access control policy mining. In: IRI. IEEE (2019)
5. Chakraborty, S., Sandhu, R., Krishnan, R.: On the feasibility of RBAC to ABAC policy mining: a formal analysis. In: Sahay, S.K., Goel, N., Patil, V., Jadliwala, M. (eds.) SKM 2019. CCIS, vol. 1186, pp. 147–163. Springer, Singapore (2020). https://doi.org/10.1007/978-981-15-3817-9_9

6. Cheng, Y., Park, J., Sandhu, R.: Attribute-aware relationship-based access control for online social networks. In: Atluri, V., Pernul, G. (eds.) DBSec 2014. LNCS, vol. 8566, pp. 292–306. Springer, Heidelberg (2014). https://doi.org/10.1007/978-3-662-43936-4_19

7. Fong, P.W., Siahaan, I.: Relationship-based access control policies and their policy languages. In: Proceedings of the 16th ACM SACMAT, pp. 51–60. ACM (2011)

8. Hu, V., et al.: Guide to Attribute Based Access Control (ABAC) definition and considerations. NIST Special Publication, pp. 162–800 (2014)

9. Rizvi, S.Z.R., Fong, P.W.L.: Efficient authorization of graph-database queries in an attribute-supporting REBAC model. ACM Trans. Priv. Secur. **23**(4), 1–33 (2020)

Author Index

Printed in the United States
by Baker & Taylor Publisher Services

Printed in the United States
by Baker & Taylor Publisher Services